Drought Frontiers in Rice

Crop Improvement for Increased Rainfed Production

Drought Frontiers in Rice: Crop Improvement for Increased Rainfed Production

Edited by R. Serraj, J. Bennett, and B. Hardy

2009

NEW JERSEY • LONDON • SINGAPORE • BEIJING • SHANGHAI • HONG KONG • TAIPEI • CHENNAI

Published by

World Scientific Publishing Co. Pte. Ltd.
5 Toh Tuck Link, Singapore 596224
USA office: 27 Warren Street, Suite 401-402, Hackensack, NJ 07601
UK office: 57 Shelton Street, Covent Garden, London WC2H 9HE

The International Rice Research Institute (IRRI) was established in 1960 by the Ford and Rockefeller Foundations with the help and approval of the Government of the Philippines. Today, IRRI is one of the 15 nonprofit international research centers supported in part by more than 40 donors: members of the Consultative Group on International Agricultural Research (CGIAR – www.cgiar.org), other government funding agencies, foundations, the private sector, and nongovernment organizations.

The responsibility for this publication rests with the International Rice Research Institute.

Mailing address: IRRI, DAPO Box 7777, Metro Manila, Philippines
Phone: +63 (2) 580-5600
Fax: +63 (2) 580-5699
Email: irri@cgiar.org
Web: www.irri.org.
Rice Knowledge Bank: www.knowledgebank.irri.org
Courier address: Suite 1009, Security Bank Center
6776 Ayala Avenue, Makati City, Philippines
Tel. +63 (2) 891-1236, 891-1174, 891-1258, 891-1303

Suggested citation: Serraj R, Bennett J, Hardy B, editors. 2009. Drought frontiers in rice: crop improvement for increased rainfed production. Singapore: World Scientific Publishing and Los Baños (Philippines): International Rice Research Institute. 400 p.

ISBN-13 978-981-4280-006
ISBN-10 981-4280-00-3

Cover design: Juan Lazaro IV
Page makeup and composition: Ariel Paelmo
Figures and illustrations: Ariel Paelmo

Printed in Singapore by Mainland Press Pte Ltd.

Contents

Foreword

Most regions with extensive poverty in Asia are dominated by rainfed ecologies where rice is the principal source of staple food, employment, and income for the rural population. Success has been limited in increasing productivity in rainfed rice systems. Poor people in these ecosystems lack the capacity to acquire food, even at lower prices, because of low productivity in food production and limited employment opportunities elsewhere. Among all abiotic stresses, drought is the major constraint to rice production in rainfed areas across Asia and sub-Saharan Africa. At least 23 million hectares (20% of rice area) are potentially affected in Asia alone. Frequent droughts result in enormous economic losses and have long-term destabilizing socioeconomic effects on resource-poor farmers and communities.

In the context of current and predicted water scarcity scenarios, irrigation is generally not a viable option to alleviate drought problems in rainfed rice-growing systems. It is therefore critical that genetic management strategies for drought focus on maximum extraction of available soil moisture and its efficient use in crop establishment, growth, and maximum biomass and seed yield. However, success has been limited in drought-prone rainfed systems. The rice yields in these ecosystems remain low at 1.0 to 2.5 t ha^{-1}, and tend to be unstable due to erratic and unpredictable rainfall. Drought mitigation, through improved drought-resistant rice varieties and complementary management practices, represents an important exit pathway from poverty.

Recent advances in drought genetics and physiology, together with progress in cereal functional genomics, have set the stage for an initiative focusing on the genetic enhancement of drought resistance in rice. Extensive genetic variation for drought resistance exists in rice germplasm. However, the current challenge is to decipher the complexities of drought resistance in rice and exploit all available genetic resources to produce rice varieties combining drought adaptation with high yield potential, good quality, and tolerance of biotic stresses. The aim is to develop a pipeline for elite "prebred" varieties or hybrids in which drought-resistance genes can be effectively delivered to rice farmers.

The Frontier Project on Drought-Resistant Rice will scale up gene detection and delivery for use in marker-aided breeding. The development of high-throughput, high-precision phenotyping systems will allow genes for component traits to be efficiently mapped, and their effects assessed on a range of drought-related traits, moving the most promising genes into widely-grown rice mega-varieties. To that end, IRRI will establish a drought consortium involving top scientists from both national agricultural research and extension systems and advanced research institutes, and will develop partnerships with extension services and the private sector for the development and evaluation of drought-resistant rice.

IRRI was pleased to convene a planning workshop for the Drought Frontier Project, bringing together some of the most eminent scientists from around the world, to discuss and devise an appropriate research agenda for this project, and to establish the partnership mechanisms for its implementation. The objectives of this workshop were to (1) assess the current status and future challenges facing rice cultivation in drought-prone environments; (2) review the recent progress, breakthroughs, and potential impact of drought research in rice and other tropical crops; (3) identify priority research areas and state-of-the-art methodologies and approaches to tackle drought challenges; and (4) establish a research consortium and an integrated research strategy on drought resistance in rice.

Robert S. Zeigler
Director General
IRRI

Rice drought-prone environments and coping strategies

Drought: economic costs and research implications[1]

Sushil Pandey and Humnath Bhandari

Drought is a major constraint to rice production in Asia. Drought occurs frequently and is one of the major reasons for wide fluctuations in rainfed production. The economic cost of drought estimated in this study was found to be substantial in rainfed areas of eastern India. The economic cost of drought depends largely on the frequency and coverage of drought, and the importance of rice in total farm income. Farmers deploy various coping strategies but these strategies were found to be largely unable to prevent a reduction in income and consumption in rainfed areas of eastern India. As a result, a large number of people fall back into poverty during drought years. The overall implications of these results for research, technology design, and policy interventions for a long-term mitigation of drought are discussed.

Drought is a recurrent phenomenon and an important constraint to rainfed rice production in Asia. Frequent major shortfalls in rice production—the staple crop of Asia—in this vast drought-prone area threaten food security, human health, and livelihood of millions of poor. At least 23 million ha of rice area (20% of the total rice area) in Asia are subject to drought of different intensities (Table 1). Drought is one of the major factors contributing to low and unstable rice production in the region (Fig. 1).

Drought can cause great harm in terms of human suffering, economic loss, and adverse environmental impact. The effect of drought in terms of production losses and consequent human misery is well publicized during years of crop failure. However, losses to drought of milder intensity, although not so visible, can also be substantial. Agricultural production losses, which are often used as a measure of the impact of drought, are only a part of the overall socioeconomic impact. Severe droughts can result in starvation and even death of the affected population. However, different types of economic impact such as production shortfall, price rise, employment and income fall, food insecurity, poor health, and so on arise before such severe consequences

[1]This paper draws heavily from the book *Economic costs of drought and rice farmers' coping mechanisms* edited by S. Pandey, H. Bhandari, and B. Hardy (2007).

Table 1. Drought-prone rice area in Asia (million ha).

Country	Rice area[a]		Drought-prone rice area	
	UR	RL	UR[b]	RL[c]
India	6.3	16	6.3	7.30
Bangladesh	0.9	6	0.9	0.80
Sri Lanka	0.06	0.2	–	na
Nepal	0.1	1.0	0.1	0.27
Myanmar	0.3	2.5	0.3	0.28
Thailand	0.05	8	–	3.1
Laos	0.2	0.4	0.2	0.09
Cambodia	–	1.7	–	0.20
Vietnam	0.5	3	0.5	0.30
Indonesia	1.1	4	1.1	0.14
China	0.6	2	0.6	0.50
Philippines	0.07	1.2	–	0.24
Total	10	46	10	13

[a]Source: IRRI (1997). [b]Assuming all upland rice (UR) area as drought-prone. [c]Source: Mackill et al (1996). Rainfed lowland (RL) rice area is classified as drought-prone and drought- and submergence-prone. The numbers represented in the table provide lower-bound estimates as the drought-prone and submergence-prone areas are excluded. na = not available.

occur. Because of market failures, farmers attempt to "self-insure" by making costly adjustments in their production practices and adopting conservative practices to reduce the negative impact during drought years. Although these adjustments reduce direct production losses, they do entail some economic costs in terms of opportunities for income gains lost during good years.

In rural areas where agricultural production is a major source of income and employment, a decrease in agricultural production will set off second-round effects through forward and backward linkages of agriculture with other sectors. A decrease in agricultural income will reduce the demand for products of the agro-processing industries that cater to local markets. This will lead to a reduction in income and employment in this sector. Similarly, the income of rural households engaged in providing agricultural inputs will also decrease. This reduction in household income will set off further "knock-on" effects. By the time these effects have been fully played out, the overall economic loss from drought may turn out to be several times more than what is indicated by the loss in production of agricultural output alone. The loss in household income can result in a loss in consumption of the poor, whose consumption levels are already low. Farmers may attempt to cope with this loss by liquidating productive assets, pulling children out of school, migrating to distant places in search of employment, and going deeper into debt. The economic and social impact of all these consequences can indeed be enormous.

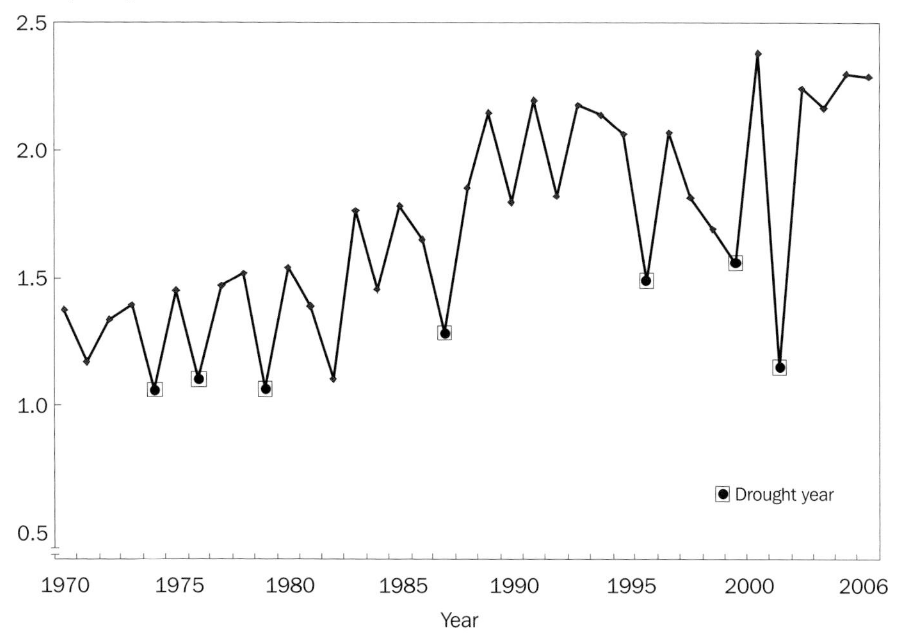

Fig. 1. Trends in rice yield and major drought years, eastern India (Orissa), 1970-2006.

This paper synthesizes the major findings of a recent cross-country comparative study of the impact of drought and farmers' coping mechanisms (see Pandey et al 2007). The countries included in the study were China, India, and Thailand. These countries vary in climatic conditions, level of economic development, rice yields, and institutional and policy contexts of rice farming. The specific regions selected for the study were southern China, eastern India, and northeastern Thailand.

Drought: definition, coping strategies, and consequences[2]

Conceptually, drought is considered to describe a situation of limited rainfall that is substantially below what has been established to be a "normal" value for the area concerned, leading to adverse consequences on human welfare. Although drought is a climatically induced phenomenon, its impact depends on the social and economic context as well. Hence, in addition to climate, economic and social parameters should also be taken into account in defining drought. This makes developing a universally applicable definition of drought impractical. Three generally used definitions of drought

[2]Details of different types of drought and farmers' coping mechanisms are presented in Pandey et al (2007).

are based on meteorological, hydrological, and agricultural perspectives (Wilhite and Glantz 1985).

Risk-coping strategies can be classified into ex ante and ex post depending upon whether they help to reduce risk or reduce the impact of risk after a production shortfall has occurred. Because of a lack of efficient market-based mechanisms for diffusing risk, farmers modify their production practices to provide "self-insurance" so that the likely impact of adverse consequences is reduced to an acceptable level. These ex ante strategies help reduce fluctuations in income and are also referred to as income-smoothing strategies. These strategies can, however, be costly in terms of forgone opportunities for income gains as farmers select safer but low-return activities.

Ex ante strategies can be grouped into two categories: those that reduce risk by diversification and those that do so by imparting greater flexibility in decision making. Diversification is simply captured in the principle of not putting "all eggs in one basket." The risk of income shortfall is reduced by growing several crops that have negatively or weakly correlated returns. This principle is used in different types of diversification common in rural societies. Examples include spatial diversification of farms, diversification of agricultural enterprises, and diversification from farm to nonfarm activities.

Maintaining flexibility is an adaptive strategy that allows farmers to switch between activities as the situation demands. Flexibility in decision making permits farmers not only to reduce the chances of low income but also to capture income-increasing opportunities when they do arise. Examples are using split doses of fertilizers, temporally adjusting input use to crop conditions, and adjusting the area allocated to a crop depending on the climatic conditions. Although postponing agricultural decisions until uncertainties are reduced can help lower potential losses, such a strategy can also be costly in terms of income forgone if operations are delayed beyond the optimal biological window. Other ex ante strategies include maintaining stocks of food, fodder, and cash.

Ex post strategies are designed to prevent a shortfall in consumption when the income drops below what is necessary for maintaining consumption at its normal level. Ex post strategies are also referred to as consumption-smoothing strategies as they help reduce fluctuations in consumption. These include migration, consumption loans, asset liquidation, and charity. A consumption shortfall can occur despite these ex post strategies if the drop in income is substantial.

Farmers who are exposed to risk use these strategies in different combinations. Over a long period of time, some of these strategies are incorporated into the nature of the farming system and are often not easily identifiable as risk-coping mechanisms. Others are deployed only under certain risky situations and are easier to identify as responses to risk.

Opportunity costs associated with the deployment of various coping mechanisms can, however, be large. Climatic uncertainties often compel farmers, particularly those who are more risk-averse, to employ conservative risk management strategies that reduce the negative impact in poor years, but often at the expense of reducing the average productivity and profitability. For example, by growing drought-hardy but low-yielding traditional rice varieties, farmers may be able to minimize the drought risk but may end up sacrificing a potentially higher income in normal years. Also, poor farmers in high drought-risk environments may be reluctant to invest in seed-fertilizer technologies that could increase profitability in normal years but lead to a loss of capital investment in poor years. In addition to these opportunity costs, poor households that are compelled to sell their productive assets such as bullocks and farm implements will suffer future productivity losses as it can take them several years to reacquire those assets. A cut in medical expenses and children's education will affect future income-earning capacity of the household. Such an impact may linger on into the future generation also. The loss of income and assets can convert transient poverty into chronic poverty, making the possibility of escape from poverty more remote (Morduch 1994, Barrett 2005).

Frequency of drought and economic loss[3]

An analysis of historical rainfall data indicated that drought is a regular phenomenon in all three regions (eastern India, northeastern Thailand, and southern China). The probability of drought varies in the range of 0.1–0.4, with the probability being higher in eastern India than in southern China and northeast Thailand (Fig. 2). The probability of late-season drought was found to be higher than that of early-season drought generally. Late-season drought was also found to be spatially more covariate than early-season drought. This means that late-season drought tends to cover large areas. As rice yield is more sensitive to drought during flowering/grain-filling stages (i.e., during the late season, according to the definition used here), late-season drought is thus likely to have a larger aggregate production impact than early-season drought.

The temporal instability in rice production as measured by the de-trended coefficient of variation of rice yield was found to be high in eastern India (17%) relative to southern China (4%) and northeast Thailand (9%). The corresponding much lower coefficients of variation for southern China and northeast Thailand indicated that droughts in these regions are not as covariate spatially as in eastern India, with their effects being limited to some pockets. Given the nature of the temporal variability, the aggregate impact of drought on production is also likely to be higher in eastern India than in the other two regions.

The estimated average loss in rice production during drought years for eastern India is 5.4 million tons (Table 2). This is much higher than for northeast Thailand (less than 1 million tons) and southern China (around 1 million tons but not statistically significant). The loss (including any nonrice crops included) during drought years is thus 36% of the average value of production in eastern India. This indeed represents a massive loss during drought years (estimated at US$856 million).

---

[3]Estimation methods for various empirical results presented are described in Pandey et al (2007).

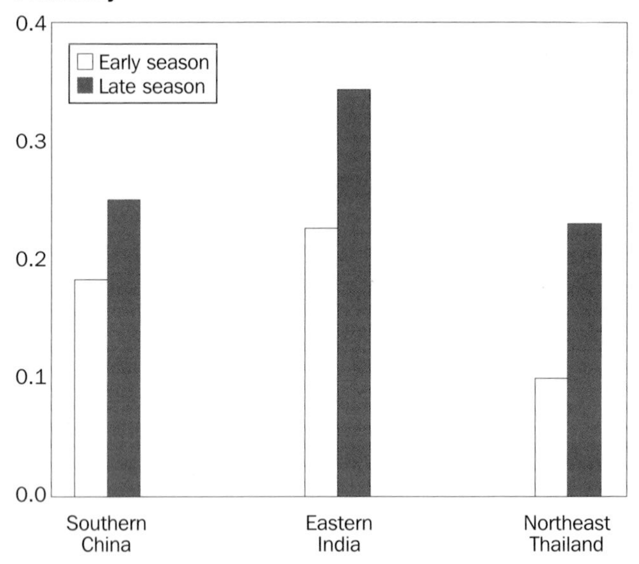

Fig. 2. Estimated probability of early- and late-season drought in southern China (1982-2001), eastern India (1970-2000), and northeast Thailand (1970-2002).

Table 2. Estimated value of crop production losses due to drought using rainfall-based drought years, 1970-2002.

Country[a]	Drought years			Annual	
	Quantity of rice production loss (million t)	Value of crop production loss[b] (million $)	Ratio of loss to average value of production (%)	Value of crop production loss[b] (million $)	Ratio of loss to average value of production (%)
Southern China	1.2	133	3	16	0.4
Eastern India	5.4	856 ***	36	162	7.0
Northeast Thailand	0.7	85 *	10	10	1.2

[a]The values are estimated based on secondary data of study provinces/states. [b]The value of production losses is estimated using both rice and nonrice crops for India while only the rice crop is used for China and Thailand. * = $P < 0.1$ and *** = $P < 0.01$.

Table 3. Percentage change in rice area, yield, and production among sample farm households in drought years compared with normal years.

Rice	Southern China	Eastern India	Northeast Thailand
Area	–19	–36	–21
Yield	–31	–54	–45
Production	–44	–71	–56

As droughts do not occur every year, the above estimate of production loss needs to be averaged over a run of drought and nondrought years to get the annual average loss estimate. Again for eastern India, this represents an annual average loss of $162 million (or 7.0% of the average value of output). For northeast Thailand and southern China, the losses were found to be much smaller and averaged less than $20 million per year (or less than 1.5% of the value of output).

The estimates thus indicate that, at the aggregate level, the production losses are much higher for eastern India than for the other two regions. Lower probability of drought, a smaller magnitude of loss during drought years, and less covariate nature of drought together have reduced production losses at the aggregate level in the other two regions relative to eastern India.

The overall economic cost of drought includes the value of production losses, the costs farmers incur in making adjustments in production systems during drought years, opportunities for gains forgone during good years by adopting ex ante coping strategies that reduce losses during drought years, the generally lower productivity of drought-prone areas due to moisture deficiency, and the costs of government programs aimed at long-term drought mitigation. The average annual cost for eastern India is in the neighborhood of $400 million (Pandey et al 2007). Overall, the cost of drought is a substantial proportion of the agricultural value added in eastern India.

Household-level consequences of drought

A detailed analysis of the household-level impact of drought was conducted using farm survey data. Drought-affected households suffered rice production losses of 44–71% (Table 3). Even in southern China and northeast Thailand, where aggregate production losses were small, production losses for the households affected by drought were substantial. Production losses resulted from both yield loss and area loss. The loss in yield, however, accounted for the major share of production losses. Across the toposequence, production losses were higher in upper fields that drain quickly than in bottom lands, which tend to have more favorable hydrological conditions.

Drought resulted in an overall income loss of 24% to 58%.[4] The drop in rice income was the main factor contributing to the total income loss. Earnings from farm labor also dropped substantially because of reduced labor demand. Farmers attempted to reduce loss in agricultural income during drought years by seeking additional employment in the nonfarm sector. This mainly included employment as wage labor in the construction sector, for which farmers often migrated to distant places. The additional earnings from nonfarm employment were clearly inadequate, however, to compensate for the loss in agricultural income.

Farmers relied on three main mechanisms to recoup this loss in total income: the sale of livestock, sale of other assets, and borrowing. These adjustment mechanisms helped recover only 6–13% of the loss in total income. Compared with normal years, households still ended up with a substantially lower level of income despite all these adjustments. Thus, all the different coping mechanisms farmers deployed were found to be inadequate to prevent a shortfall in income during drought years.

The incidence of poverty increased substantially during drought years. Almost 13 million additional people "fell back" into poverty as a result of drought (Fig. 3). This is a substantial increase in the incidence of poverty and translates into an increase in rural poverty at the national level by 1.8 percentage points. Some of the increase in poverty may be transitory, with households being able to climb out of poverty on their own. However, other households whose income and assets fall below certain threshold levels may end up joining the ranks of the chronically poor (Barrett 2005). The data collected, however, did not permit the estimation of the proportion of these two categories of households. Households with small farm sizes, with proportionately more area under drought-prone upland fields, and with a smaller number of economically active members, are more vulnerable to such adverse income consequences of drought.

In terms of crop management practices, farmers seem to have less flexibility in making management adjustments in rice cropping in relation to drought. Other than delaying crop establishment if the rains are late, replanting and resowing when suitable opportunities arise, and some reduction in fertilizer use, farmers mostly follow a standard set of practices irrespective of the occurrence of drought. This could partly be because drought mostly occurs during the late season, by which time opportunities for crop management adjustments to reduce losses are no longer available. The timing of drought (mostly late rather than early) and the lack of suitable technological options probably has limited flexibility in making tactical adjustments in crop management practices to reduce losses.

Since rice is the staple food, a loss in its production can be expected to result in major adjustments in consumption. Such adjustments could involve a reduced sale of rice, reduced quantity retained as seeds for the following year, increased amounts purchased, substitution of other crops for rice, supplementation of food deficit by other

---

[4]The household-level impact of drought presented here is based mainly on the study in eastern India. Relative to eastern India, impact in northeast Thailand and southern China was found to be quite small and, hence, is not discussed here.

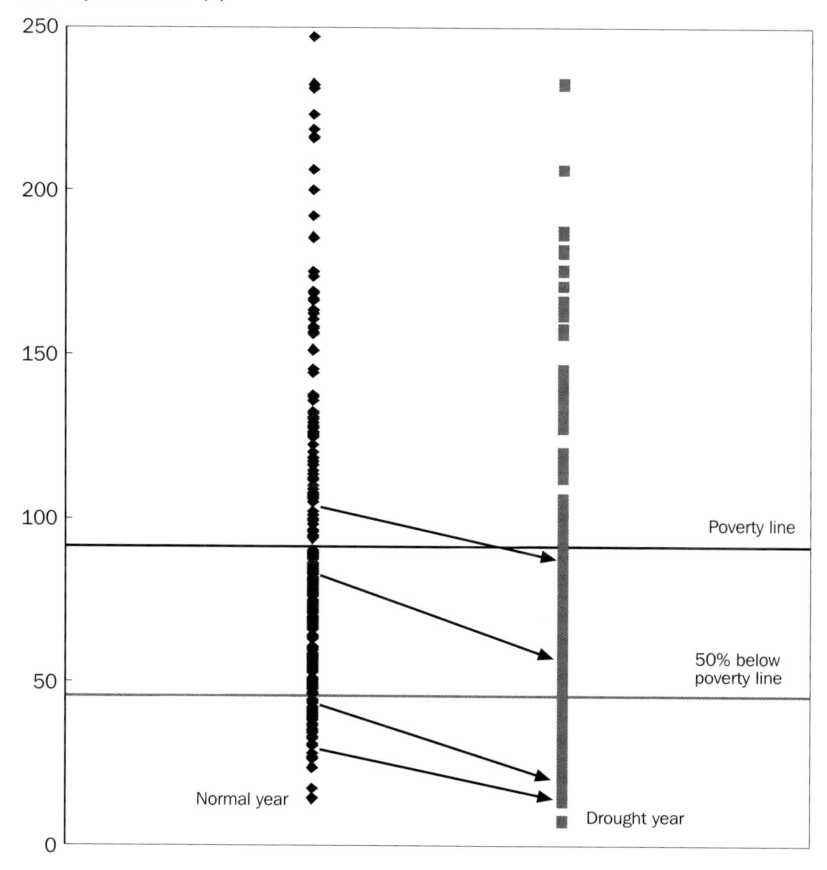

Fig. 3. Effect of drought on incidence and severity of poverty, Jharkhand, India (each dot refers to a household).

types of food not normally consumed, and, in the worst-case scenario, a reduction in consumption.

Farmers made all these types of adjustments to a varying degree. Despite these various adjustments, most farmers were unable to maintain consumption at the pre-drought level. They reduced both the number of meals eaten per day and the quantity consumed per meal. As a result, the average number of meals eaten per day dropped from close to three to close to two, with 10–30% of the households reducing their frequency of food intake to one meal per day. A large proportion (60–70%) of the households also reduced the quantity of food consumed per meal. In addition, households consumed other "inferior" food items that were not normally consumed.

The interruption and/or discontinuation of children's education is a disinvestment in human capital, which will most definitely reduce their future earning potential in most cases. An important pathway for escape from poverty may be foreclosed as a result of drought. More than 50% of the farmers reported curtailing children's education.

Relative to eastern India, the economic costs in southern China and northeast Thailand were found to be small, in both absolute and relative terms. Production losses at the aggregate level in these two regions were relatively small because of a lower frequency and less covariate nature of drought. In addition, rice accounted for a smaller proportion of household income because of a more diversified income structure. The differences in rice production systems, the level of income diversification, and the nature of drought in these latter two regions are hence the major factors determining the relative magnitudes of economic losses.

Implications

Agricultural research

Improved rice technologies that help reduce losses to drought can play an important role in long-term drought mitigation. Important scientific progress is being made in understanding the physiological mechanisms that impart tolerance of drought (Blum 2005, Lafitte et al 2006). Similarly, progress is being made in developing drought-tolerant rice germplasm through conventional breeding and the use of molecular tools (Bennett 1995, Atlin et al 2006, Serraj 2005). The probability of success in developing rice germplasm that is tolerant of drought is likely to be substantially higher now than what it was a decade ago. Complementary crop management research to manipulate crop establishment, fertilization, and general crop care for avoiding drought stress, better use of available soil moisture, and enhancing the plant's ability to recover rapidly from drought can similarly help reduce losses.

Despite the potential role of improved technologies in drought mitigation, the level of agricultural research in developing countries is generally low. Although industrialized countries invest about 2.6% of their agricultural GDP in research, the research intensity (or the ratio of research expenditure to agricultural GDP) for developing countries has been estimated to be around 0.62% (Pal and Byerlee 2003). For China and India, research intensities are only 0.43% and 0.29%, respectively. Clearly, agricultural research in the developing countries of Asia remains underinvested. The total agricultural research investment in India in 1998-99 was about $430 million (Pal and Byerlee 2003). The economic losses from drought alone as estimated in this study by considering just rainfed rice-growing areas are close to this figure.

The allocation of research resources to rainfed areas and specifically to address abiotic constraints such as drought and submergence is even lower relative to the size of losses resulting from these constraints. A recent study from India illustrates the case in point. It has been found that the allocation of rice research resources to rainfed areas in India is disproportionately small relative to the potential contribution of these areas in making efficiency and equity impacts (Pandey and Pal 2007). The share of even this limited amount of resources targeted to address abiotic constraints such as drought and submergence is less than 10%.

It has been established that the marginal productivity of research resources may now be higher in rainfed environments than in irrigated environments and that agricultural research in unfavorable (rainfed) environments can generate a substantial poverty impact (Fan et al 2005). There is a strong justification for increasing research intensity in agriculture and allocating a larger proportionate share to rainfed areas to address drought and submergence, which are the dominant constraints to productivity growth.

Technology design considerations

Several design features need to be considered when developing improved technologies for effective drought mitigation. An important design criterion is that the technologies should improve flexibility in the decision regarding crop choices, the timing and method of crop establishment, and the timing and quantity of various inputs to be used. Flexibility in agricultural technologies permits farmers not only to reduce the chances of low income but also to adaptively capture income-increasing opportunities when they do arise. Technologies that lock farmers into a fixed set of practices and timetables do not permit effective management of risk in agriculture. In fact, the empirical analyses presented in this report indicate that farmers do not seem to have much flexibility in making management adjustments in rice cropping in relation to drought. Other than delaying crop establishment if rains are late, replanting and resowing when suitable opportunities arise, and some reduction in fertilizer use, farmers mostly follow a standard set of practices irrespective of the occurrence of drought. The timing of drought (mostly late rather than early) and the lack of suitable technological options have probably limited flexibility in making tactical adjustments in crop management practices to reduce losses. Examples of technologies that provide greater flexibility are varieties that are not adversely affected by delayed transplanting caused by early-season drought, varieties that perform equally well under both direct seeding and transplanting, and crop management practices that can be implemented over a wider time window.

Losses in agricultural production and income are important factors that contribute to increases in poverty during drought years. Technologies that reduce yield losses during drought years can avoid such adverse impacts on poverty even if there may be some associated trade-offs in yield during favorable years. Hence, in terms of poverty impact, higher priority should be accorded to research focused on lopping off the lower tail of the yield distribution than to raising average yield by improving performance during normal years, if there are trade-offs involved in achieving both simultaneously.

Late-season drought is more frequent and tends to have more serious economic consequences for poor farmers than early-season drought. In addition to having to deal with the consequences of low or no harvest, farmers also lose their investments in seed, fertilizer, and labor if the crop is damaged by late-season drought. Although early-season drought may prevent planting completely, farmers can switch early to other coping strategies such as wage labor and migration to reduce income losses in such years. Thus, the poverty impact of technology is likely to be higher if research focuses on late-season drought if tolerance of early- and late-season drought cannot be achieved simultaneously.

In rainfed areas, the land endowment of farmers typically consists of fields across the toposequence that have different hydrological conditions. Fields in the upper part of the toposequence are typically more drought-prone than those in the lower part. Farmers use such a hydrologically diversified portfolio of land by growing different varieties of rice that match field hydrological features. In addition, farmers grow a range of varieties for other reasons such as staggering of labor demand, grain quality, taste, and suitability to various uses. Breeding programs that produce a wider choice of plant materials with different characteristics and varying responses to drought that correspond with field hydrological features can play an important role in effective protection from drought.

Crop diversification is an important drought-coping mechanism of farmers. Rice technologies that promote but do not constrain such diversification are therefore needed. In rainfed areas, shorter-duration rice varieties can facilitate planting of a second crop using residual moisture. Similarly, rice technologies that increase not just yield but also labor productivity will facilitate crop and income diversification. Higher labor productivity in rice production will help relax any labor constraint to diversification that may exist. Examples of such technologies are selective mechanization, direct seeding, and chemical weed control.

Complementary options

The development of water resources is an important area that is emphasized in all three countries for providing protection against drought. Opportunities for large-scale development of irrigation schemes that were the hallmark of the Green Revolution are limited now because of high costs and increasing environmental concerns (Rosegrant et al 2002). However, there are still substantial opportunities to provide some protection from drought through small and minor irrigation schemes and through land-use approaches that generally enhance soil moisture and water retention (Shah 2001, Moench 2002). Similarly, watershed-based approaches that are implemented in drought-prone areas of India provide opportunities for achieving long-term drought proofing by improving overall moisture retention within the watersheds (Rao 2000).

In all three countries studied, a major response to drought has been to provide relief to the affected population. Although the provision of relief is essential to reduce the incidence of hunger and starvation, the major problems with relief programs are slow response, poor targeting of beneficiaries, and limited coverage due to budgetary constraints. A “fire-fighting” approach that underlies the provision of drought relief cannot provide long-term drought proofing despite the large amount spent during drought years (Rao 2000, Hirway 2001). It is important that the provision of relief during drought years be complemented by a long-term strategy of investing in soil and water conservation and use, policy support, and infrastructure development to promote crop and income diversification in drought-prone areas (Rao 2000).

The scientific advances in meteorology and informatics have made it possible now to forecast drought with reasonable degrees of accuracy and reliability. Various indicators such as the Southern Oscillation Index (SOI) are now routinely used in several countries to make drought forecasts (Wilhite et al 2000, Meinke and Stone 2005). Suitable refinements and adaptations of these forecasting systems are needed to enhance drought preparedness at the national level as well as to assist farmers in making more efficient decisions regarding the choice of crops and cropping practices (Abedullah and Pandey 1998). Improvements in drought forecasting systems, the identification of efficient agricultural management practices to reduce the impact of drought, and the provision of timely advice to farmers are activities that can help reduce the overall economic costs of drought and improve preparedness to manage drought risk effectively.

Although technological interventions can be critical in some cases, this is not the only option for improving the management of drought. A whole gamut of policy interventions can improve farmers' capacity to manage drought through more effective income- and consumption-smoothing mechanisms. Improvements in rural infrastructure and marketing that allow farmers to diversify their income sources can play an important role in reducing overall income risk. Investment in rural education can similarly help diversify income. In addition, such investments contribute directly to income growth that will further increase farmers' capacity to cope with various forms of agricultural risks. Widening and deepening of rural financial markets will also be a critical factor for reducing fluctuations in both income and consumption over time (Barrett 2005). Although the conventional forms of crop insurance are unlikely to be successful because of problems such as moral hazard and adverse selection (Hazell et al 1986), innovative approaches such as rainfall derivatives and international re-insurance of agricultural risks can provide promising opportunities (Skees et al 2001, Glauber 2004). However, these alternative schemes have not yet been adequately evaluated. More work is needed for developing and pilot testing new types of insurance products and schemes suited to hundreds of millions of small farmers of Asia who grow rice primarily for subsistence.

Concluding remarks

The socioeconomic impacts of drought are enormous even in subhumid rice-growing areas. Drought causes huge economic costs, in terms of both actual economic losses during drought years and losses arising from the opportunities for economic gains forgone. The provision of relief has been the main form of drought management of the government. Although important in reducing the hunger and hardship of the affected people, the provision of relief alone is clearly inadequate and may even be an inefficient response for achieving longer-term drought mitigation. Given the clear linkage between drought and poverty, it is critically important to include drought mitigation as an integral part of a rural development strategy. Policies that in general increase income growth and encourage income diversification also serve to protect farmers from the adverse consequences of risk, including that of drought.

The scientific progress made in understanding the physiology of drought and in the development of biotechnology tools has opened up promising opportunities for making a significant impact on drought mitigation through improved technology. However, agricultural research in general remains grossly underinvested in the developing countries of Asia. This is a cause for concern, not only for drought mitigation but also for promoting overall agricultural development.

References

Abedullah, Pandey S. 1998. Risk and the value of rainfall forecast for rainfed rice in the Philippines. Philipp. J. Crop Sci. 23(3):159-165.

Atlin GN, Lafitte HR, Tao D, Laza M, Amante M, Courtois B. 2006. Developing rice cultivars for high-fertility upland systems in the Asian tropics. Field Crops Res. 97(1):43-52.

Barrett CB. 2005. Rural poverty dynamics: development policy implications. Agric. Econ. 32(1):45-60.

Bennett J. 1995. Biotechnology and the future of rice production. GeoJournal 35(3):335-337.

Blum A. 2005. Drought resistance, water-use efficiency, and yield potential: are they compatible, dissonant, or mutually exclusive? Aust. J. Agric. Res. 56(11):1159-1168.

Fan S, Chan-Kang C, Qian K, Krishnaiah K. 2005. National and international agricultural research and rural poverty: the case of rice research in India and China. Agric. Econ. 33(3):369-379.

Glauber JW. 2004. Crop insurance reconsidered. Am. J. Agric. Econ. 86(5):1179-1195.

Hazell P, Pomerada C, Valdes A. 1986. Crop insurance for agricultural development. Baltimore, Md. (USA): Johns Hopkins University Press.

Hirway I. 2001. Vicious circle of droughts and scarcity works: why not break it? Ind. J. Agric. Econ. 56(4):708-721.

Lafitte HR, Yongsheng G, Yan S, Li ZK. 2006. Whole plant responses, key processes, and adaptation to drought stress: the case of rice. J. Exp. Bot. 58(2):169-175.

Mackill DJ, Coffman WR, Garrity DP. 1996. Rainfed lowland rice improvement. Manila (Philippines): International Rice Research Institute. 242 p.

Meinke H, Stone RC. 2005. Seasonal and inter-annual climate forecasting: the new tool for increasing preparedness to climate variability and change in agricultural planning and operations. Climatic Change 70(1-2):221-253.

Moench M. 2002. Groundwater and poverty: exploring the connections. In: Llamas R, Custodio E, editors. Intensive use of groundwater: challenges and opportunities. Abingdon, UK.

Morduch J. 1994. Poverty and vulnerability. Am. Econ. Rev. 84(2):221-225.

Pal S, Byerlee D. 2003. The funding and organization of agricultural research in India: evolution and emerging policy issues. National Center for Agricultural Economics and Policy Research (NCAP) Policy Paper 16. New Delhi (India): Indian Council for Agricultural Research (ICAR). 53 p.

Pandey S, Bhandari H, Hardy B, editors. 2007. Economic costs of drought and rice farmers' coping mechanisms: a cross-country comparative analysis from Asia. Los Baños (Philippines): International Rice Research Institute. 203 p.

Pandey S, Pal S. 2007. Are less-favored environments over-invested? The case of rice research in India. Food Policy 32(5-6):606-623.

Rao CHH. 2000. Watershed development in India: recent experiences and emerging issues. Econ. Polit. Weekly, 4 November 2004.

Rosegrant MW, Cai X, Cline S. 2002. World water and food to 2025: dealing with scarcity. Washington, D.C. (USA): International Food Policy Research Institute and Colombo (Sri Lanka): International Water Management Institute.
Serraj R. 2005. Genetic and management options to enhance drought resistance and water use efficiency in dryland agriculture. Selected talk. Interdrought-II: coping with drought. The 2nd International Conference on Integrated Approaches to Sustain and Improve Plant Production under Drought Stress. Rome, Italy, 24-28 September 2005. University of Rome La Sapienza, Rome, Italy. Online at www.plantstress.com.
Shah T. 2001. Wells and welfare in the Ganga basin: public policy and private initiative in eastern Uttar Pradesh, India. Research Report 54. Colombo (Sri Lanka): International Water Management Institute.
Skees JR, Gober S, Varangis P, Lester R, Kalavakonda V. 2001. Developing rainfall-based index insurance in Morocco. World Bank Policy Research Working Paper 2577. Washington, D.C. (USA): The World Bank.
Wilhite DA, Glantz MH. 1985. Understanding the drought phenomenon: the role of definitions. Water Int. 10:111-120.
Wilhite DA, Shivakumar MKV, Wood DA, editors. 2000. Early warning systems for drought preparedness and drought management. Proceedings of an expert Group Meeting in Lisbon, Portugal, 5-7 September. Geneva (Switzerland): World Meteorological Organization.

Notes

Authors' address: International Rice Research Institute, DAPO Box 7777, Metro Manila, Philippines.

Modeling spatial and temporal variation of drought in rice production

Robert J. Hijmans and Rachid Serraj

We present a preliminary crop growth simulation model-based characterization of the spatial and temporal distribution of drought stress in rice production. The main objectives of this approach are to assist in estimating the potential benefits of drought-tolerant rice varieties, and to help select target areas for evaluation and dissemination of these varieties. The simulation model results provide a simple way to reduce daily weather data to a single or a few indices, to be used as predictors in characterization or data-mining modeling methods. We emphasize the need to refine the simulation modeling methods and to integrate the simulation results with census data and those obtained from studies of farmer behavior in response to drought, and the effect of drought on rice yield in farmers' fields.

Rice evolved in semiaquatic environments and is particularly sensitive to drought stress (O'Toole 2004). Drought is typically defined as a rainfall shortage compared with a normal average for a region. However, drought occurrence and effects on rice productivity often depend more on rainfall distribution than on total seasonal rainfall. A typical case is what happened in a recent experiment at IRRI (Los Baños, Philippines) during the wet season of 2006. Seasonal rainfall exceeded 1,200 mm, including a downpour, during a major typhoon (Milenyo), of 320 mm in a single day. Yet, a short dry spell that coincided with the flowering stage of the crop resulted in a dramatic decrease in grain yield and harvest index compared with those of the irrigated control (Serraj et al, unpublished).

An obstacle to the estimation of potential impacts of drought-tolerant varieties is that the effect, and hence adoption and impact, of these technologies is highly site- and time-specific. That is, their utility depends strongly on spatially and temporally variable environmental conditions, particularly rainfall, as well as on social and economic circumstances (Pandey and Bhandari, this volume). In this paper, we discuss only the aspects of environmental variation, aiming at estimating the potential yield benefits of drought-tolerant rice varieties and selecting target areas for evaluation and dissemination.

Defining drought

The meaning of the term "drought" often depends on a disciplinary outlook, and this includes meteorological, hydrological, and agricultural perspectives. Agricultural drought occurs when soil moisture is insufficient to meet crop water requirements, resulting in reduced crop growth and yield losses. Depending on timing, duration, and severity, this can result in catastrophic, chronic, or inherent drought stress, which would require different coping mechanisms, adaptation strategies, and breeding objectives. The 2002 drought in India could be described as typical for a catastrophic event, as it affected 55% of the country's crop area and 300 million people. Rice production was 20% below the trend values (Pandey et al 2007). Similarly, the 2004 drought in Thailand affected more than 8 million people in almost all provinces. Severe droughts generally result in impoverishment of the affected population, with dramatic, and often long-term, socioeconomic consequences (Pandey and Bhandari, this volume).

Production losses to drought of milder intensity, although not so alarming, can be substantial. The average rice yield in rainfed eastern India during "normal" years still varies between 2,000 and 2,500 kg ha^{-1}, far below achievable yield potential. Chronic dry spells of relatively short duration can often result in substantial yield losses, especially if they occur around flowering stage. In addition, drought risk reduces productivity even during favorable years in drought-prone areas because farmers avoid investing in inputs when there is large uncertainty about the attainable yield (Pandey et al 2007).

Inherent drought is associated with the increasing problem of water scarcity, even in traditionally irrigated areas, due to rising demand and competition for water uses. This is, for instance, the case in China, where the increasing shortage of water for rice production is a major concern, although rice production is mostly irrigated (Ding et al 2005).

Systems analysis and simulation

Breeding strategies for improved drought tolerance could benefit from detailed and precise characterization of the target population of environments (TPEs). One approach to characterization is the classification of rainfall patterns in relation to crop phenology (e.g., Saleh et al 2000). Although this can be very useful, it is rather difficult to do objectively, particularly for larger areas. An alternative approach involves the use of crop growth simulation models. Crop growth models encapsulate knowledge of eco-physiological processes and allow simulation of crop yield for specific varieties and locations. In this way, complex location data, such as daily weather data, can be summarized with an easy-to-interpret index such as crop yield.

For example, Heinemann et al (2008) recently used a crop simulation model to determine the patterns of drought stress for short- and medium-duration upland rice across 12 locations in Brazil. This study allowed the characterization of drought-prone TPE and confirmed the greater yield impact of drought stress when it occurred around flowering and early grain-filling.

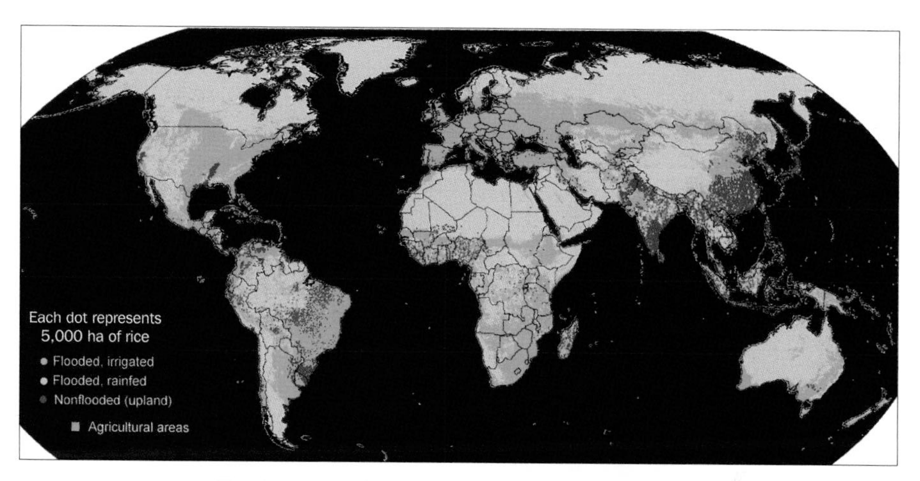

Fig. 1. Global rice area by major rice ecosystems.

Simulation models can also provide a tool to assist in understanding, and incorporating, genotype-by-environment interaction, by combining mechanistic understanding of a drought (Chapman 2008). Given a historical record of weather for a location, the probability of a yield increase (and maybe a decrease) resulting from the incorporation of any trait into the crop can be simulated. Combining the probabilities for yield change with the farmers' adversity to risk gives a strong indication to a breeder of the desirability of incorporating a particular drought trait for cultivars to be grown in a specific location. System analysis can hence allow breeding for specific drought-adaptive traits to be targeted to those geographical regions where their benefit will be largest (Sinclair and Muchow 2001). However, in the case of rice, most simulation efforts have focused on irrigated environments, and an improved rice model needs to be developed or adapted specifically for the drought-prone rainfed systems, based on better physiological understanding of rice interaction with the environment under water deficits.

Distribution of rice production systems and rainfall

Worldwide, there are more than 100 million ha of rice, with 89% in Asia. About 45% of the rice area is rainfed, of which 25% is never flooded (upland). Asia has large areas of rainfed rice in eastern India and Bangladesh, northeast Thailand, Cambodia, and the island of Sumatra in Indonesia (Fig. 1). The majority of rice production in Africa is rainfed (Balasubramanian et al 2007).

It is not a surprise that rainfed rice is not produced much in very dry areas. In Asia, about 11% of irrigated rice is produced in areas with less than 750 mm of average annual rainfall, versus 0.5% of rainfed rice. About 23% of irrigated rice is in areas with less than 1,000 mm of rainfall versus 4% of rainfed rice (Fig. 2). Rainfed rice in very dry areas is either a misclassification or it is planted in atypical humid locations on the landscape, such as valley bottoms and marshes.

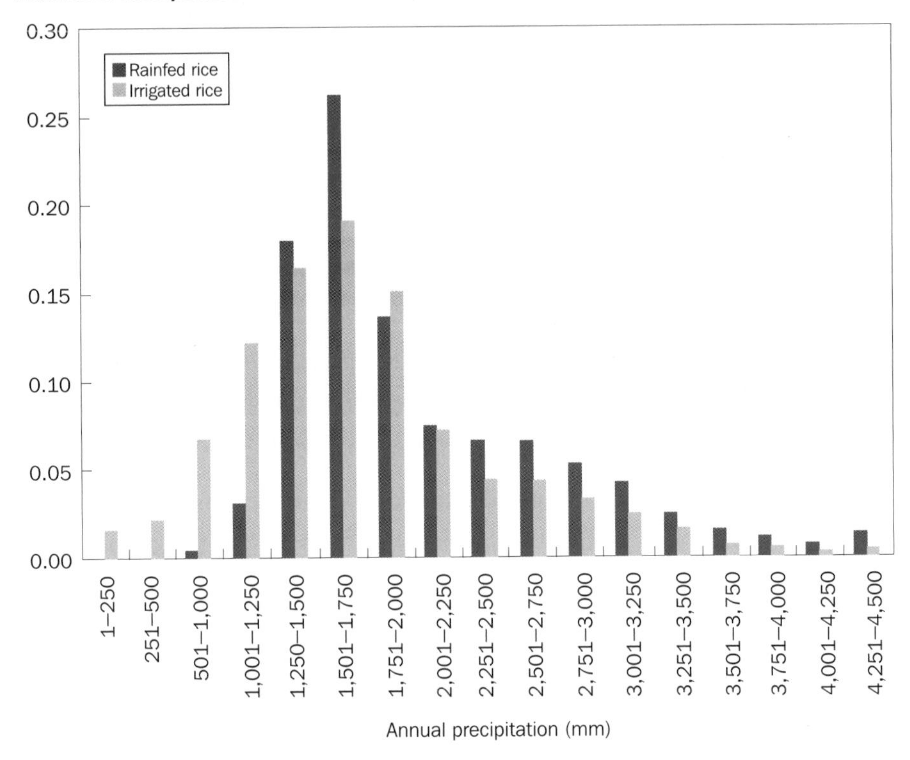

Fig. 2. Distribution of rainfed and irrigated rice area by annual precipitation.

There is also a large amount of irrigated rice in very humid areas. This is in part because irrigation in many cases provides only a part of the water required, if and when necessary, for example, during a dry spell. It is also because irrigation allows for the production of a second or third rice crop in the dry season, and in some of these areas irrigated rice during the rainy season is in most years equivalent to rainfed rice.

Water as a yield-limiting factor

We used the ORYZA2000 model (Bouman et al 2003) calibrated for variety IR72. We ran the model for 1 degree grid cells with 9 years of daily weather data estimated from satellite observations by NASA (data available at http://earth-www.larc.nasa.gov/cgi-bin/cgiwrap/solar/agro.cgi). Rainfed rice can be produced on land that never gets flooded (upland rice) or in fields that can get flooded when there is sufficient water and bunds are present (rainfed lowland rice). The source of water in these fields can be local rainfall or water flowing laterally into the fields. Here, we show only results for rainfed lowland rice (flat, bunded fields). A single rainfed crop was considered for each cell. Planting time was estimated by first simulating rainfed rice crops that

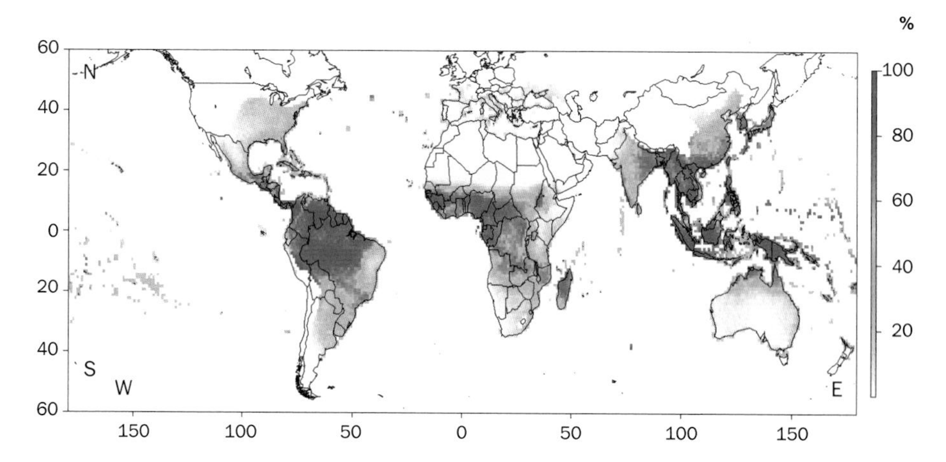

Fig. 3. Simulated rainy-season rice yield of rainfed lowland rice (flat, flooded fields) relative to irrigated conditions (%). Computed with the ORYZA2000 simulation model for variety IR72.

were planted at 2-week intervals throughout the year. We then selected the fortnight that most frequently (across the 9 years) gave the highest yield. We subsequently used that planting period for all 9 years to compute yields for rainfed and for irrigated conditions.

Figure 3 shows the simulated yield of rainfed lowland rice relative to the simulated yield with full irrigation. We refer to this as the "relative rice yield." We use relative yield because that makes it easier to compare drought effects across sites, as it is not influenced by the potential yield, which depends on temperature and radiation, but only by the (model prediction of) yield limitation due to drought.

Although our simulations do not capture many known sources of variation, such as local hydrological processes and differences in soil types, we believe that the results nevertheless show some basic facts about drought in rainfed rice. First of all, there are some places where you cannot grow much rice without irrigation. This does not necessarily mean that water stress is an important problem there. In fact, some of the most productive irrigated rice areas are found here, including the Punjab in India and the Nile Valley in Egypt. On the other hand, if water becomes scarce in these regions ("inherent drought")—as is happening in many areas—water-saving irrigation technologies and appropriate varieties would be very useful.

Figure 4 shows relative yield as a function of rainfall during the growing season, computed across all grid cells where rice can grow. It shows that, when rainfall is below 450 mm, rice production is virtually impossible. Only at 750–850 mm does the median simulated yield pass 50% of irrigated yield during that season. But, as we have seen (Fig. 4), very few farmers choose to plant rainfed rice under these conditions, probably because it is very risky.

Variation in simulated yield between sites is highest at the intermediate relative yields. When rainfall is very low or very high, the distribution of rainfall in the

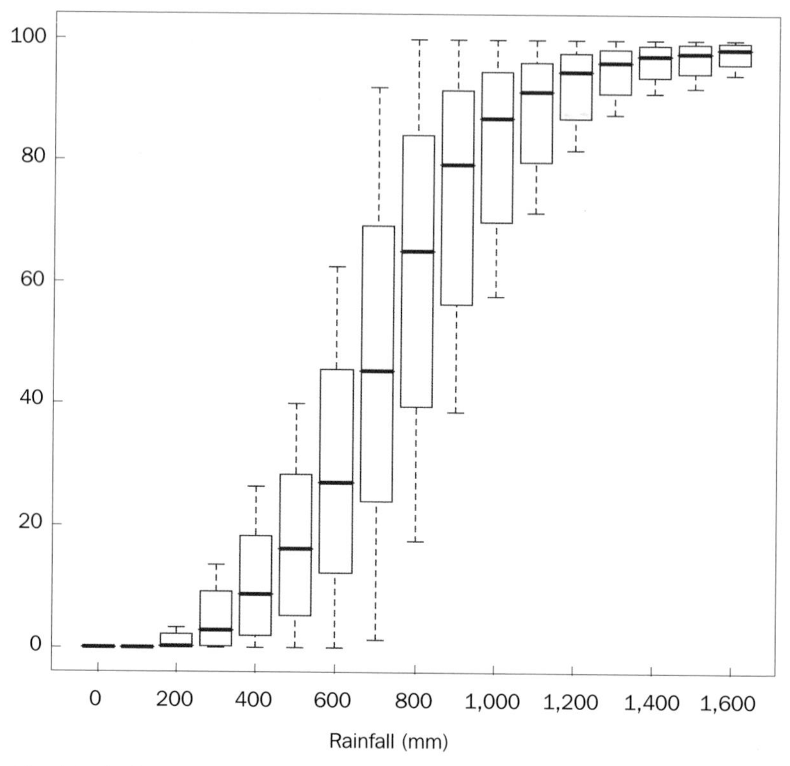

Fig. 4. Simulated rainy-season rice yield of rainfed lowland rice (flat, flooded fields) relative to irrigated conditions (%), by the amount of rain during the growing season. Box and whisker plot showing quartiles and median values. Computed with the ORYZA2000 simulation model for variety IR72.

growing season, or the effect of other climate variables (the atmospheric evaporative demand), does not matter much. But, at 600 to 800 mm during the growing season, relative yields range between 0 and 100%.

Figure 4 summarizes variation in relative yield between years, whereas Figure 5 shows variation between years within sites, expressed as the coefficient of variation. Note that, at an average seasonal rainfall of around 1,100 mm, the median coefficient of variation is still rather high at 25%. This appears to be an important property of rainfed rice production: even if the expected (median) yield is good, there can be a high frequency of years with poorer yields. The variation we found is probably a bit exaggerated because we did not allow for adaptive planting times, depending on the onset of rainfall. However, this accounts for only a part of the variation in yield reduction through drought, so even with further refining of our modeling approach, the between-year variation will likely remain high.

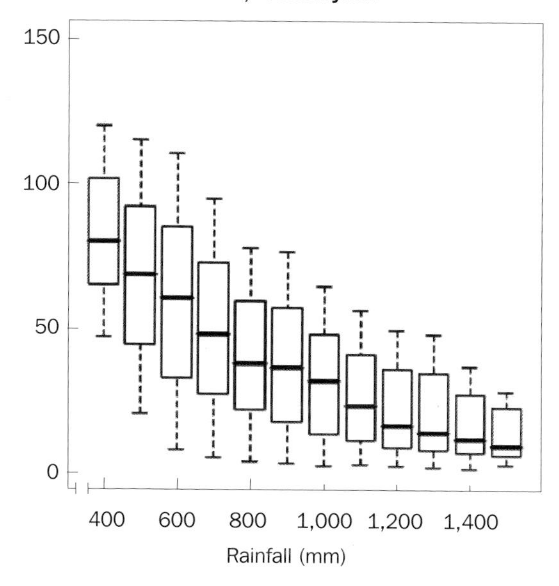

Fig. 5. Simulated coefficient of variation (over 9 years, 1993-2001) of rainy-season rice yield of rainfed lowland rice (flat, flooded fields), by the amount of rain during the growing season. Box and whisker plot showing quartiles and median values. Computed with the ORYZA2000 simulation model for variety IR72.

Drought-tolerant varieties will allow farmers to obtain higher yields without adjusting their cropping practices. However, they could also respond by adjusting cropping practices and shifting rainfed rice production to drier areas or seasons. Shifting rice production out of the main rainy season can be attractive because, if there is no drought stress, yields can often be higher when the dry season is associated with higher solar radiation (fewer clouds) and lower temperatures (longer growing season, fewer respiratory losses). In addition, early planting and harvesting may allow for double cropping of rice. Such shifts appear to be particularly relevant in eastern India, Bangladesh, and Southeast Asia (data not shown). Although drought tolerance alone can probably not do much in this context, particularly as it cannot help in planting in dry fields before the rains start, it could be very useful in combination with irrigation (pumps) to get the crop started, but with minimal additional irrigation to save on water and fuel costs.

The potential benefits of drought tolerance

Current research aims to improve the data used to run the models, and by running models for different rice ecosystems and for different varieties to contrast existing versus new drought-tolerant varieties, and to contrast current cropping practices with water-saving technologies. Simulation models are in principle very useful for estimating the benefit (in terms of yield) of drought tolerance and other traits. This could be achieved by comparing a standard variety with a variety with increased tolerance (e.g., Hijmans et al 2003). However, right now, the ORYZA2000 model has not been calibrated for any of the more drought-tolerant lines that the IRRI breeding program has developed. If the physiological mechanisms that make these new lines more drought tolerant were known, they could be incorporated into a hypothetical variety for simulation.

Because of the current knowledge gaps in understanding drought-tolerance mechanisms (Serraj et al, this volume), we decided to express drought not as a physiological trait (water demand), but rather as an environmental supply in terms of available water. The assumption is that increasing water availability to a standard rice variety is equivalent to some types of drought tolerance. We implemented this by increasing the amount of rain, on each rainy day, by 10%. Although at this point we cannot relate that to existing varieties, it does serve as an indicator of how much and where drought tolerance could be beneficial. Moreover, preliminary research findings at IRRI suggest that an important characteristic of some of our new drought-tolerant varieties is that they are able to extract more water from the soil, on the order of 7% more water (Bernier et al 2008).

Figure 6 shows the results of these simulations. As expected, drought tolerance is not useful in extremely dry or in very wet areas. Areas with rainfall between 550 and 1,050 mm during the growing season would have the most benefits, typically on the order of 500 kg ha^{-1} (on average across years). These are the areas with moderate to high relative yields (Fig. 4). When increasing water availability by 25%, the yield effects generally doubled relative to a 10% increase (data not shown).

Figure 7 shows the yield effects of drought tolerance. It could clearly be very important for large tracts in Africa. We also looked at the benefit of drought tolerance in terms of impact on total production. This was computed by multiplying the yield gain by the area under rainfed rice for each grid cell. Different regions then come out as most important (Fig. 8). Drought tolerance could be particularly important to boost production in eastern India and Thailand, where the combination of a huge area with a considerable yield gain makes for very large predicted increases in production. The simulated total global annual production increase in rainfed rice areas due to an increase in water availability of 10% is about 18 million tons of rough rice. Because of production constraints other than water, which were not considered in the simulations, this production increase would not likely be achieved by drought tolerance alone. However, this bias may be compensated for if increased drought tolerance leads to higher investments in, for example, fertilizers.

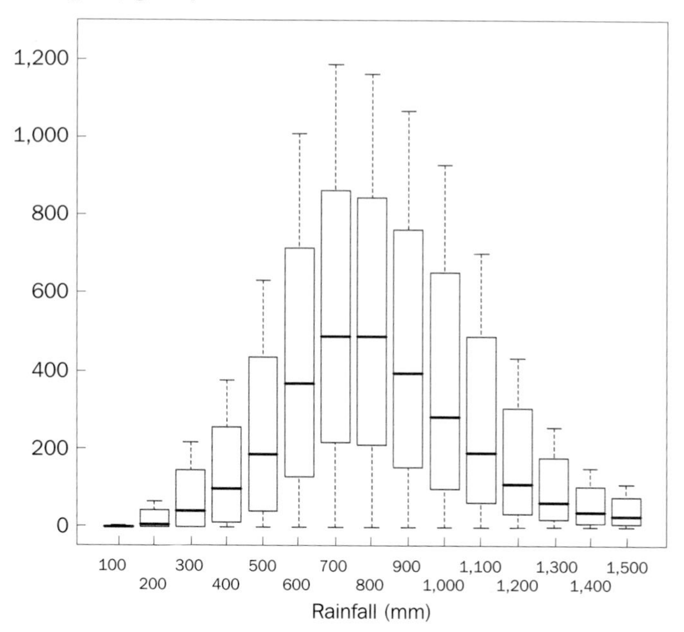

Fig. 6. Yield gain of rainfed lowland rice due to a hypothetical 10% increase in rainfall, by current rainfall. Box and whisker plot showing quartiles and median values.

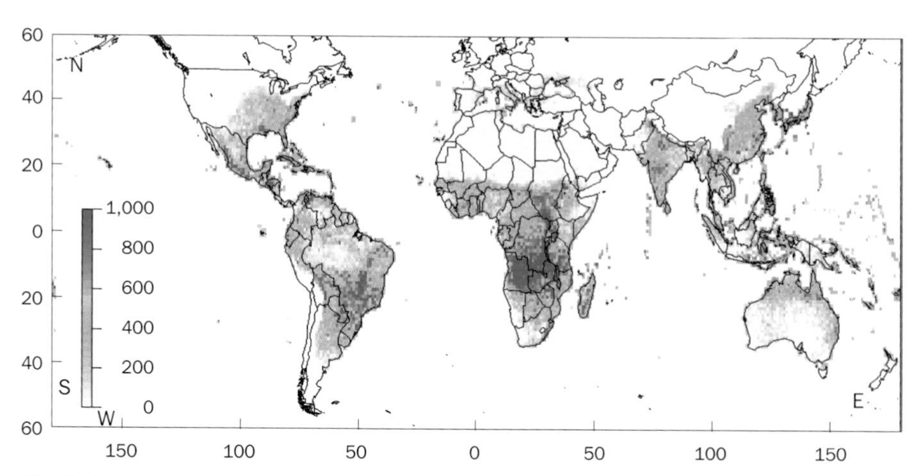

Fig. 7. Spatial distribution of yield gains (kg ha^{-1}) in rainfed rice with an increase in water availability of 10%.

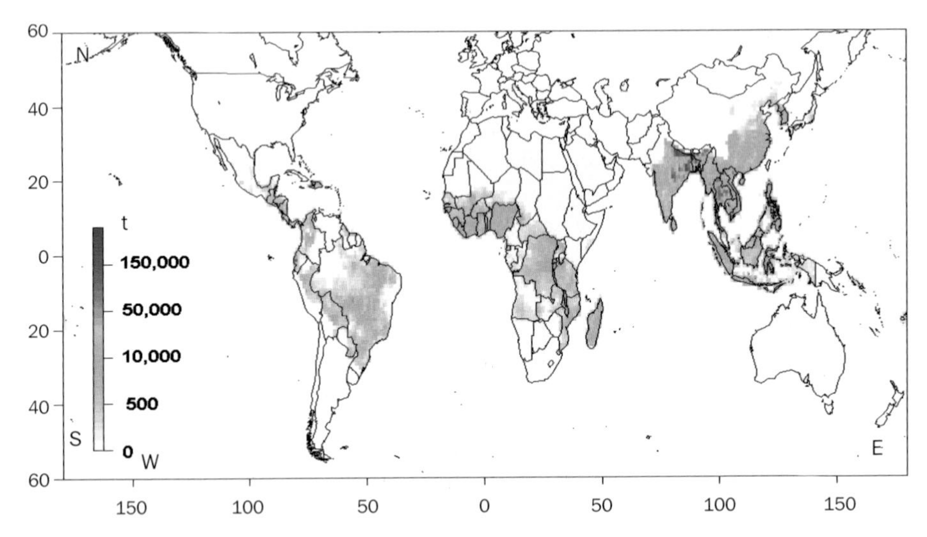

Fig. 8. Spatial distribution of production gains (tons) in rainfed rice with an increase in water availability of 10%.

Discussion

We have presented a preliminary analysis of drought stress in rainfed rice production. The modeling approach can, and will, be improved in many different ways. For example, we are working on incorporating insights on farmer behavior to improve our crop calendars (Shinji et al 2008) and allow for more adaptive variation to create distribution functions for management practices within each cell. The amount of detail that can be put in a broad regional study like this one is limited. Ideally, the regional work would help select sites for more detailed analysis (e.g., Fukai et al 2001), which could then be used to refine the regional modeling. In future work, we intend to look in more detail at several additional aspects, including different drought-coping mechanisms such as escape and daylength sensitivity.

In this study, we considered only rainfed lowland rice production. The hydrology of rainfed flooded rice fields does not only depend on rainfall but is also dependent on their position in the landscape (Haefele and Bouman, this volume). At the scale of our analysis, this probably does not matter that much. However, we do not really know, and in the future we would like to use hydrological models linked to the rice model to look at local variation in drought stress and yield. The highly unstable dynamics of hydrology, with frequent shifting between flooded and aerobic conditions within a paddy, impose a high amount of environmental variability and result in strong impact of the spatial variation in the toposequence on crop growth (Cooper et al 1999). We also ignored upland rice production systems that are more sensitive to drought as they do not have as much in field water storage capacity that can serve as a buffer.

Although we are continuing to refine our simulation modeling methods, it is equally, if not more, important to improve our simulation modeling method by contrasting, and perhaps integrating, the simulation modeling results with census data and the results obtained from studies of farmer behavior and households (e.g., Pandey and Bhandari, this volume). The simulation model results provide an easy way to reduce daily weather data to a single or a few indices. Rather than using them as truth, they could be used as predictors in a regression or data-mining modeling approach such as is commonly done in ecology ("ecological niche modeling," Elith et al 2006). However, to do so, we also need to develop broad-scale spatial data sets on farmer behavior in response to drought, and the effect of drought on yield in farmers' fields.

This study reported progress in using simulation models to characterize drought in rainfed rice. The results are preliminary and many more integrative studies are available. For example, drought-prone rainfed rice ecosystems were classified based on toposequence and water regime by Garrity et al (1986) and upland systems were described by Courtois and Lafitte (1999). Several studies have also previously discussed the biophysical characteristics of the rainfed lowland ecosystem and their implications in breeding (Mackill et al 1996, Fukai et al 2001, Wade et al 1999).

It would seem that our results somewhat underestimate drought risk. For example, in northwest Bangladesh, the average annual rainfall varies between 1,500 and 2,000 mm, with more than 200 mm of rainfall per month during the monsoon period (June to September), when transplanted aman rice (T. aman) is grown mostly under rainfed conditions. However, the erratic rainfall distribution causes drought frequently in this region, and results in yield losses that are generally higher than the damage caused by flooding and submergence (Towfiqul Islam 2008). A recent characterization and modeling study showed that the recurrence interval of drought is around 2–3 years, especially during the latest part of T. aman, generally recognized as terminal drought (Towfiqul Islam 2008). Short-duration varieties such as BRRI dhan 39 are generally used to escape terminal drought in this region. However, the risk of early droughts is also very serious, with a return period of 10 mm of rainfall deficit as high as 1.3 years in some districts, which requires a new set of drought-adapted T. aman rice varieties.

References

Balasubramanian V, Sie M, Hijmans RJ, Otsuka K. 2007. Increasing rice production in sub-Saharan Africa: challenges and opportunities. Adv. Agron. 94:55-133.

Bernier J, Serraj R, Impa S, Gowda V, Owane R, Bennett J, Kumar A, Spaner D, Atlin G. 2008. Increased water uptake explains the effect of qtl12.1, a large-effect drought-resistance QTL in upland rice. Field Crops Res. (In press.)

Bouman BAM, Kropff MJ, Tuong TP, Wopereis MCS, ten Berge HFM, van Laar HH. 2003. ORYZA2000: modeling lowland rice. IRRI. Available at www.knowledgebank.irri.org/oryza2000.

Chapman SC. 2008. Use of crop models to understand genotype by environment interactions for drought in real-world and simulated plant breeding trials. Euphytica 161:195-208.

Cooper M, Rajatasereekul S, Immark S, Fukai S, Basnayake J. 1999. Rainfed lowland rice breeding strategies for Northeast Thailand. I. Genotypic variation and genotype × environment interactions for grain yield. Field Crops Res. 64:131-151.

Courtois B, Lafitte R. 1999. Improving rice for drought-prone environments. In: Ito O, O'Toole J, Hardy B, editors. Genetic improvement of rice for water-limited environments. Los Baños (Philippines): International Rice Research Institute. p 35-56.

Ding S, Pandey S, Chen C, Bhandari H. 2005. Drought and rice farmers' coping mechanisms in the rice production systems of southern China. A final research report submitted to the Rockefeller Foundation. Department of Agricultural Economics, Huazhong Agricultural University, Hubei, China.

Elith J, Graham CH, Anderson RP, Dudík M, Ferrier S, Guisan A, Hijmans RJ, Huettmann F, Leathwick J, Lehmann A, Li J, Lohmann LG, Loiselle B, Manion G, Moritz C, Nakamura M, Nakazawa Y, Overton J McC, Peterson AT, Phillips S, Richardson K, Scachetti-Pereira R, Schapire R, Soberón J, Williams S, Wisz M, Zimmerman N. 2006. Novel methods improve prediction of species' distributions from occurrence data. Ecography 29:129-151.

Fukai S, Basnayake J, Cooper M. 2001. Modeling water availability, crop growth, and yield of rainfed lowland rice genotypes in northeast Thailand. In: Tuong TP, Kam SP, Wade L, Pandey S, Bouman BAM, Hardy B, editors. Proceedings of the International Workshop on Characterizing and Understanding Rainfed Environments, Bali, Indonesia, 5-9 December 1999. Los Baños (Philippines): International Rice Research Institute. p 111-130.

Garrity DP, Oldeman LR, Morris RA. 1986. Rainfed lowland rice ecosystems: characterization and distribution. In: Progress in rainfed lowland rice. Los Baños (Philippines): International Rice Research Institute. p 3-23.

Heinemann AB, Dingkuhn M, Luquet D, Combres JC, Chapman S. 2008. Characterization of drought stress environments for upland rice and maize in central Brazil. Euphytica 162(3):395-410.

Hijmans RJ, Cameron SE, Parra JL, Jones PG, Jarvis A. 2005. Very high resolution interpolated climate surfaces for global land areas. Int. Climatol. 25:1965-1978.

Hijmans RJ, Condori B, Carillo R, Kropff MJ. 2003. A quantitative and constraint-specific method to assess the potential impact of new agricultural technology: the case of frost resistant potato for the Altiplano (Peru and Bolivia). Agric. Syst. 76:895-911.

Mackill, DJ, Coffman WR, Garrity DP. 1996. Rainfed lowland rice improvement. Los Baños (Philippines): International Rice Research Institute. 242 p.

O'Toole JC. 2004. Rice and water: the final frontier. First International Conference on Rice for the Future, 31August-2 Sept. 2004, Bangkok, Thailand.

Pandey S, Bhandari H, Hardy B, editors. 2007. Economic costs of drought and rice farmers' coping mechanisms: a cross country comparative analysis from Asia. Los Baños (Philippines): International Rice Research Institute. 203 p.

Saleh AFM, Mazid MA, Bhuiyan SI. 2000. Agrohydrologic and drought risk analyses of rainfed cultivation in northwest Bangladesh. In: Tuong TP, Kam SP, Wade L, Pandey S, Bouman BAM, Hardy B, editors. Characterizing and understanding rainfed environments. Los Baños (Philippines): International Rice Research Institute. p 233-244.

Shinji S, Hasegawa T, Goto S, Konghakote P, Polthanee A, Ishigooka Y, Kuwagata T, Hitoshi Toritan H. 2008. Modeling the dependence of the crop calendar for rain-fed rice on precipitation in Northeast Thailand. Paddy Water Environ. 6:83-90.

Sinclair TR, Muchow RC. 2001. System analysis of plant traits to increase grain yield on limited water supplies. Agron. J. 93:263-270.

Towfiqul Islam M. 2008. Modeling of drought for Aman rice in the Northwest region of Bangladesh. PhD thesis. Bangladesh Agricultural University, Mymensingh. 194 p.
Wade LJ, Fukai S, Samson BK, Ali A, Mazid MA. 1999. Rainfed lowland rice: physical environment and cultivar requirements. Field Crops Res. 64:3-12.

Notes

Authors' address: International Rice Research Institute, Los Baños, Philippines.

Recent progress in breeding and genetics of drought resistance

Rice germplasm development for drought-prone environments: progress made in breeding and genetic analysis at the International Rice Research Institute

G.N. Atlin, R. Venuprasad, J. Bernier, D. Zhao, P. Virk, and A. Kumar

Drought is the most important constraint affecting yield in rainfed rice. Drought effects are most severe in unbunded upland fields and upper-toposequence bunded fields that infrequently accumulate standing water. Drought reduces productivity both through direct effects on biomass production and grain set and through delaying or rendering impossible crop management operations such as transplanting, fertilizer application, and weeding. There is substantial genetic variation for tolerance of drought stress, direct seeding in dry soil, and delayed transplanting. In areas where transplanting is likely to remain the major establishment system, varieties with tolerance of both drought and delayed transplanting are needed to reduce risk and increase productivity. In light-textured soils, direct seeding of drought-tolerant varieties in dry, unpuddled fields has the potential to eliminate the risk of transplanting failure and to advance maturity sufficiently to permit the production of a postrice crop. However, varieties for use in this establishment system must be highly weed competitive and have a high degree of tolerance of drought at the reproductive stage. The IRRI breeding program now routinely screens lines targeted at dry direct-seeding systems for rapid early biomass accumulation under aerobic conditions, a trait that has been shown to be closely associated with weed competitiveness. Lines targeted at transplanted systems are screened for yield under transplanting as 60-day-old seedlings. For both systems, advanced breeding lines are screened both for yield potential and for yield under continuous recurring drought stress after maximum tillering. IRRI research has confirmed that yield under drought stress has both a moderate heritability and a moderate positive correlation with yield potential, permitting the development of varieties combining high yield potential with stress tolerance. IRRI research has shown that direct selection for yield under drought stress, combined with selection for yield potential under favorable conditions, is an effective way to develop such cultivars. It has also been shown that hybrids are higher yielding than pure lines, on average, under both moderate lowland stress and delayed transplanting. Lines and hybrids combining high yield potential with yield of more than 2 t ha^{-1} under severe lowland stress, and more than 1.5 t ha^{-1} under severe upland stress, have been identified. Such varieties have the potential to reduce risk and increase overall productivity in drought-prone environments. Recent research indicates that, in many crosses, approximately

20–50% of the genetic variation for yield under severe stress is controlled by factors that also affect yield potential. The remainder appears to be affected by relatively few genes with large effects that are detectable only in drought-stressed environments. Such genes may be used to increase the drought tolerance of widely grown varieties via marker-assisted backcrossing.

Drought stress occurs frequently in rice ecosystems that are either rainfed or rely on impounded surface water, affecting about 20–25 million ha worldwide. The eastern Indo-Gangetic Plain, with more than 17 million ha of rainfed rice area, is the worst affected region (Huke and Huke 1997). In this area, drought losses are most severe in the key rice-producing states of eastern India, as well as in neighboring areas of Nepal. Northeastern Thailand and Laos, with more than 3 million ha of drought-prone rainfed rice area, is the other severely drought-affected area in Asia (Pandey et al 2005). Drought also affects production on millions of hectares in irrigated areas dependent on surface irrigation, where river flows and water impounded in ponds, tanks, and reservoirs may be insufficient to irrigate the crop in dry years (Maclean et al 2002). In water-short areas, drought risk reduces productivity even in favorable years because farmers avoid investing in inputs when they fear crop loss (Pandey et al 2005).

Although water shortage is one of the most severe constraints to rice yield, limited effort has been devoted to the development of rice cultivars with improved drought tolerance. Breeding for drought tolerance is complicated by the intermittent occurrence of natural stress, the strong relationship between plant phenology and sensitivity to stress (Fukai et al 1999, Pantuwan et al 2002), and the specificity of tolerance mechanisms for particular soil hydrological environments. Genetic analyses of traits related to drought tolerance in rice conducted to date have usually reported the detection of many QTLs with relatively small effects (e.g., Babu et al 2003, Lanceras et al 2004), indicating that grain yield under drought stress in rice is a complex trait affected by many loci. Despite this complexity, substantial genetic variability for yield under drought stress has been documented in many types of field screen, and single QTLs with large effects on drought performance have recently been documented (Bernier et al 2007, Kumar et al 2007). Genetic variability has been detected in trials in which water was withheld at a defined phenological stage (e.g., Lafitte and Courtois 2002), populations were screened in dry seasons by stopping irrigation to the entire field on one date (Babu et al 2003, Venuprasad et al 2006, 2007), and populations were screened under wet-season stress imposed by draining paddies (IRRI, unpublished data; Kumar et al 2007). Variability exists both for drought tolerance per se (e.g., Pantuwan et al 2002, Lafitte and Courtois 2002) and for traits that confer adaptation to water-short environments, such as seedling vigor and weed competitiveness (Zhao et al 2006c). There is strong evidence that this variability can be exploited by screening lines for high yield under managed drought stress. Rice cultivars that combine improved yield under stress with high yield potential can be obtained by screening breeding lines for both yield potential in favorable environments and yield under managed stress. This approach has been successful in improving drought tolerance in several other crop species, notably maize (Bänziger et al 2006), but has been little used in rice. Recent evidence also indicates that there are alleles with large additive effects on yield under stress in rice that can be mobilized via marker-assisted selection (Bernier et al 2007). However, because the nature and timing of drought stress differ greatly among production environments, drought breeding efforts, whether based on phenotypic selection alone or incorporating molecular methods, must be tailored to meet the needs of farmers in specific water-short regions, land types, and production systems. The main objectives of this paper are therefore

- To identify particular drought-prone target environments and management systems.
- To clarify the physiological and agronomic effects of water shortage in drought-prone target environments.
- To outline screening and breeding strategies that can be used to develop drought-tolerant cultivars.
- To assess progress in identifying QTL alleles with large effects on yield under water stress, and assess their potential contribution to the development of drought-tolerant rice cultivars through marker-aided selection (MAS).

Target environments for drought germplasm improvement

Four major hydrological environments for rice production can be defined in terms of toposequence position, or the relative elevation of a rice field within a watershed consisting of terraced fields that drain into each other. Within distances of several hundred meters, the toposequence may include

1. Unbunded uplands that never retain standing water.
2. Bunded but drought-prone upper fields that retain standing water only briefly after a rainfall or irrigation.
3. Well-drained mid-toposequence fields that receive a reliable supply of water from fields higher in the watershed, but that rarely experience stagnant flooding.
4. Poorly drained lower fields in which water accumulates to depths of 1 m or more during the rainy season.

All four of these hydrological environments are often found within a small area in rainfed ecosystems. The latter three may also often be found within a single irrigation command area. Water shortage is mainly observed in unbunded uplands and bunded upper-toposequence fields. Drought stress in these environments varies in severity across years due to variability in the amount and distribution of rainfall, but occurs with predictable frequency in a given field, based on its toposequence position and soil texture. Yield variability under stress can be great even within a single field because of its variability in soil texture and levelness. This micro-scale variability among and within fields results in very large estimates of genotype × environment and residual error in the analysis of rainfed rice trials, thus complicating selection (Cooper et al 1999).

Over time, rice farmers develop a deep understanding of the hydrological behavior of their fields, allowing them to target varieties and management techniques to specific fields. In unbunded fields at the top of a toposequence, farmers grow short-duration, drought-tolerant upland rice varieties established via direct seeding. Varieties used in these systems are usually tall, unimproved, and of the *aus* (in South Asia) or tropical japonica (in Southeast Asia and West Africa) varietal groups. In upper bunded fields, farmers tend to grow short-duration, photoperiod-insensitive modern varieties that flower before the withdrawal of the monsoon, escaping late-season drought stress. In well-drained mid-toposequence fields, farmers usually grow semidwarf varieties developed for irrigated systems because of their high yield potential, and usually establish their crops via transplanting. In lower and flood-prone fields, farmers usually direct-sow tall, photoperiod-sensitive varieties that flower as the rains cease and thus stagnant water begins to decrease (Mackill et al 1996). Individual farmers often have fields at several toposequence levels, and thus often grow several varieties, each adapted to a particular hydrological environment.

The principal target environments requiring germplasm with improved drought tolerance are unbunded uplands and bunded upper fields at the top of a toposequence; drought occasionally occurs in lower fields but is relatively rare because these fields benefit from runoff and seepage from upper fields and usually remain saturated long after upper fields are dry. Unbunded uplands are highly drought-prone, but make up a relatively small and decreasing part of the South Asian rice area. Bunded upper fields are the most important target environment for drought tolerance breeding, because of both their extent and their potential for improved productivity. Rainfed rice breeding programs need to develop varieties with the duration, plant type, and stress tolerances required for this environment, which covers millions of hectares in most rice-growing areas. Rice crops in such fields were originally established via variants of the direct-seeding system known in eastern India as *beusani, biasi,* or *beushening* (Singh et al 1994), wherein dry seed is broadcast on moist soil; the fields are then re-plowed after seedlings are established and standing water has accumulated in the field, uprooting both weeds and rice seedlings. The rice seedlings are then re-rooted by hand or by running a plank over the field. Although the beusani system is still widely used in some areas, notably in the Indian state of Chhattisgarh, transplanting has spread extensively throughout eastern India in rainfed upper-toposequence fields since the general adoption of high-yielding semidwarf varieties in the 1970s. This new establishment technique is very risky due to the frequent failure of these upper-toposequence fields to accumulate sufficient standing water for timely transplanting, and because of frequent occurrence of drought stress after transplanting. This is an example of a change in crop management that has rendered the production system more sensitive to drought. A major research effort is now required to develop a more stable production system for upper bunded fields.

Physiological and agronomic effects of drought, and implications for germplasm improvement

Direct effects of water shortage on growth and yield

Effects of drought stress at flowering. The direct effects of water shortage on growth and yield can be acute, occurring at critical crop stages, or they may result from continually recurring nonsaturated conditions that reduce biomass accumulation and tillering over many weeks. The acute effects of drought immediately before and during flowering (Atlin et al 2006, Ekanayake et al 1990, Garrity and O'Toole 1994) are severe, so tolerance at this stage is particularly critical. This is especially true in upland rice, where the lack of standing water makes the crop vulnerable to brief periods of drought around flowering, possibly leading to near-complete spikelet sterility. For this reason, much research on drought tolerance has focused on tolerance of stress at the flowering stage. Substantial genetic variation exists within *Oryza sativa* for the trait (Atlin et al 2006). Some varieties have a high degree of tolerance of short periods of stress around flowering, whereas others experience markedly reduced seed set and harvest index. A set of varieties was evaluated at IRRI under rainfed upland conditions in the wet seasons of 2004 and 2005. In both seasons, drought at flowering resulted in severe stress between panicle initiation and anthesis. For a subset of lines with similar days to flower under nonstress conditions, mean yield and harvest index are presented in Table 1. In this set, yields ranged from 0.7 to 2.3 t ha^{-1}. Nearly all of the variation in yield was explained by variation in harvest index; lines that are high-yielding under stress, such as IR71525-19-1-1 and CT 6510-24-1-2, were able to maintain a high amount of seed set under stress at flowering. The physiological basis for this differential tolerance is unknown. Root architecture and root depth vary greatly among upland rice cultivars (e.g., Price et al 1997, Venuprasad et al 2002), but some deep-rooted upland cultivars, such as the traditional Philippine tropical japonica cultivar Azucena, are highly susceptible to dry soil conditions at flowering (unpublished data). Similar susceptibility to acute stress around flowering is observed in lowland rice, although stress may take longer to develop in a lowland field. In a lowland rice experiment repeated over two seasons in the mapping population CT9993/IR62266, stress at the flowering stage reduced yield by an average of 80% relative to a nonstressed control in a set of approximately 100 recombinant inbred lines (RILs). In this experiment, the relationship between yield under stress and maintenance of HI was very high, with a genetic correlation of 0.94 The range in tolerance of lowland stress in this population was also great; the highest-yielding line produced a mean yield of 1.39 t ha^{-1} over two years, nearly three times the trial mean (Kumar et al 2007).

Effects of intermittent stress throughout the season. Much less attention has been paid to the effects of growth reduction due to intermittent soil drying throughout the season in upper fields than to the acute effects of water shortage around flowering, but the former likely causes similar or greater overall losses, particularly in bunded upper fields that are managed by farmers as lowland (i.e., puddled and transplanted), but that do not maintain standing water. In puddled fields, relatively few rice roots penetrate the hardpan of the puddled soil layer, and most roots occur within the top

Table 1. Mean yield and harvest index of rice cultivars exposed to severe reproductive-stage stress under upland conditions: IRRI, WS 2004.

Designation	Harvest index	Yield (t ha^{-1})
IR71525-19-1-1	0.22	2.3
CT6510-24-1-2	0.19	2.0
UPL RI 7	0.16	1.9
Apo	0.18	1.7
IR77298-12-7	0.17	1.2
IR71700-247-1-1-2	0.16	1.1
IR77298-14-1-2	0.12	0.9
PR26406-4-B-B-2	0.09	0.8
PSBRc 82	0.11	0.7
IR72875-94-3-3-2	0.11	0.7
$LSD_{0.05}$	0.06	0.7

15 cm or less (Pantuwan et al 1997, Samson and Wade 1998). Therefore, when puddled fields dry at the surface, rice roots cannot access water that is deeper in the soil profile, and stress may develop quickly. Rice yields in such fields are closely related to the number of days in the growing season in which soil is saturated (Boling et al 2004, Haefele et al 2004). The ability to maintain biomass accumulation and seed set in relatively dry soils, and to acquire water from deeper soil, is therefore a key feature required in drought-tolerant varieties. Intermittent soil drying substantially reduces biomass production and therefore total yield potential. IRRI research has shown that there is substantial genetic variation in the ability of upland and lowland rice cultivars to maintain biomass accumulation in unsaturated water conditions. For example, in a set of lowland cultivars evaluated at IRRI under intermittently drained conditions in the wet season of 2005, yields averaged 1.6 t ha^{-1}, a reduction of more than 50% relative to the fully irrigated control. In this trial, there was a range in total biomass among cultivars of 4.1 to 7.4 t ha^{-1}. Variation in biomass was more closely related to final grain yield than was harvest index in this trial (Table 2).

Screening cultivars for tolerance of acute stress at flowering versus intermittent stress

Because crop phenological stages differ in their sensitivity to drought, researchers have devoted considerable efforts to the development of screening techniques that permit genotypes of different growth durations to be evaluated in common experiments at equivalent levels of stress at key stages such as flowering. These include techniques such as line-source irrigation (Lanceras et al 2004), which subjects cultivars to a constant stress gradient throughout the season, and field designs that permit each genotype to be irrigated independently (Lafitte and Courtois 2002), allowing stress to be targeted to a specific phenological stage for each cultivar in the trial. However, these methods are not practical in a breeding program that must screen hundreds

Table 2. Cultivar differences in yield, harvest index, and biomass production in an intermittently-dried lowland field: IRRI, WS 2005.

Designation	Harvest index	Biomass ($t\ ha^{-1}$)	Yield ($t\ ha^{-1}$)
IR70213-10-CPA 4-2-2-2	0.28	7.6	2.1
IR79670-125-1-1-3	0.26	7.3	1.9
PSBRc 80	0.30	5.2	1.6
PSBRc 14	0.34	4.9	1.7
IR36	0.31	4.2	1.3
PSBRc 82	0.40	4.1	1.6
$LSD_{0.05}$	0.13	2.5	0.4

of lines. The IRRI breeding program screens for drought tolerance using protocols (described below) in which stress is repeatedly imposed on a large nursery or trial on a uniform date, shortly after transplanting in lowland rice and at around maximum tillering in upland rice, with cycles of stress and re-irrigation repeated until harvest. Variety means in screens of this type are highly correlated with means from trials in which stress is precisely applied at the sensitive flowering stage (IRRI, unpublished data). They are also at least as repeatable as means from nonstress trials (Venuprasad et al 2007, Bernier et al 2007).

Effects of drought on crop management and agronomic practices

Land preparation, transplanting, fertilizer application, and weed control in lowland rice production are all dependent on the presence of a standing water layer in the paddy. If standing water is not present, these operations may be delayed or omitted, resulting in large yield losses, even though plants may not have suffered physiological water stress. Losses from these management disruptions may be as great as those from direct drought damage. Cultivars differ in their sensitivity to these management disruptions. These differences can be exploited in the development of more resilient varieties for drought-prone environments.

Transplanting delay. Transplanting is the management step that is most vulnerable to water shortage. The optimum age of seedlings at transplanting is 2 to 4 weeks old, but rainfed farmers must often delay transplanting due to water shortage, and therefore plant seedlings that are much older than optimum. Farmers cannot transplant until sufficient water accumulates in fields to permit puddling (usually 400–500 mm of rainfall); often, this may not occur until seedlings are 60 to 80 days old. Such delays result in large yield losses because of reductions in both panicle number and weight. In experiments conducted at IRRI in 2005, transplanting 65-day-old as opposed to 22-day-old seedlings resulted in a yield reduction of more than 50%, averaged across 125 cultivars. Yield reductions due to delayed transplanting were experienced on this scale in large areas of eastern India in 2004, and in the Nepali *terai* and adjoining regions of Uttar Pradesh in 2006. Even high-rainfall regions that are not truly drought-prone, such as southern Cambodia, may experience severe losses due to delayed transplanting resulting from an early-season pause in the monsoon.

Weed management. Water shortage also affects weed management. Standing water in lowland fields after sowing or transplanting suppresses the germination of weeds. Under the nonflooded, aerobic conditions characteristic of upland or drought-affected lowland fields, weeds germinate freely. Most upland weed species grow more quickly than rice in nonsaturated soils, resulting in greater competition from weeds under drought conditions. The widespread indigenous eastern Indian rainfed lowland establishment and weed-management practice of *beushening* (described above) (Singh et al 1994) is also highly sensitive to early drought stress, as the uprooting and replanting process requires the presence of standing water in the field. Early drought therefore results in a failure of weed control in this system. Extensive genetic variation among rice cultivars with respect to weed competitiveness has been documented both for upland (Zhao et al 2006a) and lowland (Haefele et al 2004) systems, but little effort has been made to exploit this variation in the development of cultivars for water-short environments. Recently, however, Zhao et al (2006b) showed that weed-suppressive ability and weed competitiveness under upland conditions are strongly associated with rapid seedling growth in the first 4 weeks after sowing, a trait that is easily scored and for which substantial variability exists within and among the major rice germplasm groups (Zhao et al 2006c). Thus, selection for rapid early vegetative growth can be relatively easily incorporated into breeding programs that aim to develop cultivars for aerobic or direct-seeded systems.

Cultivar development for drought-prone environments

The analysis of drought hydrological environments and production systems described above indicates that reducing drought risk in rainfed production environments will require the development of two different types of germplasm:

- In both unbunded uplands and in the uppermost bunded fields, which almost never accumulate standing water, varieties are required that combine adaptation to dry direct seedling, the ability to maintain biomass production at a high rate in soils that are usually below field capacity, tolerance of severe drought stress at flowering, and high yield potential under favorable conditions. These varieties, which differ from traditional upland rice varieties in their input responsiveness and yield potential that allow them to achieve yields of 4–5 t ha^{-1} under favorable conditions, are often referred to as aerobic rice.
- On slightly lower fields, where transplanting is usually possible, but may be delayed because of water shortage, and where stress may occur at any point after establishment, varieties are needed that combine high yield potential with tolerance of delayed transplanting, and the ability to maintain biomass production and seed set in soils that are frequently unsaturated and usually below field capacity.

These two cultivar types are being developed by IRRI's aerobic rice and drought-prone lowland breeding programs. Overall, the most effective strategy for improving drought tolerance for these environments has proven to be *direct selection* for grain yield under water stress (Venuprasad et al 2006, 2007). Screening and breeding strategies are described below.

Developing cultivars with improved lowland drought tolerance for bunded upper terraces

The most drought-affected lowland fields are upper-toposequence bunded fields that are established by transplanting or traditional broadcasting methods. Critical traits for these fields include the ability to maintain biomass accumulation in intermittently dry fields, tolerance of severe stress at flowering, tolerance of delayed transplanting, and responsiveness to favorable conditions when they occur.

Screening for drought tolerance. Bunded (lowland) fields regularly affected by drought are usually upper-toposequence fields with light to medium soil texture. These fields are without standing water for much of the growing season, and may dry out repeatedly. Screening of cultivars targeted at this environment should mimic these intermittently dry conditions. Effective screening for lowland drought tolerance can be done even in the wet season in trials situated in upper light-textured fields that can easily be drained; in such wet-season screens, planting may be delayed to increase the possibility that the monsoon will withdraw before flowering, increasing the chances of imposing severe flowering-stage drought stress (e.g., Kumar et al 2007). Care should be taken to ensure that the field used is at the top of the toposequence, and that there is no higher field from which water flows into the drought-screening site. Because the objective of screening is to identify cultivars with improved yield under stress, such screening is conducted at IRRI in replicated trials in plots 5 m in length to achieve adequate precision. Seedlings are transplanted into puddled soil, and then the trial should be drained 7 days after transplanting. The field is allowed to dry until the soil cracks and/or the surface is completely dry. The field is not irrigated again until the local check variety is wilting severely, and the water table is at least 1 m below the surface. If tensiometers are installed, the field is irrigated when soil water tension = 40–70 kPa at a depth of 20 cm. When these conditions are achieved (the time needed for this to occur varies with soil texture, rainfall, and evapotranspiration), the field is then re-irrigated by flash flooding. One day after re-irrigation, the field is drained again. The cycle of stress followed by re-irrigation and drainage is repeated until the field is finally drained for harvest. Drought tolerance is expressed simply as the yield produced by a cultivar under stress.

Screening under this type of managed stress has identified large differences among lowland breeding lines and mega-varieties in yield under stress at IRRI (Table 3). Several lines (e.g., the IR64-derived line IR77298-14-1-2 and the hybrid IR80228H) have been identified that are comparable in yield potential with current elite irrigated varieties under nonstress conditions, but that outyield them substantially under drought stress. Screening for grain yield under drought stress has now been incorporated as a routine cultivar evaluation step by IRRI and by several Indian

Table 3. Days to flower, harvest index, and yield of medium-duration varieties and breeding lines under severe intermittent lowland drought stress and full irrigation: IRRI, 2006 dry season.

Designation	Days to flowering		Harvest index		Yield (t ha^{-1})	
	Stress	Nonstress	Stress	Nonstress	Stress	Nonstress
IR77298-14-1-2	94	85	0.21	0.40	1.2	3.3
IR80461-B-7-1	95	84	0.22	0.37	1.1	3.7
IR80228 H	101	85	0.27	0.46	0.9	5.8
PSBRc 82	104	91	0.10	0.36	0.3	2.6
Trial mean	100	88	0.10	0.34	0.4	2.2
$LSD_{0.05}$	8	2	0.10	0.16	0.4	0.9

breeding programs in collaboration with the IRRI-India Drought Breeding Network, a collaborative network serving drought-prone rainfed environments. In 2005, this network tested a number of breeding lines developed at IRRI as well as at different national research institutes in India for their performance under drought. These lines were screened in alpha lattice designs with three replications under fully irrigated conditions and two levels of stress. In one stress level, fields were drained just after transplanting, water from rains was never allowed to stand, and the trial was never irrigated. These experiments generally experienced severe stress resulting in at least a 70% reduction in mean yield as compared with control yields. In the second stress level, fields were drained 35–40 days after transplanting with the aim of screening the lines for tolerance of reproductive-stage stress. The mean yield reduction in these moderately-stressed experiments ranged from 30% to 60%. Screening under severe drought, moderate drought, and flooded control at Raipur identified breeding lines of 100–120 days' duration that had yield potential of 4.0–5.2 t ha^{-1} under nonstress conditions and produced grain yields of 1.7–2.1 t ha^{-1} under severe drought stress (Table 4). In the group of 120–140 days' duration, breeding lines with yield potential of 6.3 t ha^{-1} under flooded control and yields of up to 1.9 t ha^{-1} under severe drought stress were identified (Table 5). The screening also showed that the widely grown rainfed variety Swarna was moderately tolerant of lowland drought, whereas the related variety Sambha Mahsuri was extremely susceptible to drought stress.

Screening for tolerance of delayed transplanting. Delayed transplanting due to drought is probably the main cause of yield loss in most rainfed lowand systems, but tolerance of delayed transplanting has rarely been systematically evaluated or incorporated as a rice breeding objective. Variability for tolerance of delayed transplanting appears to be large, even in photoperiod-insensitive germplasm. In an evaluation in the 2005 wet season of 125 photoperiod-insensitive varieties with medium duration transplanted when seedlings were 65 days old, cultivar mean yields ranged from 0.3 to 3.3 t ha^{-1}. Some elite breeding lines and cultivars yielded well when normally

Table 4. Grain yield, days to flower, and harvest index of medium-duration (120–140 days) varieties and breeding lines under three levels of water stress: IRRI-India Drought Breeding Network, 2005 wet season. Mean of cultivars over 7, 3, and 1 trial in southern and eastern India for nonstressed, moderately stressed, and severely stressed trials, respectively.

Designation	Days to flowering Stress level			Harvest index Stress level			Yield (t ha^{-1}) Stress level		
	None	Moderate	Severe	None	Moderate	Severe	None	Moderate	Severe
Tolerant lines and varieties									
Baranideep	82	87	87	0.42	0.40	0.38	5.5	3.9	1.4
CB00-15-24	81	83	82	0.40	0.40	0.36	5.0	3.1	1.4
IR74371-3-1-1	83	83	88	0.41	0.42	0.34	5.0	3.9	1,2
Widely grown varieties									
MTU 1010	86	92	91	0.28	0.21	0.13	2.9	1.9	0.6
IR64	87	90	90	0.41	0.35	0.17	5.2	2.9	0.5
IR36	85	97	94	0.41	0.27	0.04	4.2	2.0	0.1
Trial mean	84	89	91	0.38	0.32	0.22	4.6	2.8	0.8
$LSD_{0.05}$	4	5	3	0.05	0.08	0.11	0.8	0.9	0.4

Table 5. Grain yield, harvest index, and days to flowering of 120–140 days' duration entries from the IRRI-India Drought Breeding Network: Raipur, WS 2005.

Designation	Harvest index			Days to flowering			Yield (t ha^{-1})		
	Control	Moderate stress	Severe stress	Control	Moderate stress	Severe stress	Control	Moderate stress	Severe stress
ARB 6	0.37	0.43	0.4	79	78	81	6.7	4.3	1.9
IRMBP-2	0.38	0.32	0.35	82	84	85	6.1	3.2	1.3
Mahamaya	0.34	0.19	0.14	92	93	96	6.5	1.9	0.6
PSBRc-9	0.42	0.42	0.37	90	89	91	5.8	4.3	1.6
Sambha Mahsuri	0.41	0.09	0.02	103	111	–	6.7	0.8	0.0
Swarna	0.38	0.25	0.34	103	110	126	6.0	2.1	1.3
Swarna/IR42253-54	0.42	0.33	0.38	83	85	80	6.4	2.8	1.7
$LSD_{0.05}$	0.04	0.06	0.07	1	1	5	0.7	0.6	0.4

Table 6. Agronomic performance of 10 hybrids versus 115 pure lines when transplanted at 22 or 65 days after sowing: IRRI, 2005 wet season.

Cultivar type	Days to flowering		Height (cm)		Harvest index		Yield (t ha^{-1})	
	Seedling age at transplanting (d)							
	22	65	22	65	22	65	22	65
Hybrid	85	114	115	90	0.41	0.38	5.0	2.7
Inbred	82	113	119	92	0.37	0.28	3.4	1.5
Pr > F.	ns[a]	ns	ns	ns	0.0012	<0.0001	<0.0001	<0.0001

[a]ns = nonsignificant.

transplanted, but poorly when delay-transplanted. A notable example is IR77298-14-1-2, a tungro-resistant derivative of IR64, which yielded 4.0 t ha^{-1} under normal transplanting, but only 1.8 t ha^{-1} under delayed transplanting. In contrast, a hybrid, IR80642H, yielded 4.4 t ha^{-1} under normal transplanting and 3.3 t ha^{-1} under delayed transplanting. In general, hybrids were found to be more tolerant of delayed transplanting than were inbreds (Table 6).

Developing cultivars with improved drought tolerance for unbunded uplands

Upland rice is grown as a subsistence crop in unbunded upper fields by some of the poorest farmers in Asia. Upland rice growers use few improved varieties and, because of risk of crop loss due to drought or weed pressure, apply only small amounts of fertilizer to their fields. Recently, studies in traditional upland rice-growing areas of Yunnan (Atlin et al 2006) and Laos (Saito et al 2006) demonstrated that improved upland rice varieties have at least 50% higher yield potential than traditional cultivars, and can serve as the basis for more productive and sustainable upland rice-based cropping systems. However, since upland systems are almost exclusively rainfed, adoption of such systems will depend on the development of varieties that combine high yield potential with high levels of drought tolerance and weed competitiveness.

Screening for tolerance of upland drought stress. Strategies for drought-tolerance screening under upland conditions are similar to those described above for lowland management. Most upland varieties are photoperiod-insensitive, so, if temperatures permit, dry-season screening is the preferred option for reliably imposing stress. Many upland varieties have a moderate degree of vegetative drought tolerance, but are often highly susceptible to stress around flowering. For this reason, screening protocols should emphasize tolerance of stress at flowering. At IRRI, drought screening is conducted in replicated yield trials of fixed lines that have been previously selected for yield potential and disease resistance. Screening is conducted in an unbunded well-drained field at the top of a toposequence. No irrigated or flooded trials are planted above the drought-screening site, and lines are screened in trials with at least two replicates. Trials are direct-sown into dry soil. The field is irrigated to maintain soil water potential near field capacity until canopy closure, or for about 50 days after seeding (DAS), and the frequency of irrigation is then reduced until harvest. Irrigation is withheld until the soil surface is completely dry, susceptible check varieties are severely wilted, and soil water tension reaches 50 to –70 kPa at a depth of 30 cm. When the target level of soil dryness and plant stress is reached, the field is liberally irrigated to saturate the root zone. Per irrigation, this requires around 40–60 mm of water.

There is evidence that differences in drought tolerance measured in this screen are predictive of differences observed under natural stress in the target population of environments. For example, 30 varieties were screened under severe upland stress artificially imposed at IRRI in the dry season (DS) of 2005. These same varieties were screened under rainfed upland conditions at IRRI in the wet season (WS) of 2004 and WS of 2005. In both of these years, severe drought stress occurred at flowering during the wet season. The mean correlation between variety means for grain yield in the dry-season stress screen and under natural stress in the wet season was 0.87, indicating that the ability of the artificial drought screen to predict performance under natural stress was high (IRRI, unpublished data).

Selection of breeding lines under artificial stress has been shown to result in gains under natural stress in wet seasons. Venuprasad et al (2007) screened several hundred lines from the crosses Apo/IR64 and Vandana/IR64 in the DS of 2003. The lines were evaluated for grain yield under both severe upland stress and irrigated control conditions. Selected lines from both the stress and the irrigated control screens were then evaluated under natural stress at IRRI in the WS of 2004 and 2005. Yield gains under natural stress were greater in the subset of lines selected under artificial stress than under fully irrigated conditions. Selection under stress gave no gains under nonstress conditions nor did it reduce yield potential.

Screening for weed competitiveness. Upland rice cultivars that compete well against weeds are often thought to be tall, rapid in early growth, and have droopy leaves and high specific leaf area. These traits have been linked to low yield potential in some studies (Jennings and Aquino 1968, Kawano et al 1974), but not in others (Garrity et al 1992, Ni et al 2000, Fischer et al 2001). More recently, Zhao et al (2006b) have shown that differences in cultivar weed competitiveness in direct-sown rice are largely determined by differences in the rate of seedling biomass accumulation in the first 4 weeks after sowing. They observed that, averaged over 3 years, there was a twofold difference between the most and least competitive cultivars in weed biomass at 9 weeks in plots that were hand-weeded once at 3 weeks after sowing, and that there was no trade-off between yield potential and weed competitiveness. Improved weed competitiveness can be selected for in replicated trials by visually rating advanced breeding lines for total biomass at 4 weeks after sowing (Zhao et al 2006b). Screening for seedling biomass accumulation has been incorporated as a routine screening step in the IRRI rainfed and aerobic rice breeding programs. Cultivars with high seedling biomass accumulation tend to be erect, moderately drought-tolerant, and derived from the indica and *aus* germplasm groups.

Direct seeding to reduce drought risk in drought-prone upper fields

As noted above, rice establishment either by transplanting or the traditional beushening/biasi practice in bunded upper fields frequently leads to heavy crop yield loss because of delayed transplanting, exposure of the transplanted seedlings to early drought, or heavy weed pressure. In crops where establishment has been delayed due to lack of standing water in fields, the risk of drought occurring during the reproductive stage or grain filling is also increased. Direct seeding of unsprouted seed in dry soil, with herbicide-based weed control, may be a useful alternative to transplanting or beushening in areas where early-season drought is frequent.

Direct seeding can be undertaken in dry or moist soil starting with the earliest rains, and therefore allows establishment to take place 4 to 6 weeks earlier than is possible in puddled transplanted systems. Early establishment reduces drought risk during flowering and grain filling associated with early withdrawal of the monsoon, and, because direct-sown crops mature approximately 10–14 days earlier than transplanted crops seeded on the same date, increases the probability of successfully establishing a postrice rainfed crop. Direct-seeded establishment also eliminates the risk associated with delayed transplanting, which occurs when rainfall is insufficient for main-field puddling by the time seedlings are ready to be removed from the nursery bed; planting overaged seedlings due to early-season drought is a major cause of yield reduction in light soils and upper rainfed terraces.

Cultivars differ substantially in their adaptation to dry direct-seeded establishment in nonsaturated soils. Component traits include weed competitiveness, seedling vigor, ability to maintain biomass development in intermittently dry fields, and tolerance of late-season drought. The development of adapted cultivars with these traits is therefore an important element in the design of successful direct-seeding establishment systems in rainfed upland and shallow lowland systems. Such cultivars are often referred to as *aerobic rice*, and are also potentially useful in irrigated rice systems where water availability is limited (Bouman et al 2006).

A new generation of aerobic-adapted varieties for direct-seeded systems has been identified with yield potential of 4–5 t ha^{-1} but that produce yields of more than 1 t ha^{-1} when subjected to severe intermittent stress bracketing the entire reproductive period. The yield potential of these materials is not greater than that of current elite aerobic adapted variety Apo or the lowland variety PSBRc 80, but yields under moderate drought stress are three- to fourfold higher (Table 7).

Designing cultivar development programs that can combine drought tolerance with yield potential

For the drought-prone target environments described above, breeding programs must combine selection under stressful conditions with selection for yield potential because farmers want cultivars that are both drought-tolerant and have high yield potential in favorable years. It can be useful to think of the breeder's task as raising both the "ceiling" of yield potential that can be achieved in favorable years and the "floor" yield that can be protected under drought conditions.

Table 7. Grain yield of elite aerobic-adapted varieties and a lowland-adapted check (PSBRc 80) evaluated under aerobic management with severe intermittent stress applied following maximum tillering: IRRI, DS 2005.

Designation	Nonstress yield (t ha^{-1})	Stress yield (t ha^{-1})	Days to flowering (nonstress)
IR78875-190-B-1-3	4.6	0.8	81
IR71525-19-1-1	4.2	1.4	85
IR78875-131-B-1-3	4.1	1.0	85
IR78875-131-B-1-2	4.0	1.0	79
IR74371-54-1-1	4.0	1.1	75
Apo	3.4	0.2	80
PSBRc 80	1.0	0.1	78
$LSD_{0.05}$	1.1	0.3	

Substantial evidence indicates that these goals are not mutually exclusive. Most studies in which large populations of unselected lines have been screened under both stress and nonstress conditions show that there is a moderate to large positive correlation between yield under drought stress and yield potential under favorable conditions. Atlin et al (2004) surveyed 10 experiments in which populations of random recombinant inbred or doubled haploid lines were evaluated under both water stress and control, with a mean reduction of 65% due to water stress. They reported genetic correlations for yield across stress levels that ranged from 0.35 to 0.91, averaging 0.67.

Even when reductions due to stress are extreme, genetic correlations for yield across stress levels tend to be positive and often quite high. Kumar et al (2007), in an experiment involving a population of doubled-haploid lines from the cross CT9993-5-10-1-M/IR62266-42-6-2 evaluated over 2 years under lowland conditions in a stress regime that reduced yield by 80% relative to a well-watered control, observed a genetic correlation of 0.8 across stress levels for yield, indicating that two-thirds of the genetic variation for yield under stress involved factors that also affected yield potential. Venuprasad et al (2007) evaluated five large populations under upland drought stress and lowland nonstress conditions and found that, on average, stress reduced yield by more than 64% but still the genetic correlation between yields in stress and nonstress was 0.48. In an upland experiment involving the Vandana/Way Rarem population, in which mean yield reduction due to water stress over 2 years was 88%, the genetic correlation for yield across stress levels was 0.44 (Bernier et al 2007). In general, even under extremely stressful conditions, perhaps 30–50% of the genetic variance for yield under drought in random mapping populations is due to factors that also affect yield potential, such as partitioning of biomass to grain. The remaining 50–80% of genetic variation for yield under severe stress is due to factors that affect only drought tolerance rather than yield potential. To ensure that these factors are screened for during the selection process, it is important that the yield of managed drought stress trials be reduced by at least 50% relative to nonstress controls (Venuprasad et al 2007, Pantuwan et al 2002).

The moderate positive correlation between yield under optimal conditions and yield under severe drought stress in mapping populations, which are sets of unselected lines, should not be taken as evidence that selection for yield under stress is unnecessary in the development of drought-tolerant cultivars. On the contrary, most elite cultivars developed for irrigated or favorable rainfed systems have very poor drought tolerance (e.g., Table 4). The moderate positive correlation is evidence, however, that it is feasible to produce cultivars combining high yield potential and improved drought tolerance.

How should a breeding program that aims to produce such cultivars be organized? The key feature of a successful drought breeding program is the incorporation of a managed-stress screening step early in the selection process, preferably at the initial replicated testing stage (because the heritability of yield under stress, like yield under well-watered conditions, is relatively low, only replicated screening should be used as a basis for selection; selection for yield under stress or nonstress conditions in unreplicated nurseries or on a single-plant basis is likely to be ineffective). In a well-conducted managed-stress drought screen, the repeatability of genotype yield estimates is usually similar to or only slightly less than in well-watered trials (at IRRI, trials conducted under severe drought stress often have higher repeatability than well-watered trials). Therefore, selection of lines for advancement can be based on means over stress and nonstress trials. Plot yield measurements within stress levels should be standardized (i.e., divided by their within-trial standard deviation) before analysis for such selection, so that means from nonstress trials, which may be three- to fivefold higher than means in stress trials, do not overwhelm the information from the stress trials; selection on the basis of raw means over stress levels would be heavily weighted in favor of performance under nonstress conditions. Selection on the basis of mean performance over stress levels may not, however, be appropriate in situations where selection for yield potential is done in the wet season and selection for stress tolerance is conducted only in the dry season. In this case, it may be appropriate to screen first for yield potential in the wet season, subjecting only those lines with high yield potential to drought tolerance screening in the dry season. Choice of parents is a critical step in designing crosses for drought tolerance breeding. To maximize the prospects for selecting progeny combining high yield potential with improved drought tolerance, at least one parent in the cross should be known to be drought-tolerant or to produce drought-tolerant offspring. Relatively little information is available on such potential donors. Experience at IRRI has shown that donors of upland drought tolerance are not necessarily useful donors for lowland drought tolerance; donors conferring a form of tolerance appropriate to the target environment should be used. A list of such donors, as well as some highly susceptible check varieties, is presented in Table 8.

The use of drought-tolerant parents and application of the screening methods described above have resulted in the development or identification of lines combining improved stress tolerance with high yield potential at IRRI. Table 9 presents partial results from IRRI's 2006 dry-season trials of advanced rainfed lowland breeding lines

Table 8. Drought tolerance donors and susceptible checks identified through testing at IRRI.

Genotype	Adaptation	Drought tolerance level	Yield potential	Notes
IR71525-19-1-1	Upland	Highly tolerant	Low	Improved japonica type with high vegetative- and reproductive-stage drought tolerance, medium duration
IR55419-04	Upland	Moderately tolerant	Moderate	Improved indica type with excellent upland adaptation and early vigor, medium duration
IR55423-01 (also PSBRc 9, Apo)	Upland and lowland	Moderately tolerant under upland and lowland conditions	High	High yield potential under both favorable upland and lowland conditions, long duration
IR74371-46-1-1	Upland	Moderately tolerant	Moderate	Aerobic-adapted rice with moderate drought tolerance, medium duration
Vandana	Upland	Highly tolerant	Low	Improved eastern Indian upland rice derived from an aus/japonica cross, short duration
IR71524-44-1-1	Upland	Highly tolerant	Low	Improved japonica type with high vegetative- and reproductive-stage drought tolerance, medium duration
UPLRI-5	Upland	Highly susceptible	Moderate	Highly susceptible upland variety
IR77298-14-1-2	Lowland	Moderately tolerant	Moderate	IR64 derivative, tungro-resistant in addition to drought-tolerant
IR81047-B-106-4-3	Lowland	Moderately tolerant	High	Medium-duration indica line combining moderate drought tolerance with high yield potential
IR77843H	Lowland	Moderately tolerant	High	Drought-tolerant hybrid
IR80228H	Lowland	Moderately tolerant	High	Drought-tolerant hybrid
IR36	Lowland	Highly susceptible	Moderate	Widely grown short-duration lowland variety, highly susceptible
IR64	Lowland	Highly susceptible	Moderate	Widely grown short-duration lowland variety, highly susceptible

Table 9. Yield under lowland drought stress and fully irrigated conditions of medium-duration lines selected either under nonstress conditions only or under both stress and nonstress conditions: IRRI, dry season 2006.

Designation	Selection history	Stress yield	Nonstress yield
		(t ha^{-1})	
IR80461-B-79-3	Stress and nonstress	1.1	5.0
IR72	Nonstress only	0.7	3.4
PSBRc 82	Nonstress only	0.3	3.3
IR80461-B-7-1	Stress and nonstress	1.3	4.5
$LSD_{0.05}$		0.4	1.2

in the medium-duration group. Some lines that had been selected under both stress and nonstress conditions significantly outyielded, under both stress and nonstress conditions, elite irrigated varieties selected only under optimal conditions.

Hybrid rice varieties: an option for drought-prone lowland fields

Hybrid varieties appear to offer a route to combining improved tolerance of drought stress with high yield potential, particularly in drought-prone lowland fields or fields where transplanting is often delayed. In replicated field experiments conducted at IRRI during the dry seasons of 2003 through 2006, seven hybrids not previously selected for drought tolerance were compared with elite pure lines (also not selected for drought tolerance) from the IRRI irrigated (n = 31) and aerobic (n = 4) breeding programs under (1) full irrigation; (2) a nonstress alternate wetting-and-drying irrigation protocol, with the water table maintained within 15 cm of the soil surface; and (3) the intermittent lowland drought stress protocol described above. Mean yields of the three treatments over two years were 6.3, 5.6, and 1.8 t ha^{-1}, a 71% yield reduction for the stress protocol relative to full irrigation (Table 10). The hybrids outyielded pure lines from both the irrigated and aerobic breeding programs under all three irrigation regimes; under the stress protocol, the mean yield advantage of the hybrids relative to the pure lines selected under similar irrigated management was 1.2 t ha^{-1}. The advantage of hybrids is both proportionately and absolutely greater under moderate stress than under fully irrigated conditions. The tolerance of hybrids of moderate water stress and, as noted earlier, delayed transplanting, combined with their high yield potential in favorable environments, may have contributed to their rapid adoption in eastern India, where they have been introduced by the commercial seed sector over the past five years. Particularly in the drought-prone shallow lowland areas of the poorest states in the region, including Jarkhand, Bihar, Uttar Pradesh, and Chhattisgarh, smallholders have been eager to replace short-duration but drought-susceptible varieties such as IR64 and IR36 with hybrids.

Table 10. Agronomic trait means under full irrigation, alternate wetting and drying (AWD), and severe water stress of hybrids, pure lines selected under full irrigation, and pure lines selected under upland management: IRRI, 2003-04.

Variety type	Grain yield (kg ha^{-1})			Days to flower			Height (cm)		
	Full irrigation	AWD	Severe stress	Full irrigation	AWD	Severe stress	Full irrigation	AWD	Severe stress
Hybrids	7,321*	6,348*	2,753**	84**	85**	85*	97	86	76
Lowland lines	6,185	5,527	1,514	91	92	96	104	96	81
Upland lines	5,751	5,043	2,356**	82**	83**	82**	106	96	86
Mean	6,330	5,616	1,794	89	90	93	103	94	81

*,** Single df contrast with mean of lowland lines significant at P = 0.05 and P = 0.01, respectively.

Prospects for marker-aided selection for drought tolerance in rice

A relatively few improved varieties, including Swarna, Samba Mahsuri, IR36, IR64, BR11, and MTU 1010, sometimes referred to as "mega-varieties" (Mackill 2006), together now account for much of South Asian rainfed rice production. Most of these varieties are valued for their quality, marketability, and yield potential under favorable conditions. Extensive multienvironment testing by the IRRI-India Drought Breeding Network has shown that most of these important varieties are highly susceptible to even moderate drought stress (e.g., Tables 4 and 5). However, these rainfed mega-varieties will be very difficult to be replaced by more drought-tolerant genotypes unless they are matched in terms of quality and agronomic performance in favorable years.

Prospects for the adoption of drought-tolerant varieties will be improved if yield under stress can be enhanced through the development of mega-varieties introgressed with a small number of genes for drought tolerance via marker-assisted selection (MAS), leaving the rest of the desirable recurrent-parent genotype largely intact, a strategy that has been highly successful in rice for abiotic stresses such as submergence (Xu et al 2006). Until recently, however, the possibility of finding this type of gene in rice appeared to be slight. Genetic analyses of traits related to drought tolerance in rice conducted to date have usually reported the detection of many QTLs with relatively small effects (e.g., Babu et al 2003, Lanceras et al 2004), leading most researchers working in the field to conclude that grain yield under drought stress in rice is a highly complex trait affected by many loci with small effects, making progress from MAS unlikely. Efforts to introgress chromosomal regions with small or moderate effects on secondary root traits thought to be related to drought tolerance have not succeeded in significantly improving yield under stress (Shen et al 2001, Steele et al 2006), further increasing skepticism about the potential for MAS-based approaches. However, the bulk of the data on which these conclusions were based were derived from only two mapping populations, Azucena/IR64 and CT9993-5-10-1-M/IR62266-42-6-2. These experiments also attempted to introgress large chromosome segments carrying a putative QTL, rather than a small fine-mapped region, and therefore are likely to have been affected by linkage drag. Use of fine-mapped targets could be more successful.

Recently, IRRI initiated a broader survey to systematically identify genes or oligogenic combinations with large effects on yield under drought stress, both in donors known to have high yield under stress and in random donors. Evidence is now accumulating that a relatively small number of genes can have a large effect. Lafitte et al (2006) reported that, in backcross populations derived from the susceptible recurrent parent IR64 crossed to unselected donors and mass-selected in the BC_2F_2 for yield under stress, lines were selected that significantly outyielded IR64 under moderate stress. In another study, BC_3-derived sister lines from the cross IR77298, developed with IR64 as a recurrent parent and the tungro-resistant pure-line variety Aday Selection as a donor, were shown to differ substantially in yield under severe lowland stress, despite sharing a coefficient of co-parentage of more than 0.9 (Venuprasad et al 2007, IRRI, unpublished data). In an upland rice population derived from the cross Vandana/Way Rarem, a single QTL accounted for more than 50% of genetic variation for yield under severe upland stress over 2 years, but had no effect under nonstress conditions (Bernier et al 2007). The allele conferring improved tolerance more than doubles the mean yield of homozygotes under stress (from approximately 0.2 to 0.6 t ha^{-1}). In lowland screening of a population derived from the cross CT9993-5-10-1-M/IR62266-42-6-2, a single QTL located near the *sd-1* locus on chromosome 1 accounted for more than 30% of genetic variation for yield under severe reproductive-stage stress, but only 4% under nonstress conditions (Kumar et al 2007). It should be recalled from the discussion in the section "Defining cultivar development programs" that approximately half of the genetic variance for yield under stress is due to variation in yield potential expressed under both stress and nonstress conditions. Thus, single genes explaining 30–50% of the genetic variance for yield under stress are actually accounting for the bulk of yield variation under stress that is not associated with variation in yield potential. In many populations, it therefore seems that genetic variation for yield under severe stress is under oligogenic rather than polygenic control. If confirmed, these results would indicate that large improvements in the performance of mega-varieties under drought stress may result from the marker-assisted introgression of a small number of genes.

Conclusions

Drought is a severe risk for rice producers who farm upper terraces with light soils, under both upland and lowland management, affecting productivity in both stress and favorable years, due to direct yield losses and underinvestment in inputs, respectively. Many widely grown varieties in rainfed rice-producing areas are highly susceptible to drought. More drought-tolerant cultivars are needed to replace them, but they are unlikely to be adopted if the quality and yield potential of current varieties are not maintained.

Several adaptations are required to increase productivity and reduce risk of crop loss due to drought on these lands, which comprise perhaps 20% of the rice area of Asia. These adaptations include increased ability to maintain vegetative biomass growth in intermittently dry soils, increased weed competitiveness, tolerance of delayed transplanting, and tolerance of severe drought stress at flowering. Specific adaptation to dry direct seeding in nonpuddled soils is also required for dry direct-seeding systems, which could allow farmers in drought-prone upper fields to establish their crops earlier to reduce the risk of drought during the critical flowering and grain-filling periods. The substantial genetic variation for all these traits can be exploited by rice breeding programs by incorporating screens for yield under appropriate timings and levels of drought stress for the target environment, and by incorporating screens for traits that are relevant to drought-prone rice systems, such as tolerance of delayed transplanting or dry direct seeding. High-yielding cultivars tolerant of lowland drought stress and delayed transplanting have been developed. Hybrids are particularly promising for drought-prone lowland fields because of their tolerance of moderate drying during vegetative growth and of delayed transplanting. Drought-tolerant cultivars adapted to direct seeding under nonpuddled aerobic conditions that combine yield potential of more than 5.0 t ha^{-1}, yields of at least 1.5 t ha^{-1} under severe stress, and a high amount of weed competitiveness have been developed by IRRI and collaborators. These cultivars are ready for evaluation as the basis for intensified management systems for drought-prone rainfed lowland rice environments. Further improvements in drought tolerance, particularly for lowland conditions, await better characterization of potential donors. Currently, IRRI is screening a large sample of lines from its core germplasm collection for tolerance of lowland stress; promising donors are being identified and will be made available to rice breeders.

In the future, marker-assisted backcrossing holds considerable promise for improving the drought tolerance of Asian rainfed mega-varieties. Genetic control over variation for yield under severe stress appears to be oligogenic, rather than polygenic, in many crosses. QTLs with large effects on yield under stress have been identified in several populations, and may not be infrequent in the rice germplasm. A major effort should be mounted to identify and characterize such genes for use in crop improvement.

References

Atlin GN, Lafitte HR, Tao D, Laza M, Amante M, Courtois B. 2006. Developing rice cultivars for high-fertility upland systems in the Asian tropics. Field Crops Res. 97:43-52.

Atlin GN, Lafitte HR, Venuprasad R, Kumar R. 2004. Heritability of rice yield under reproductive-stage drought stress, correlations across stress levels, and effects of selection: implications for drought tolerance breeding. In: Resilient crops for water-limited environments. Abstracts of an international workshop, 24-28 May 2004, Cuernavaca, Mexico.

Bänziger M, Setimela PS, Hodson D, Vivek B. 2006. Breeding for improved abiotic stress tolerance in maize adapted to southern Africa. Agric. Water Manage. 80:212-224.

Babu RC, Nguyen BD, Chamarerk V, Shanmugasundaram P, Chezhian P, Jeyaprakash P, Ganesh SK, Palchamy A, Sadasivam S, Sarkarung S, Wade LJ, Nguyen HT. 2003. Genetic analysis of drought resistance in rice by molecular markers: association between secondary traits and field performance. Crop Sci. 43:1457-1469.

Bernier J, Kumar A, Venuprasad R, Spaner D, Atlin G. 2007. A large-effect QTL for grain yield under reproductive-stage drought stress in upland rice. Crop Sci. 47:505-517.

Boling A.,Tuong TP, Jatmiko SY, Burac MA. 2004. Yield constraints of rainfed lowland rice in Central Java, Indonesia. Field Crops Res. 90:351-360.

Bouman BAM, Yang XG, Wang HQ, Wang ZM, Zhao JF, Chen B. 2006. Performance of aerobic rice varieties under irrigated conditions in North China. Field Crops Res. 97:53-65.

Cooper M, Rajatasereekul S, Immark S, Fukai S, Basnayake J. 1999. Rainfed lowland rice breeding strategies for northeast Thailand. I. Genotypic variation and genotype × environment interactions for grain yield. Field Crops Res. 64:131-151.

Ekanayake IJ, Steponkus PL, De Datta SK. 1990. Sensitivity of pollination to water deficits at anthesis in upland rice. Crop Sci. 30:310-315.

Fischer AJ, Ramirez HV, Gibson KD, Pinheiro BDS. 2001. Competitiveness of semi-dwarf upland rice cultivars against palisadegrass (*Brachiaria brizantha*) and signalgrass (*B. decumbens*). Agron. J. 93:967-973.

Fukai S, Pantuwan G, Jongdee B, Cooper M. 1999. Screening for drought resistance in rainfed lowland rice. Field Crops Res. 64:61-74.

Garrity DP, Movillon M, Moody K. 1992. Different weed suppression ability in upland rice cultivars. Agron. J. 84:586-591.

Garrity DP, O'Toole JC. 1994. Screening for drought resistance at the reproductive phase. Field Crops Res. 39:99-110.

Haefele SM, Johnson DM, Bodj, DM, Wopereis, MCS, Miezan KM. 2004. Field screening of diverse rice genotypes for weed competitiveness in irrigated lowland ecosystems. Field Crops Res. 88:39-56.

Huke RE, Huke EH. 1997. Rice area by type of culture: South, Southeast, and East Asia. Los Baños (Philippines): International Rice Research Institute.

Jennings PR, Aguino RC. 1968. Studies on competition in rice. III. The mechanism of competition among phenotypes. Evolution 22:529-542.

Kawano K, Gonalez H, Lucena M. 1974. Intraspecific competition with weeds, and spacing response in rice. Crop Sci. 14:841-845.

Kumar R, Venuprasad R, Atlin GN, 2007. Genetic analysis of rainfed lowland rice drought tolerance under naturally-occurring stress in eastern India: heritability and QTL effects. Field Crops Res. 103:42-52.

Lafitte HR, Courtois B. 2002. Interpreting cultivar × environment interactions for yield in upland rice: assigning value to drought-adaptive traits. Crop Sci. 42:1409-1420.

Lafitte HR, Li ZK, Vijayakumar CHM, Gao YM, Shi Y, Xu JL, Fu BY, Yu SB, Ali AJ, Domingo J, Maghirang R, Torres R, Mackill DJ. 2006. Improvements of rice drought tolerance through backcross breeding: evaluation of donors and selection in drought nurseries. Field Crops Res. 97:77-86.

Lanceras JC, Pantuwan GP, Jongdee B, Toojinda T. 2004. Quantitative trait loci associated with drought tolerance at reproductive stage in rice. Plant Physiol. 135:384-399.

Mackill DJ, Coffman WR, Garrity DP. 1996. Rainfed lowland rice improvement. Manila (Philippines): International Rice Research Institute. p 45-46.

Mackill DJ. 2006. Breeding for tolerance to abiotic stresses in rice: the value of quantitative trait loci. In: Lamkey KR, Lee M, editors. Plant breeding: The Arnel R. Hallauer International Symposium. Ames, Iowa (USA): Blackwell Pub. p 201-212

Maclean JL, Dawe DC, Hardy B, Hettel GP, editors. 2002. Rice almanac. Los Baños (Philippines): International Rice Research Institute, Bouaké (Côte d'Ivoire): West Africa Rice Development Association, Cali (Colombia): International Center for Tropical Agriculture, Rome (Italy): Food and Agriculture Organization. 253 p.

Ni H, Moody K, Robles RP, Paller EC, Lales JS. 2000. *Oryza sativa* plant traits conferring competitive ability against weeds. Weed Sci. 48:200-204.

Pandey S, Bhandari H, Sharan R, Naik D, Taunk SK, Sastri ASRAS. 2005. Economic costs of drought and rainfed rice farmers' coping mechanisms in eastern India. Final project report. Los Baños (Philippines): International Rice Research Institute.

Pantuwan G, Fukai S, Cooper M, O'Toole JC, Sarkarung S. 1997. Root traits to increase drought resistance in rainfed lowland rice. In: Fukai S, Cooper M, Salisbury J, editors. Breeding strategies for rainfed lowland rice in drought-prone environments. Proc. No. 77. Canberra (Australia): Australian Centre for International Agricultural Research. p 170-179.

Pantuwan G, Fukai S, Cooper M, Rajatasereekul S, O'Toole JC. 2002. Yield response of rice (*Oryza sativa* L.) genotypes to different types of drought under rainfed lowlands. Part 1. Grain yield and yield components. Field Crops Res. 73:153-168.

Price AH, Tomos AD, Virk, DJ. 1997. Genetic dissection of root growth in rice (*Oryza sativa* L.). I. A hydroponic screen. Theor. Appl. Genet. 95:132-142.

Saito K, Linquist B, Atlin GN, Phanthaboon K, Shiraiwa T, Horie T. 2006. Response of traditional and improved upland rice cultivars to N and P fertilizer in northern Laos. Field Crops Res. 96:216-223.
Samson BK, Wade LJ. 1998. Soil physical constraints affecting root growth, water extraction, and nutrient uptake in rainfed lowland rice. In: Ladha JK et al, editors. Rainfed lowland rice: advances in nutrient management research. Manila (Philippines): International Rice Research Institute. p 231-244.
Shen L, Courtois B, McNally KL, Robin S, Li Z. 2001. Evaluation of near-isogenic lines of rice introgressed with QTLs for root depth through marker-aided selection. Theor. Appl. Genet. 103:75-83.
Singh RK, Singh VP, Singh CV. 1994. Agronomic assessment of beushening in rainfed lowland rice cultivation in Bihar, India. Agric. Ecosyst. Environ. 51:271-280.
Steele KA, Price AH, Shashidar HE, Witcombe JR. 2006. Marker-assisted selection to introgress rice QTLs controlling root traits into an Indian upland rice variety. Theor. Appl. Genet. 112:208-221.
Venuprasad R, Lafitte R, Atlin GN. 2006. Response to direct selection for grain yield under drought stress in rice. Crop Sci. 47:285-293.
Venuprasad R, Shashidhar HE, Hittalmani S, Hemamalini GS. 2002. Tagging quantitative trait loci associated with grain yield and root morphological traits in rice (*Oryza sativa* L.) under contrasting moisture regimes. Euphytica 128:293-300.
Venuprasad R, Zenna N, Choi IR, Amante M, Virk PS, Kumar A, Atlin GN. 2007. Identification of marker loci associated with tungro and drought tolerance in near-isogenic rice lines derived from IR64/Aday Sel[4]. Intl. Rice Res. Notes 32:27-29.
Xu K, Xia X, Fukao T, Canlas P, Maghirang-Rodriguez R, Heuer S, Ismail AI, Bailey-Serres J, Ronald PC, Mackill DJ. 2006. *Sub1A* is an ethylene response factor-like gene that confers submergence tolerance to rice. Nature 442:705-708.
Zhao DL, Atlin GN, Bastiaans L, Spiertz, HJ. 2006a. Cultivar weed-competitiveness in aerobic rice: heritability, correlated traits, and the potential for indirect selection in weed-free environments. Crop Sci. 46(1):372-380.
Zhao DL, Atlin GN, Bastiaans L, Spiertz JHJ. 2006b. Developing selection protocols for weed competitiveness in aerobic rice. Field Crops Res. 97:272-285.
Zhao DL, Atlin GN, Bastiaans L, Spiertz JHJ. 2006c. Comparing rice germplasm groups for growth, grain yield and weed-suppressive ability under aerobic soil conditions. Weed Res. 46:444-452.

Notes

Authors' address: International Rice Research Institute, DAPO Box 7777, Metro Manila, Philippines.

Drought research at WARDA: current situation and prospects

M. Sié, K. Futakuchi, H. Gridley, S. Mande, B. Manneh, M.N. Ndjiondjop, A. Efisue, S.A. Ogunbayo, M. Moussa, H. Tsunematsu, and H. Samejima

Drought is one of the major constraints to rice production in the rainfed ecology in sub-Saharan Africa (SSA). It occurs not only in uplands but also in lowlands; for example, 70% of lowland rice farmers experience a drought problem at the reproductive stage, which can reduce yield more severely than at the vegetative stage. At WARDA, therefore, drought is one priority area to be addressed. In WARDA's interspecific breeding between *Oryza sativa* and *O. glaberrima* for the rainfed ecology, one of the most important characteristics is short duration to evade drought that often occurs in the latter days of the cropping season. Several interspecific lines (NERICA: New Rice for Africa) with growth duration of 90–100 days have been developed. Apart from this escape type, NERICAs showing better drought resistance than local *O. sativa* checks have also been identified. Several *O. glaberrima* landraces showing high resistance were identified after collection from the flood plains of the Niger River and crossing of these landraces with *O. sativa* and existing NERICAs with high agronomic performance has already been done. Several *O. sativa* varieties were also screened for drought resistance at both vegetative and reproductive phases. Promising *O. sativa* varieties with good drought resistance have been used in crosses with susceptible genotypes for creating breeding and mapping populations. The populations being created are expected to segregate for root characteristics and/or osmotic adjustment, both of which are important drought-resistance traits in rice. A WARDA-JIRCAS joint project also aims to develop drought-resistant varieties. The project narrowed target characters down to root penetration into the deeper soil layers, which is very effective for growth maintenance in certain drought situations. Highly promising *O. sativa* lines have already been identified, and the next step is to identify genes/QTLs for this trait for use in marker-assisted selection. Past drought research at WARDA mostly concentrated on varietal improvement. However, an agronomic approach is also within our purview. We inventory farmers' existing cultural practices to minimize the risks of yield reduction by drought and test their true usefulness. Integrated drought management options for rainfed rice combining resistant varieties and cultivation practices will also be developed and evaluated.

Rice has been cultivated in West and Central Africa for centuries and is now one of the region's staple foods. This is a unique crop adaptable to various ecologies ranging from free-draining upland soils to inundated and irrigated lowland soils. However, rice is more sensitive to water deficiency than other field crops such as cowpea and maize. Therefore, drought is one of the major constraints to rice production in rainfed environments (Cruz and O'Toole 1984). It occurs not only in uplands but also in lowlands; for example, 70% of lowland rice farmers experience a drought problem at the reproductive stage, which can reduce yield more severely than at the vegetative stage, according to a recent survey conducted at Sikasso, Mali, using the participatory rural appraisal (PRA) approach. Recently, in order to solve the problem, interspecific hybrids were developed by the Africa Rice Center by crossing *Oryza glaberrima* and *O. sativa* (WARDA 2000, Ishii 2003, Futakuchi et al 2003).

One of the most important characteristics of these interspecifics is short duration to evade drought that often occurs in the latter days of the cropping season. Several interspecifics (NERICA: New Rice for Africa) with growth duration of 90–100 days have been developed. As shown in the section on "Evaluation of breeding lines developed," NERICAs showing better drought resistance than local *O. sativa* checks have also been identified along with drought-escape types.

Lilley and Fukai (1994), Kobata et al (1996), and Fujii and Horie (2001) showed that high dry matter production by drought-resistant cultivars of rice (*O. sativa*) is caused by superior ability to gather soil water. Also, molecular tools have been used to facilitate the identification and genomic locations of genes controlling traits related to drought resistance (Lanceras et al 2004). Several researchers have reported genetic variation in rice for drought resistance. Several traits implicated in drought resistance of rice are deep rooting ability, osmotic adjustment, anther dehiscence, leaf rolling or nonrolling, recovery ability, early vigor, death of leaves, delay of heading, deformed rachids, and grain weight (Chang et al 1974). *O. sativa* indica varieties of rice are reported to have higher osmotic adjustment than *O. sativa* japonica varieties. On the other hand, japonica varieties generally have deeper root systems than indica varieties.

As new varieties are arguably a technology readily adopted by small farmers, the development of varieties combining improved drought resistance with good yield potential would help stabilize on-farm yield while increasing productivity and production. To develop such varieties requires the identification and transfer of genes from drought-resistant sources into high-yielding well-adapted varieties. QTLs associated with several important traits in drought resistance have already been identified; for example, osmotic adjustment (Zhang et al 1999, Robin et al 2003), relative water content (Babu et al 2003), canopy temperature (Babu et al 2003), flowering date (Lafitte et al 2004), maximum root length (Li et al 2005), deep root mass (Kamoshita et al 2002), root dry weight (Zhang et al 1999, Li et al 2005), root-pulling force (Zhang et al 1999), seedling vigor (Zhang et al 2005), and grain yield (Lanceras et al 2004, Bernier et al 2007).

In view of the importance of drought as a production constraint of rice in sub-Saharan Africa, it is be desirable to adopt an integrated approach to mitigate drought problems. However, the major component of the drought research at WARDA has been varietal development. Several special projects aiming to develop drought-resistant varieties, for example, a Rockefeller-funded project and a joint WARDA-JIRCAS project, have been implemented at WARDA. In this report, WARDA's drought-related activities are reviewed and future prospects are described.

Identification of genetic sources for drought resistance

To identify sources for the development of drought-resistant varieties, several field trials to screen *O. sativa*, *O. glaberrima*, and NERICA have been conducted.

Screening of *O. sativa*

This trial mostly focused on *O. sativa*, although some *O. glaberrima* and NERICAs were included in the entries. One hundred and twenty genotypes (Table 1) inclusive of *O. sativa* indica, *O. sativa* japonica, *O. glaberrima*, and NERICA (interspecific *O. sativa* × *O. glaberrima* progeny), which were sourced from WARDA, CIAT, and IRRI, were screened for drought resistance at the Togoudo research station (Benin) between 2005 and 2007. In this research, the drought screening protocol involved imposing a 21-day drought stress at 45 days after sowing (DAS), which coincides with the vegetative/reproductive phase of crop development. Two trials were conducted during the main dry season (Dec. 2005-March 2006 and Dec. 2006-March 2007) and one during the short dry season (July-August 2006). The trials were laid out in a split-plot design with irrigation regime (the plot of full irrigation throughout growth as a control and the drought treatment plot) as the main plot factor and genotype as the subplot factor. Within each subplot, the genotypes were randomized using an alpha lattice design. Data were collected following the standard evaluation system (SES) of IRRI (1996), in which applicable data collected were plant height, tiller number, leaf greenness rating (using a SPAD meter), leaf rolling, leaf drying, recovery ability, flowering date, leaf temperature, fresh and dry weights of organs before and after drought stress, number and length of leaves, number and weight of panicles, and grain yield per plant.

Over the two seasons of screening, grain yield under drought was found to be positively correlated with yield under continuous irrigation, implying that it is possible to breed drought-resistant rice genotypes with high yield potential.

Significant phenotypic correlations were detected between grain yield and several morphological and physiological traits (Table 2). Leaf greenness rating (SPAD reading), leaf width, and leaf length consistently had positive correlations with grain yield under drought conditions, whereas significant correlations were detected between grain yield and tiller number, days to 50% flowering, and leaf temperature, but the signs differed between the years. In 2005-06, grain yield was positively correlated with tiller number and leaf temperature and negatively correlated with days to 50% flowering, whereas grain yield in 2006-07 was negatively correlated with tiller number and leaf temperature and positively correlated with days to 50% flowering. It is noteworthy that all traits with significant correlations with grain yield under drought stress were only weakly correlated with grain yield (correlations below 50%). Hence, breeding

Table 1. One hundred and twenty lines and breeding lines used for drought screening at Cotonou.

No.	Line	No.	Line	No.	Line
1	Aliança	41	IR62266-42-6-2	81	RAM 134
2	Araure 4	42	IR64	82	RAM 152
3	B6144F-MR-6-0-0	43	IR74371-54-1-1	83	RAM 24
4	Bala	44	IRAT 104	84	RAM 3
5	Black Gora	45	IRAT 109	85	RAM 55
6	CAIAPO	46	IRAT 13 × OS6-AL-1CM-1JN	86	RAM 25
7	Carolino Blanco	47	IRAT 216	87	RHS 107-2-1-2TB-1JM
8	CG17	48	IRAT 13	88	RHS 107-2-2-1TB-1JM
9	CG14	49	ITA 186	89	Salumpikit
10	CG20	50	ITA 212	90	Short Grain
11	CO39	51	M 17	91	TGR 68
12	CT 6510-24-1-2	52	MGL 2	92	Tog5681
13	CT6946-6-2-2P-1X	53	Morobérékan	93	TOX 1011-4-1
14	CT7201-16-5P	54	NERICA 1	94	TOX 1012-12-3-1
15	CT7203-6-5P	55	NERICA 2	95	TOX 1177-17-16-8-1CH-2P
16	CT7415-6-5-2-2X	56	NERICA 3	96	TOX 1177-17-16-B-1CH-1P
17	CT7415-6-5-3-1X	57	NERICA 4	97	TOX 1779-3-3-201-1B
18	CT9993-5-10-1-M	58	NERICA 5	98	TOX 1840-3-2-3X
19	Dourado	59	NERICA 6	99	TOX 1857-3-2-201-1
20	Dourado Precoce	60	NERICA 7	100	TOX 1871-38-1
21	FONAIAP 2000	61	NERICA 8	101	TOX 718-AL-11-1CM-1JU
22	IAC 164	62	NERICA 9	102	TOX 718-AL-20-1CM-1JN
23	IAC 165	63	NERICA 10	103	TOX 718-AL-27-1CM-1JN
24	IAC 25	64	NERICA 11	104	TOX 891-212-2-102-2-101-1
25	IAC 47	65	NERICA 12	105	Vandana
26	ICC 004 Azucena	66	Ngovie	106	Vermelho Comun
27	ICC 124 Lac 23	67	OS6	107	WAB181-18
28	ICC 134 Kinandang Patong	68	P 5589-1-1-2P	108	WAB56-125
29	ICC 137 Ma Hae	69	P. Resistente Sequia	109	WAB96-1-1
30	ICC 208 Trembese	70	Paga Divida	110	WAB450-6-2-9-MB-HB
31	IDSA10	71	Palawan	111	WAB502-12-2-1
32	IDSA6	72	Perola	112	WAB56-104
33	IR55419-04	73	Pratao	113	WAB56-50
34	IR55423-01	74	Pratao Precoce	114	WAB56-57
35	IR58821-23-1-3-1	75	PSBRC 9	115	WAB638-1
36	IR71525-19-1-1	76	PSBRC 80	116	WAB706-35-K1-KB
37	IR74371-3-1-1	77	RAM 100	117	WAB880-1-38-19-26-P2-HB
38	IR78875-131-B-1-3	78	RAM 120	118	WAB96-1-1
39	IR78905-105-1-2-2	79	RAM 13	119	WAB96-3
40	IR52561-UBN-1-1-2	80	RAM 131	120	Zhen Shan 97

Table 2. Means of traits measured during and after 21 days of drought stress on a diverse population of rice genotypes under irrigated and drought-stressed conditions at Togoudo Research Station, Benin (n = 97) in 2005-06. Correlation of traits measured under stress with yield under stress is also included.

Trait[a]	Fully irrigated	Drought stress	Correlation with stress yield	S.E.D.
Height 64	90	75	0.05 n.s.	2.59
Tiller no. 60	19	12	0.163*	0.62
Tiller no. 92	22	19	0.170*	2.64
Leaf greenness 92	43.30	43.10	0.155*	0.52
Leaf no. 74	5	4	0.180*	0.39
Leaf length 74	42	34	0.128*	3.39
Leaf temp. 59	31	33	0.158*	0.22
Leaf drying 67	–	2.5	–0.153*	–
Leaf rolling 80	–	2.00	–0.157*	–
Leaf drying 80	–	1.70	–0.185**	–
Biomass 70 (g)	35.36	11.14	0.325**	–
Moisture content 70 (g)	107.13	31.09	0.220*	–
Biomass during stress	29.43	5.00	0.215*	–
50% flowering (days)	79	91	–0.196**	0.85
Fertile panicle no.	13	8	0.366**	1.26
Fertile panicle wt.	14.28	6.10	0.559**	2.30
Final biomass	62.26	49.49	0.212*	4.65
Grain yield per plant	12.35	5.03	–	2.31

[a]Numbers following trait names indicate the DAS on which the trait was measured. ** = significant at $P < 0.001$; * = significant at $P < 0.05$; n.s. = not significant. S.E.D. = standard error of the difference between mean trait value under fully irrigated and under drought stress conditions.

for drought resistance should employ complex crosses aiming at pyramiding drought-resistance alleles in adapted backgrounds.

Screening of *O. glaberrima*

The main target of this screening was *O. glaberrima*, though some *O. sativa* lines were tested too. The genetic material in the study comprised 75 genotypes composing nine *O. sativa* lines, eight drought-resistant chromosome segment substitution lines (CSSL) from CIAT, and 58 *O. glaberrima* accessions that included the RAM series (Riz Africain du Mali) from the Institut d'Economie Rurale in Mali. The material was evaluated in three trials under two drought treatments consisting of 28 days of water stress initiated at 21 days (trial A) and 42 days (trial B) after sowing (DAS), with a third treatment, the control (trial C), with no water stress imposed. Each trial was laid out as an alpha lattice with three replicates and each plot had three plants. Soil water status in the water-stress treatments was measured in three 20-cm layers of soil from the surface to 60 cm. Data collected on individual plants were number of tillers, plant height, number and length of leaves, leaf rolling, seedling vigor, number and mass of panicles, and seed yield. The ultimate priority is the identification and development of genotypes with good yield potential under drought stress. The plots in the trials were too small to obtain a good estimate of the yield potential per hectare under drought stress. However, important yield components recorded—number of tillers and seed yield per plant—provide a good indicator.

A drought-stress effect was evident from 15 days after withholding water. The first plant response was leaf rolling and the RAM series of *O. glaberrima* accessions were the first to attain a leaf-rolling score of 9 and they remained in this state until resumption of watering. Other genotypes had the capacity to recover overnight, which was particularly marked for the two *O. glaberrima* accessions, CG14 and CG17. All phenotypic traits measured showed the effect of drought stress at 28 days following initiation of the stress at 21 DAS compared with those in the nonstressed trial. The same length of drought stress initiated at 42 DAS reduced aboveground biomass through leaf wilting and drying. Regardless of the timing of drought stress, the *O. glaberrima* accessions had better recovery ability than the *O. sativa* and CSSL lines, which may reflect the late tillering ability of *O. glaberrima* accessions, as this was not evident for the *O. sativa* cultivars and CSSL lines.

Forty-nine genotypes exhibiting good performance under drought stress were selected from the set of 75 described above to compare their root traits with those of Moroberekan, a well-studied drought-resistant *O. sativa* variety from West Africa.

WARDA-JIRCAS Drought Project

This project narrowed target characters down to root penetration into the deeper soil layers, which is very effective for growth maintenance in certain drought situations. The screening started in 2004 with 600 lines (*O. sativa* and *O. glaberrima*) in Bamako, Mali, in relation to the target trait. Evaluation of the promising lines identified continued in Ibadan, Nigeria, in 2005 and 2006 (Table 3). In the intensive screening, Khao Dam and Malagkit Pirurutong were identified as deep-root varieties and Ma Hae, Trembese, and Chau as shallow-root varieties. To identify QTLs associated with deep root, crosses were made between the deep- and shallow-root varieties and populations are being developed. In 2007, F_2 (Ma Hae × Khao Dam, Chau × Khao Dam) and F_3 (Ma Hae × Malagkit Pirurutong, Trembese × Malagkit Pirurutong, Chau × Malagkit Pirurutong) populations were available. Evaluation of populations in relation to root depth starts from the F_3 generation.

Generation of breeding and mapping populations

Interspecific BC_2F_2 populations derived from a top-cross

BC_1F_2 and BC_1F_4 interspecific progeny were developed from the crosses of CG14 (*O. glaberrima*) with elite *O. sativa* lines WAB56-104 and WAB638-1. These interspecific progeny were top-crossed to Morobérékan to develop three-way-cross BC_2F_1 populations from which individuals were genotyped with 51 microsatellite markers to assess

Table 3. Root depth of promising *O. sativa* entries identified at Ibadan.

Item	Ranking in root depth[a]	Accession	Root depth (cm)	Shoot dry weight (g)
Top 7	1 (4)	Malagkit Pirurutong	23.8 ± 6.4	34.2 ± 13.7
	2 (29)	Dam Ngo	22.8 ± 6.3	25.5 ± 8.3
	3 (58)	Godawee	22.7 ± 5.0	23.4 ± 9.5
	4 (54)	Dharial	22.0 ± 6.7	21.9 ± 8.6
	5 (87)	Arang	21.8 ± 9.4	35.7 ± 11.5
	6 (90)	DA 1	21.4 ± 9.4	35.7 ± 13.4
	7 (25)	Rathal	20.8 ± 4.5	32.1 ± 11.7
Average			21.2 ± 6.3	28.8 ± 12.0
O. glaberrima	41 (83)	TOG 5495	17.4 ± 3.6	29.1 ± 9.3
	49 (14)	TOG 5484	16.9 ± 3.3	28.5 ± 9.3
	56 (77)	TOG 5979	16.7 ± 3.1	34.1 ± 12.6
	68 (38)	TOG 5556	16.1 ± 4.6	25.4 ± 11.0
	75 (88)	TOG 5725	15.4 ± 5.0	29.3 ± 8.3
	94 (63)	TOG 6639	13.2 ± 3.4	23.2 ± 8.3
	98 (34)	TOG 5675	12.1 ± 3.1	32.5 ± 11.7
Average			15.4 ± 4.2	28.9 ± 10.5

[a]Numbers in parentheses are the rank in the screening at Bamako, Mali.

the contribution of the *O. glaberrima* accessions to their genetic makeup. Selfing of superior BC_2F_1 individuals generated nine BC_2F_2 populations for further selection.

About 10,000 BC_2F_2 individuals were screened for drought resistance by subjecting them to prolonged water stress of 35 days, starting from 42 DAS. Watering resumed at 77 DAS, allowing recovery. At maturity, selection based on growth duration, resistance to shattering, resistance to lodging, fertility, and drought resistance identified 3,000 individuals with the desired parameters for these traits.

One hundred and twenty individuals with a short cycle (<100 days), representing 4% of the 3,000 individuals, were advanced to BC_2F_3, from which the following two sets of progeny are being developed:

- Seventy-four BC_2F_4 progeny from BC_2F_3 heterogeneous progeny selected for seedling vigor, high effective tiller number (weed competitiveness), panicle size, and grain number per panicle (>300 seeds per panicle).
- Fifty-four BC_2F_4 progeny from homogeneous BC_2F_3 progeny for yield evaluation in plots of 8 m^2.

The remaining 96% of the 3,000 BC_2F_2 individuals matured 100 to 125 DAS, and selection in heterogeneous BC_2F_3 progeny with confirmed drought resistance is based on tiller number, stay-green ability, appearance, and grain yield.

Breeding populations developed from *O. glaberrima* RAM accessions

To further exploit the drought resistance detected in the *O. glaberrima* RAM accessions, a series of crosses and backcrosses involving the drought-resistant RAM accessions and elite high-yielding but drought-susceptible *O. sativa* and interspecific lines have been undertaken. Forty-five crosses generated 531 BC_1F_1 seeds and BC_2F_1 populations are being developed for subsequent selfing, selection for drought-resistance and important agronomic traits, and distribution to NARES breeding programs. Additionally, to exploit and explore the *O. glaberrima* genome in greater detail than is possible in a backcross program, a large number of F_1 crosses between drought-resistant *O. glaberrima* RAM accessions and interspecific (*O. sativa* × *O. glaberrima*) lines were selfed. Although the intergenomic sterility barriers in crosses between *O. glaberrima* and *O. sativa* reduced the number of F_2 seeds to 5 or fewer in some crosses, 42 F_1 crosses yielded 155 F_2 seeds. The F_2 and subsequent generations provide an opportunity to develop a series of new interspecific recombinant progeny segregating for a wide range of genes/QTLs from the *O. glaberrima* genome.

Twelve BC_1F_1 and BC_2F_1 populations, developed from crosses between drought-resistant *O. sativa* lines and elite high-yielding but drought-susceptible *O. sativa* and from interspecific (*O. sativa* × *O. glaberrima*) crosses are being selfed to quickly develop a range of progeny to screen for drought resistance and yield.

Generation of mapping populations

IR64 and ITA212, lowland drought-susceptible lines, are being crossed, respectively, with 18 and 17 drought-resistant donor lines to develop mapping populations. Recombinant inbred lines and doubled haploids will be developed from the progeny derived from these crosses and will be used in mapping QTLs for drought resistance.

Additionally, drought-resistant upland lines NERICA 1 to 7, WAB56-104, IAC 165, and WAB96-3 are involved in 55 crosses with sources of drought-resistance traits to develop breeding populations for selection.

As a specific aspect of the project is to exploit the drought resistance detected in the *O. glaberrima* RAM accessions from Mali, CSSL populations are being developed from crosses of 18 accessions of *O. glaberrima* RAM accessions with two *O. sativa* varieties, namely, Morobérékan and WABC165. Twenty-two cross combinations generated the F_1 generation, and BC_1 and BC_2 populations are being developed.

Evaluation of breeding lines developed

Upland lines developed have been routinely subjected to evaluation for drought resistance. When the WARDA research station was in M'bé, Côte d'Ivoire, before the Ivorian crisis, some of the lines were tested in relation to their response to various soil water conditions using a sprinkler method. As an example, data from a trial in the dry season of 2000-01 are reported.

Three rows of sprinklers were installed in an upland rice field at an interval of 15 m and rice plots in four replicates were established between them. The plots were irrigated for 2 hours daily for 3 weeks from seeding, after which only the central row of sprinklers was operated. The trial was conducted for two seasons with the same varietal entries. It was expected that the amount of irrigated water would become smaller with the distance from the central row of sprinklers. Thus, five treatments of soil water conditions were used: 0–3 m (1st plot); 3–6 m (2nd plot); 6–9 m (3rd plot);

Table 4. Yield of NERICA lines and check varieties in various soil moisture conditions (2000-01).

Line/variety	Yield (t ha^{-1})				
	1st plot[a]	2nd plot	3rd plot	4th plot	5th plot
CG 14	2.34	1.68	2.33	0.62	0.04
WAB56-104	1.87	1.18	1.75	1.29	0.08
Bouaké 189	1.43	0.70	1.57	0.62	0.00
WAB450-24-3-2-P18-HB	2.06	1.29	1.97	0.83	0.27
Moroberekan	1.86	0.78	1.45	0.58	0.10
45 NERICA lines					
Mean	2.23	1.56	2.02	1.00	0.05
S.E.	0.52	0.30	0.44	0.22	0.08
Min.	1.04	1.01	1.06	0.43	0.00
Max.	3.08	2.24	3.17	1.51	0.55
LSD (5%)	0.96	0.83	1.14	0.78	0.25

[a]The distance of each plot, from 1st to 5th, was increasing from the central sprinkler row.

9–12 m (4th plot); and 12–15 m (5th plot) for the distances from the central row of sprinklers. Forty-five upland NERICA lines (WAB878, WAB880, and WAB881 series) of the second generation were tested with the following five check varieties: Moroberekan (a tolerant check, a traditional *O. sativa* japonica variety); WAB56-104 (an improved *O. sativa* japonica variety); Bouaké 189 (a susceptible check, an improved *O. sativa* indica variety); CG14 (*O. glaberrima*); and WAB450-24-3-2-P18-HB (one of the first-generation NERICA lines).

On average, the new-generation NERICA outyielded the resistant check, Morobérékan, in all water treatments other than the severest drought plot (the 5th plot) (Table 4). In the 5th plot, however, one NERICA line, WAB878-6-37-8-1-P1-HB, significantly outyielded (0.55 t ha^{-1}) all the check varieties. The yield of this new NERICA line was 1.63, 1.48, 2.45, 1.36, and 0.55 t ha^{-1} in the 1st, 2nd, 3rd, 4th, and 5th plots, respectively. This line produced a comparatively high yield (1.36 t ha^{-1}) in the 4th plot and is promising for drought-prone upland conditions.

The prime cause of yield reduction in the soil-water-deficit plots was the occurrence of a large number of unfilled grains in these conditions. The percentage ratio of the ripened grain of WAB878-6-37-8-1-P1-HB was 85.5%, 86.3%, 85.0%, 51.8%, and 16.9% in the 1st, 2nd, 3rd, 4th, and 5th plots, respectively. This line and two check varieties, WAB450-24-3-2-P18-HB and Morobérékan, had significantly higher ratios than 0 in the 5th plot with severe soil water deficit (Table 5).

Conclusions and prospects

Several *O. glaberrima* landraces showing high resistance were identified and crossed with *O. sativa* and NERICA lines with high agronomic performance. Several *O. sativa* varieties were also screened for drought resistance in both vegetative and

Table 5. Percentage ratio of filled grains in NERICA lines and check varieties in various soil moisture conditions (2000-01).

Line/variety	Ratio of filled grains (%)				
	1st plot[a]	2nd plot	3rd plot	4th plot	5th plot
CG 14	78.6	76.8	77.7	26.7	1.8
WAB56-104	82.1	78.2	80.3	52.9	3.2
Bouaké 189	38.4	18.5	51.4	16.0	0.0
WAB450-24-3-2-P18-HB	65.8	72.0	81.7	46.8	20.4
Moroberekan	68.2	60.4	70.9	44.8	15.6
45 NERICA lines					
Mean	78.9	75.9	78.7	45.7	2.8
S.E.	4.6	5.2	6.9	9.7	2.9
Min.	66.6	63.1	53.2	24.6	0.0
Max.	85.5	86.3	86.5	62.7	16.9
LSD (5%)	12.6	14.8	16.9	32.2	11.8

[a]The distance of each plot, from 1st to 5th, was increasing from the central sprinkler row.

reproductive phases. Promising *O. sativa* varieties with good drought resistance have been used in crosses with susceptible genotypes for creating breeding and mapping populations. The populations being created are expected to segregate for root characteristics and osmotic adjustment, both of which are important drought-resistance traits in rice.

WARDA's *O. glaberrima* collection has not been fully explored for drought resistance. Evaluation and screening will continue. However, sources of drought resistance will not be restricted to *O. glaberrima* and will be expanded to *O. sativa* and other wild relatives such as *O. barthii*.

QTL identification is also ongoing for some characteristics associated with drought resistance; for example, deep root is focused on in a WARDA-JIRCAS joint project at Ibadan, Nigeria.

Meanwhile, varietal improvement is a major approach at WARDA to address drought problems, and an agronomic approach is also within our purview. We will inventory farmers' existing cultural practices to minimize the risks of yield reduction by drought and test their true usefulness. Integrated drought management options for the rainfed rice ecology by combining resistant varieties and cultivation practices will also be developed and evaluated.

Geographic information systems need to focus on the methodology for assessing drought risk: the patterns of drought profile for rainfed rice and impacts of climate change on drought occurrence for rainfed rice. The historical climate data analysis of drought patterns needs to focus on the interannual variability of rainfall (start, end, length of the season), the profile of dry spells, the probability of occurrence, the water balance, and crop risk failure because of drought.

Drought profiling will be analyzed with spatial indicators using optical and thermal indicators such as normalized difference vegetation index (NDVI), land surface temperature (LST), and vegetation temperature condition index (VTCI). There are anomalous images showing areas where production deviates from long-term, average production (Fig. 1A and B).

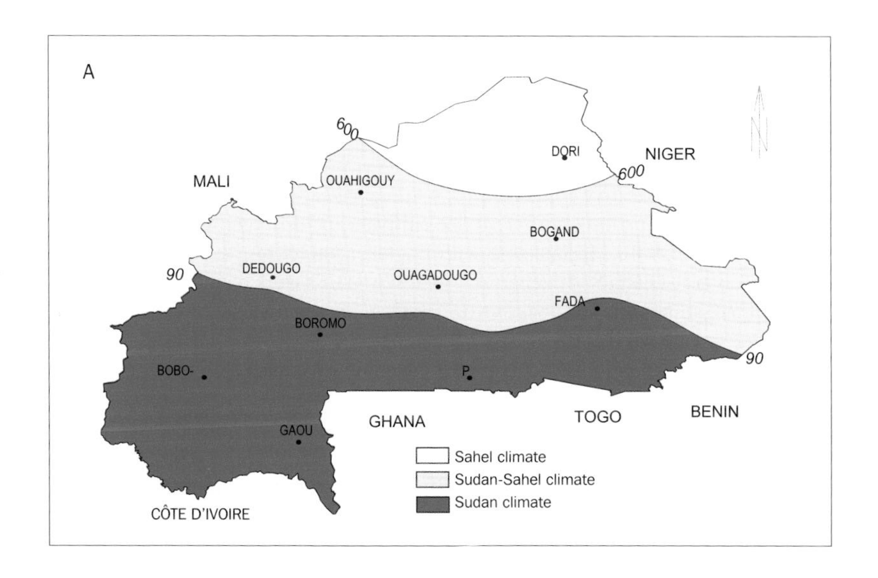

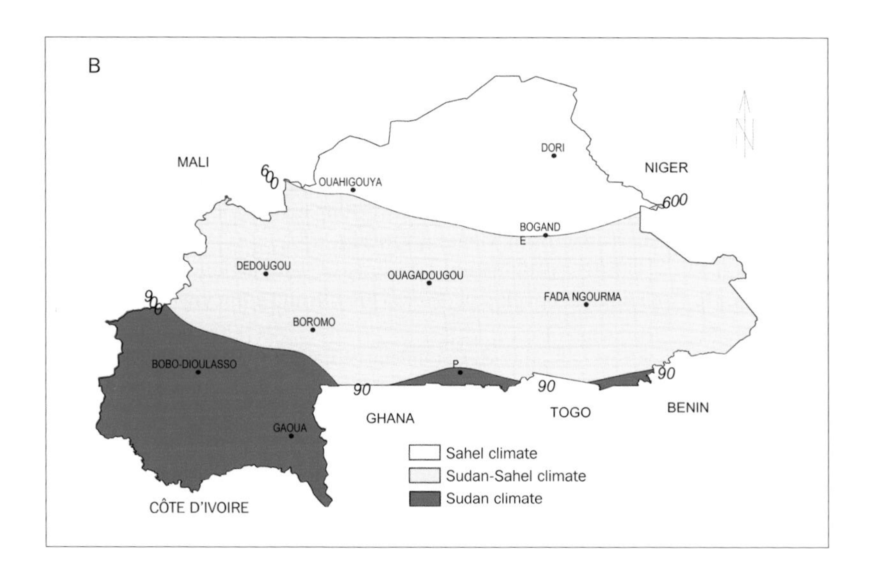

Fig. 1. Burkina Faso climatic zones, (A) 1951-80 and (B) 1971-2000.

References

Babu RC, Nguyen BD, Chamarerk V, Shanmugasundaram P, Chezhian P, Jeyaprakash P, Ganesh SK, Palchamy A, Sadasivam S, Sarkarung S, Wade LJ, Nguyen HT. 2003. Genetic analysis of drought resistance in rice by molecular markers: association between secondary traits and field performance. Crop Sci. 43:1457-1469.

Bernier J, Kumar A, Ramaiah V, Spaner D, Atlin G. 2007. A large-effect QTL for grain yield under reproductive-stage drought stress in upland rice. Crop Sci. 47:507-516.

Chang TT, Loresto GC, Tagumpay O. 1974. Screening rice germ-plasm for drought resistance. SABRAO J. 6:9-16.

Cruz RT, O'Toole JC. 1984. Dry-land rice response to an irrigation gradient at flowering stage. Agron. J. 76:178-183.

Fujii M, Horie T. 2001. Relative contributions of tolerance and avoidance to drought resistance in dry-matter production of different rice cultivars and different fertilization levels. Jpn. J. Crop Sci. 70:59-70. (In Japanese with English abstract.)

Futakuchi K, Tobita S, Diatta S, Audebert A. 2003. WARDA's work on the New Rice for Africa (NERICA)—interspecific *Oryza sativa* L. × *O. glaberrima* Steud. progenies. Jpn. J. Crop Sci. 72(Extra 1):324-325.

IRRI. 1996. Standard evaluation system for rice. 4th edition. Manila (Philippines): International Rice Research Institute. 52 p.

Ishii R. 2003. Can rice save Africa? Jpn. J. Trop. Agr. 47:35-48. (In Japanese.)

Kamoshita A, Zhang J, Siopongco J, Sarkarung S, Nguyen HT, Wade LJ. 2002. Effects of phenotyping environment on identification of quantitative trait loci for rice root morphology under anaerobic conditions. Crop Sci. 42:255-265.

Kobata T, Okuno T, Yamamoto T. 1996. Contributions of capacity for soil water extraction and water use efficiency to maintenance of dry matter production in rice subjected to drought. Jpn. J. Crop Sci. 65:652-662.

Lafitte HR, Price A, Courtois B. 2004. Yield response to water deficit in an upland rice mapping population: associations among traits and genetic markers. Theor. Appl. Genet. 109:1237-1246.

Lanceras JC, Pantuwan G, Jongdee B, Toojinda T. 2004. Quantitative trait loci associated with drought tolerance at reproductive stage in rice. Plant Physiol. 135:384-399.

Li ZC, Mu P, Li C, Zhang H, Li ZK, Gao Y, Wang X. 2005. QTL mapping of root traits in a doubled haploid population from a cross between upland and lowland *japonica* rice in three environments. Theor. Appl. Genet. 110:1244-1252.

Lilley JM, Fukai S. 1994. Effect of timing and severity of water deficit on four diverse rice cultivars. III. Phenological development, crop growth and grain yield. Field Crops Res. 37:225-234.

Robin S, Pathan MS, Courtois B, Lafitte R, Carandang S, Lanceras S, Amante M, Nguyen HT, Li Z. 2003. Mapping osmotic adjustment in an advanced back-cross inbred population of rice. Theor. Appl. Genet. 107:1288-1296.

WARDA (West Africa Rice Development Association). 2000. New Rice for Africa—with a little help from our friends. WARDA annual report for 1999. Bouaké (Côte d'Ivoire): WARDA. p 9-15.

Zhang J, Nguyen H, Blum A. 1999. Genetic analysis of osmotic adjustment in crop plants. J. Exp. Bot. 50:291-302.

Zhang ZH, Yu SB, Yu T, Huang Z, Zhu YG. 2005. Mapping quantitative trait loci (QTLs) for seedling-vigor using recombinant inbred lines of rice (*Oryza sativa* L.). Field Crops Res. 91:161-170.

Notes

Authors' address: Africa Rice Center (WARDA), 01 B.P. 2031, Cotonou, Benin. Tel.: (229) 21 35 01 88; fax: (229) 21 35 05 56; email: m.sie@cgiar.org.

Drought resistance characters and variety development for rainfed lowland rice in Southeast Asia

Shu Fukai, Jaya Basnayake, and Ouk Makara

Drought is a major problem in rainfed lowland rice in Thailand, Laos, and Cambodia. Together with national agricultural research institutes, the University of Queensland has been involved in research on breeding strategies to minimize the adverse effects of drought for the past 15 years. Several putative drought resistance characters were examined in an attempt to identify plant characters that confer drought resistance in rainfed lowland rice in the region. Our approach was to grow a population under irrigated and drained conditions to identify genotypes that yielded well under drained conditions relative to irrigated conditions, and then examine the variation in relative yield in relation to putative drought resistance characters. The common pattern that emerged from this study was that rainfed lowland rice was a drought avoider, and those genotypes that were able to maintain higher internal water potential tended to produce higher grain yield under drought that developed around flowering and thereafter. A shorter delay in flowering was an indication of a genotype's drought avoidance when drought developed just prior to flowering. However, the effectiveness of the drought resistance characters in minimizing yield losses depended on the timing of drought development; for example, a shorter delay in flowering was not effective when drought developed well before flowering. The drought environment in rainfed lowlands was characterized in these countries by determining a free water level above and below the soil surface throughout growth at several locations for several years. The most common type of drought was late-season drought, but water deficit developed also at other times and affected yield. Yield reduction due to drought was related to the free water level in the field at around anthesis. Lateral and downward water movement contributed to a large variation in available water during crop growth, and top toposequence positions lost water faster, resulting in earlier development of plant water stress. A water balance model has been developed to accommodate various water flows in rainfed lowlands.

We have screened a large number of genotypes under drained conditions in the field in the wet season. The selection was mostly based on grain yield adjusted for flowering time and potential yield, but spikelet sterility, leaf water potential, and delay in flowering were also considered for the selection of drought-tolerant genotypes. A large number of putative drought-tolerant lines were selected and crossed with elite lines with high potential yield and grain quality. The populations thus developed are available for further testing for advancing the yield levels of rainfed lowlands in these countries. The contribution of drought tolerance to overall grain yield in rainfed lowlands in these countries was not always high when drought was mild, and potential yield that was obtained under irrigated conditions contributed most to the variation in yield under drained conditions. Thus, it is important to combine the ability to produce high yield under no water limitation and the ability to tolerate drought in one variety.

In the Mekong region, including Thailand, Laos, and Cambodia, rice is grown mostly in rainfed lowlands, and the national average yield is less than that of neighboring countries, where a higher percentage of rice lands are irrigated. Drought is a major problem in rainfed lowland rice and it can affect around 50% of all rice fields in these countries.

The University of Queensland (UQ) has been involved in rice research for 15 years in the Mekong region with the general objective of improving the yield of rainfed lowland rice. One major objective was to reduce the adverse effect of drought in the rainfed lowland rice ecosystem. This work is in collaboration with the government departments responsible for rice variety development such as the Department of Agriculture (presently the Rice Department) in Thailand, the National Agriculture and Forestry Research Institute (NAFRI) in Laos, and the Cambodian Agricultural Research and Development Institute (CARDI). Funds have been provided by the Australian Centre for International Agricultural Research (ACIAR), the Rockefeller Foundation, the International Rice Research Institute (IRRI), and the national governments as well as UQ. We have had many interactions with IRRI scientists, particularly in environmental characterization in the region. Although we have been involved in cropping systems and other aspects of rice research in the region, this paper summarizes our work on genetic improvement of rice grown under drought-prone environments, particularly in the rainfed lowland ecosystem in the Mekong region.

The paper describes first how we determined drought resistance characters that helped rice crops cope with drought, and then the pattern of drought development in rainfed lowlands, particularly on sloping lands, and its effect on grain yield. Continuous efforts to develop rice varieties that produce higher yield in the drought-prone rainfed lowland ecosystem will be described, followed by the current position in our research and development on genetic improvement in the region. Our work in Cambodia and Laos is described in this paper while Pantuwan et al (this volume) describe achievements in Thailand.

Drought resistance characters

Our involvement in the identification of drought resistance characters has been reviewed in Fukai and Cooper (1995), Fukai et al (1999), and Fukai and Kamoshita (2005). Our approach was to examine yield variation for recombinant inbred lines

Table 1. Physiological and morphological characters of drought resistance under different drought types.

Type of drought	Primary/secondary characters
Terminal drought	Early flowering
	Short delay in flowering
	High leaf water potential
	Low leaf death score
Intermittent drought	Deep root system
Vegetative-stage drought	Late flowering
	Longer delay in flowering
	Large green leaf area
	Tillering after drought period

(RILs) under appropriate drought conditions, and then correlate the variation with that of putative drought resistance characters. Perhaps because of this approach, we have identified integrative characters such as reduced spikelet sterility or secondary characters such as increased plant water status to be associated with drought resistance, rather than primary characters that respond initially to reduced water availability (e.g., a genotype with a deep root system extracting more water). Integrative characters may not pinpoint the exact cause of drought resistance and are likely to involve a number of genes and QTLs. However, because they are likely to be more closely associated with grain yield under various drought conditions, they may have wider applicability in a plant breeding program. The following section briefly describes physiological and morphological characters that have been found to be useful against different types of drought (Table 1).

Terminal drought

When drought develops late in the season, early-flowering varieties often escape from terminal drought, and the use of such varieties is often the most effective way of increasing yield under terminal drought conditions.

Among genotypes of a similar maturity type, those that can maintain high leaf water potential are often advantageous in producing higher yield (Jongdee et al 2002, Pantuwan et al 2002c). The cause of variation in leaf water potential is not known although morphological traits such as larger xylem diameter have been found to be associated with this variation (Sibounheuang et al 2006). Although the RILs used in these studies showed genotypic consistency in leaf water potential, this is not always the case in other materials we have used more recently.

Favorable plant water status results in positive turgor promoting exsertion of panicles. Delay in flowering will be small with maintenance of positive turgor, and this is related to higher yield in terminal drought conditions (Pantuwan et al 2002b).

Under terminal drought, particularly that which develops rather suddenly, the maintenance of favorable water conditions for reproductive organ growth and securing adequate grain sink size are a main drought resistance mechanism, and a conservative strategy of water use has been found to be effective. Since water runs out by maturity in terminal drought, prolonging water availability to the plants is required for achieving higher yield. This may be achieved by extracting water rather slowly by a smaller root system or with a smaller shoot system with reduced water demand (Pantuwan et al 2002c).

Intermittent drought

When intermittent drought develops late around flowering stage, characters such as maintenance of favorable plant water status and short delay in flowering, which are effective against terminal drought, appear to be also useful. In contrast to terminal drought, intermittent drought is broken by rainfall events and soil water level increases, and hence there is no strong need for the conservation of water as such. A deep root system with higher root density is likely to be useful under these conditions, as association between high root length density and the amount of water extracted has been well demonstrated (Lilley and Fukai 1994).

Vegetative-stage drought

If drought develops early in the season in the main fields followed by favorable water conditions, late-flowering varieties often produce higher yield than early-flowering varieties, which would have only a short recovering period from the end of the drought period to flowering. After vegetative-stage drought, plant growth takes place, and this recovery growth will then affect sink size as well as source supply to meet the demand of the grain. Genotypes may differ in their recovery growth after vegetative-stage drought. This may be related to the amount of leaf that remains after drought (Mitchell et al 1998) or ability to tiller after drought (Lilley and Fukai 1994).

In our recent work in Cambodia, a short delay in flowering was associated with lower yield under early-season drought conditions, as later-flowering varieties had more time to recover before flowering took place. This illustrates the point that different mechanisms operate for different drought conditions, and genotypes should be selected based on criteria that match with the common drought pattern of the region concerned.

Characterizing the water environment

Because drought patterns, particularly the timing of drought occurrence, determine the effectiveness of drought resistance characters, it is important to characterize the water environment for the identification of drought patterns in rainfed lowland rice. The water environment in the rainfed lowland rice ecosystem is complex because often rice fields are located on gently sloping lands that induce lateral water movement, particularly when the soil is saturated with water.

Variation in water environment and associated variation in drought development from year to year and location to location are also a main reason for the large genotype by environment interaction for yield that has been found in rainfed lowland rice (Cooper et al 1999). This is particularly the case when there is a large variation in flowering time among genotypes tested, as the effect of drought depends on its timing in relation to the development phase of the rice plants (Rajatasereekul et al 1997).

In rainfed lowland experiments, we routinely monitor the free water level in paddies to identify the drought type. PVC tubes with 50-cm length with a number of holes in the lower part are installed in the field for this purpose. If the free water level is below the soil surface, rice plant growth is affected. The magnitude of the drought is quantified from the depth of the free water level below the soil surface and the number of days at that water level.

Our recent work indicates that the free water level for 6 weeks around anthesis (3 weeks before and 3 weeks after anthesis) is closely related to loss in grain yield in rainfed lowland rice (Ouk et al 2006), supporting our earlier simulation modeling results (Fukai et al 2001). Thus, the measurement of the free water level below the soil surface at around flowering can be used to indicate the severity of drought and its effect on grain yield. Analysis of the free water level around flowering in 23 experiments in Cambodia showed that estimated yield loss of greater than 20% occurred in only two experiments. In one case, the free water level was well below the soil surface for a long time. It should be pointed out, however, that water availability to rice plants would also depend on soil type, and therefore these estimates need to be taken with caution.

Water stress development depends not only on rainfall and evapotranspiration but also on downward and lateral water movement in the field. When there is standing water, the water losses from rainfed lowlands are greater at a higher position in the toposequence, partly because of larger lateral water flow, but also larger downward water loss that is associated with lower clay content (Tsubo et al 2007). This results in earlier disappearance of standing water in the field at higher toposequence positions (Tsubo et al 2006). Earlier maturing varieties tend to escape from the drought that would be severe at the top toposequence position, and their water productivities are generally higher than those of late-maturing genotypes. Because of the consistency in variation in drought development, toposequence positions are considered as a separate target of populations for the Thai rice breeding program (Jongdee et al 2006). Thus, upper positions with large yield losses (often 50% of potential yield) are used for screening varieties adapted for severe drought conditions, whereas middle positions with milder drought (up to 20% yield loss) are used for screening varieties adapted for mild drought conditions.

Improving yield in drought-prone environments

Screening tolerant genotypes under managed-drought environment

Widely adapted varieties that perform well in the breeding domain are commonly identified by multilocation trials. Because of a large genotype by environment interaction for yield in rainfed lowland rice, a large number of selection trials need to be conducted across locations and years before widely adapted genotypes can be determined. These adapted genotypes produce higher yield than other genotypes because they may flower at appropriate time to escape from drought, they may have higher potential yield, and/or they may possess drought-tolerance characters. Drought-tolerant genotypes are those that produce higher yield than others while they have similar flowering time and potential yield.

Drought occurrence in rainfed lowland rice is not regular in timing and severity; therefore, conducting screening trials for drought tolerance under natural conditions in the field is not effective. In our work of identifying drought-tolerant genotypes, the growing environment was manipulated to create drought that was of desired severity and timing. Drought screening can be done more readily in the dry season, but the dry-season environment does not represent growing conditions of farmers' fields in the wet season, and the variation for drought resistance among genotypes found in the dry season does not reflect that in the wet season (Pantuwan et al 2004). Nevertheless, the dry season was used for vegetative-stage screening in initial screening of a large number of genotypes in our research program. For advanced screening in which yield responses were required, managed-drought conditions were created and screening was conducted in the wet season.

In order to reduce the chance of rainfall interfering with drought development in managed-drought trials in the wet season, the experiment was planted several weeks later than the commercial practice so that the crop had a better chance of being exposed to drought, particularly for terminal drought. Another practice was to drain water from the field when drought conditions were required. If draining commenced early in the season, there was a good chance that rainfall broke drought and hence created intermittent drought (Fukai 2004).

We also conducted an identical trial under irrigation to provide a reference point for each genotype. From the pair of irrigated and drained trials, mean yield reduction was calculated for each drained trial, and this mean yield reduction was used to decide which trials should be used for identifying drought-tolerant genotypes. We aimed to have a 50% yield reduction in drained trials, but various degrees of yield reduction were obtained depending on the timing of draining commencement, rainfall pattern, and also soil type of the site. If drought was too mild, drought-tolerant genotypes could not be picked up, whereas, if drought was too severe, yield was too low to detect drought-tolerant genotypes. Yield declined between 20% and 80% in 13 out of 17 experiments conducted in 2002-05 in Cambodia and Laos (Table 2). Some experiments were not used for screening purposes because of other problems such as poor establishment, diseases, and insects.

Recent work indicates that genotypes are also not consistent in their yield response to different types of drought conditions (Pantuwan et al 2002a). Thus, the type of stress, timing in development stage, and severity of stress affect genotype performance. Therefore, the water environment of screening locations should be similar to those in the breeder's domain. We used prolonged drought from the vegetative stage as well as late-season drought for screening in Cambodia, whereas we concentrated on late-season drought in Thailand and Laos.

Table 2. Drought screening experiments in Cambodia and Laos, 2002-05.

Location	Year	No. of lines	Stress condition	Grain yield (mean)		LSD[a] (5%) for main yield comparison	Days to flower	Delay in flowering (days)	Yield reduction (days)	Leaf water potential (MPa)
				Well watered (t ha^{-1})	Drained (t ha^{-1})					
Cambodia										
CARDI	2002	75	Severe stress from flowering	2.13	0.55	0.92	112	8	74	–2.49
	2003	72	Severe stress from flowering	2.31	0.53	1.23	110	8	77	–1.36
	2004	75	Moderate stress from flowering	1.69	1.04	0.42	97	6	38	–1.50
	2005	75	Mild stress at flowering	3.10	2.97	ns	114	–1	4	–
Chrey Veal	2003	72	Severe stress from flowering	1.53	0.74	0.63	97	7	52	–2.06
	2004	58	Mild stress from flowering	1.02	0.73	0.23	92	9	28	–3.24
	2005	75	Mild stress from flowering	2.35	1.66	0.43	81	1	29	–
Laos										
ARC	2002	80	No drought	3.21	3.32	ns	106	Nil	No	–1.50
	2003	70	Severe stress from flowering	2.15	0.32	1.56	97	8	85	–2.45
	2004	68	Severe intermittent stress	1.81	0.36	1.10	98	11	80	–2.58
	2005	68	Severe stress from vegetative stage	1.60	0.86	0.68	92	3	46	–2.98

Continued on next page

Table 2 continued.

Location	Year	No. of lines	Stress condition	Grain yield (mean)		LSD[a] (5%) for main yield comparison	Days to flower	Delay in flowering (days)	Yield reduction (days)	Leaf water potential (MPa)
				Well watered ($t\ ha^{-1}$)	Drained ($t\ ha^{-1}$)					
Pakse	2002	68	Mild stress from flowering	1.85	1.53	ns	108	2	17	–1.85
Tassano	2004	68	Severe intermittent stress	2.62	1.04	1.23	98	9	60	–2.36
	2005	68	Severe stress from vegetative stage	1.91	0.90	0.98	92	7	53	–2.13
Sayabuli	2003	70	Moderate stress from vegetative stage[a]	0.54	0.37	0.13	90	5	32	–
	2004	68	Severe stress from vegetative stage	2.01	1.14	0.52	92	3	43	–
	2005	68	Mild stress at flowering stage	0.82	0.67	ns	91	1	18	–

[a]The LSD 5% values for yield comparison between well-watered and drained conditions were estimated from the combined analysis. Significant variation was found when yield reduction was above 17%. ns = nonsignificant.

Selection criteria

We used grain yield under drained conditions as the primary criterion for selecting drought-tolerant genotypes. Because yield under drained conditions was affected by the timing of draining and water stress development relative to the genotype's phenological stage, the yield was adjusted with the flowering date of the genotypes. Similarly, it was adjusted for the variation in yield achieved under irrigated conditions. These adjustments resulted in a drought response index (DRI) of Bidinger et al (1987); we modified the original calculation procedure by calculating DRI for each replication so that variation in DRI could be estimated (Ouk et al 2006).

The DRI was reasonably consistent across experiments when the drought pattern was similar and significant genotypic variation was observed. In an example of nine pairs of experiments conducted at two locations for a few years in Cambodia, heritability was 0.56 for DRI, which was slightly lower than the 0.64 obtained for grain yield under drained conditions (Ouk et al 2006). This DRI variation was not due to water availability conditions and thus was likely to be a reflection of drought tolerance of the genotypes (Ouk et al 2006).

Genotypic variation in spikelet sterility percentage under drained conditions is generally correlated with that of grain yield, but it often provides more repeatable results, partly because it does not involve the number of fertile tillers, which varies and causes variation in grain yield. Therefore, we used spikelet sterility as an additional selection criterion. We also considered other characters such as leaf water potential and delay in flowering when selecting drought-tolerant lines.

Importance of potential yield or general adaptability

Appropriate drought-tolerance characters will minimize the damage caused by drought in rainfed lowland rice, but genotypes also need to have appropriate phenology to escape from drought exposure as much as possible. The best flowering time to avoid common timing of water stress development in the breeding domain needs to be known, and genotypes with appropriate flowering time need to be selected.

Potential yield of a genotype that can be achieved under irrigation is also important in determining yield of the genotype in farmers' fields, as the latter can be considered as a function of drought tolerance, drought escape, and potential yield. When yield is reduced slightly by mild drought, genotypic ranking is similar between irrigated and droughted conditions (Pantuwan et al 2002a). Potential yield may reflect a genotype's ability to adapt to soil factors, such as nutrient deficiency or toxicity as well as any tolerance of biotic stresses. Potential yield may also differ because of genotypic variation in grain sink demand or assimilate source supply to fill grains. A good advance has been made in irrigated rice in raising potential yield, but this does not appear to have occurred at any degree in rainfed lowland rice. The degree of contribution of drought tolerance, drought escape, and potential yield to a genotype's yield under drained conditions was estimated by using multiple linear regression by Basnayake et al (2006) using nine pairs of experimental data obtained in 1998-2002 in Cambodia by Ouk et al (2006). The contribution of potential yield to a genotype's actual yield was greater than that of drought tolerance (mean 53% vs 31%)

Table 3. Mean grain yield and grain yield reduction (%) of two sets of experiments: 9 experiments with 15 genotypes in 1998-2002 and 3 experiments with 23 genotypes in 2002-03 in Cambodia under well-watered (WW) and water-stress (WS) treatments. Percentage contribution of potential yield, drought tolerance, and drought escape (days to flower) as determined from partition of sums of squares of the regression (linear or nonlinear) to yield under water-stress conditions.

Item	9 experiments in 1998-2002		3 experiments in 2002-03	
	WW	WS	WW	WS
Yield (t ha^{-1})	2.18	1.59	1.99	0.60
Yield reduction (%)	28		65	
Contribution to yield under drought				
Potential yield (%)	53		4	
Drought tolerance (%)	31		52	
Drought escape (%)	16		44	

in these experiments in which the mean yield reduction due to draining was only 28% (Table 3). The contribution of drought escape was small in these experiments, as often the effect of flowering date had no effect on grain yield, in contrast to the large effect of phenology observed in Thailand (Rajatasereekul et al 1997). In another set of selection trials in 2002-03 in Cambodia in which drought conditions were severe and mean yield reduction was 65%, potential yield contributed the least (mean 4%) and drought tolerance the most (52%), followed by drought escape (44%), indicating the increased importance of drought tolerance when drought becomes severe.

When several contrasting varieties were compared at different locations and years under rainfed lowland conditions in Thailand and Laos, genotypes with high potential yield performed better in most cases, particularly when environmental yield level was high (Fig. 1 from Fukai 2004). This confirms the importance of improving the germplasm base for potential yield, and traditional sources for parents for crossing in a rainfed lowland rice breeding program may not be most appropriate for this purpose. The challenge would be to combine characters of high potential yield and drought tolerance for genotypes that flower at the most appropriate time for the target domain of the breeding program.

Current status in research and development

We have contributed to the development of phenotyping capacity for drought resistance in rainfed lowland rice in the region. There are now scientists and technical staff members who are competent in phenotyping. Several staff members became specialized for drought screening in their higher degree studies.

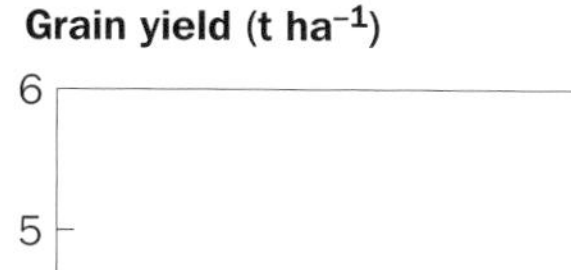

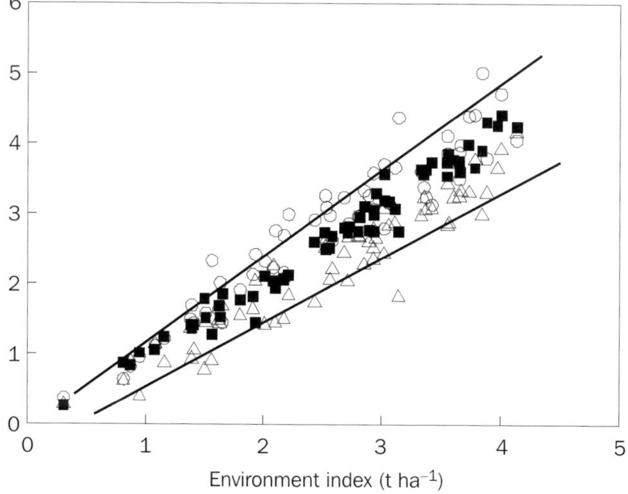

Fig. 1. Grain yield of three groups of genotypes: high-yielding (IR57514-PMI-5-B-1-2, open circle), intermediate-yielding (KDML105 and RD6, solid square), and low-yielding (Chaing Seng, open triangle) plotted against environment yield in Thailand and Laos. Thirty-five genotypes were tested in multilocation trials (Fukai et al 2004).

Field screening facilities are now available and have been used for the selection of drought-tolerant genotypes that can be used as donors in a crossing program. Drought screening trials have been repeated in locations and years in both Laos and Cambodia because of some failures of experiments in terms of achieving an appropriate stress level and also because of large genotype and drought environment interactions. This has, however, reduced the number of genotypes that have been screened in our program. Proper wet-season testing has been done only for just over 155 and 130 genotypes in Laos and Cambodia, respectively. Considering the large number of local accessions available, for example, 13,000 in Laos, the number of genotypes tested so far is very small. The effort to screen a large number of genotypes for drought resistance needs to continue in the future.

From these limited screening exercises conducted in the past few years, drought-tolerant genotypes have been identified for each maturity group (Table 4) and have been used as donor parents for the development of populations in an attempt to develop varieties that are not only drought tolerant but are also well adapted and possess high grain quality. Most drought-tolerant lines were selected because of high DRI for yield from the 2002-03 experiments, while others were selected because of high leaf water potential despite low DRI and low grain yield (e.g., Kam11 and Kam19 for Laos and Gneng for Cambodia) or better performance in other experiments. A large number of crosses have been made and most of them are being used in the breeding programs in

Table 4. **Lines selected for drought tolerance in Cambodia (7) and Laos (9). Results are means of 3 experiments conducted in 2002 and 2003 in each country. DRI = drought response index for yield or spikelet fertility, LWP = leaf water potential. Means are calculated from all entries.**

Genotype	Maturity	Days to flower	DRI-yield	DRI-fertility	Delay in flowering (days)	LWP (MPa)	Fertility (%)	Yield drought (t ha^{-1})
Cambodia								
Thmar Krem	Early	86	1.5	–0.01	–0.8	–1.77	54	1.01
Gneng	Early	96	0.06	0.23	2	–1.69	50	0.50
Neang Tai Dun	Medium	102	0.89	0.15	7	–2.14	55	0.74
Chamreak Phdau	Medium	103	0.74	0.65	7	–1.71	46	0.86
Damnoeub Krachak Sess	Medium	105	0.50	0.24	11	–1.81	51	0.82
Neang Sar Thom	Late	111	0.64	0.52	10	–1.91	48	0.80
Phcar Sla	Late	113	–0.26	0.35	12	–1.62	42	0.49
Mean		94	0	0	7	–1.81	44	0.57
LSD 5%		21	0.84	0.74	4	0.34	13	0.39

Selections	Maturity	Days to flower	DRI-yield	DRI-fertility	Delay in flowering (days)	LWP (MPa)	Fertility (%)	Yield drought (t ha^{-1})
Laos								
ANG DO	Early	77	0.50	0.44	–2	–2.44	70	0.93
Chao America	Medium	89	0.53	0.60	8	–2.34	57	0.93
Chao deng (2)	Medium	90	0.01	0.68	4	–2.36	68	0.84
Chao deng (1)	Medium	93	0.60	0.70	5	–2.29	59	0.88
Hom (1)	Medium	93	0.77	0.55	2	–2.45	52	0.47
Kam11	Medium	98	0.35	–0.16	6	–2.25	42	0.34
Kam19	Medium	98	–0.13	–0.14	7	–2.26	38	0.29
TDK21-B-24-19 -1-B	Late	102	0.01	0.11	4	–2.32	50	0.72
TDK2TDK21-1-B-6-2-1-B	Late	104	0.79	0.66	4	–2.37	55	0.96
Mean		89	0	0	6	–2.36	52	0.69
LSD 5%		19	0.76	0.69	5	0.39	16	0.31

the region; our objective here is to improve the drought tolerance of current improved varieties with generally high yield by using drought tolerance available in the traditional varieties. Table 5 shows the current stage of their development. In Cambodia, 7 drought-tolerant genotypes were crossed with 4 popular varieties, and 26 populations have been developed from these parents. They were at the F_5 stage in 2006. Eight of these populations are in a single seed descent program to create recombinant inbred lines. In Laos, similar progress has been made: 9 drought-tolerant donors were crossed with 5 popular varieties, and 23 populations have been developed. They were at the F_3-F_5 stages in 2006, and most of them are under single seed descent. These materials are available for further genetic studies as well as the development of new varieties.

Table 5. Drought-tolerant donors and recipients used in the population development program in Cambodia and Laos and the stages of the populations in 2006.

Cambodia

Drought-tolerant donors	Recipients
Thmar Krem (acc. 1703)	Phka Rumchek
Gneng (acc. 387)	CAR 4
Chamreak Phdau (acc. 1116 B)	Phka Rumduol
Damnoeub Krachak sess (acc. 1416B)	Sen Pidao
Neang Tai Dun (acc. 1273)	
Neang Sar Thom (acc.152)	
Phcar Sla (acc. 1005B)	

Stage	Population	Number of crosses
F_5	Single seed descent	8
F_5	Bulk selection	26

Laos

Drought-tolerant donors	Recipients
Ang Do	TDK1
Chao America	TDK5
Chao Deng (2)	TDK7
Chao Deng (1)	IR253-100
Hom (1)	TDK47-B-6-1-2-3-B
Kam11	
TDK-21-B-24-19-1-B	
Hang Yi	

Stage	Population	Number of crosses
F_5	Single seed descent	11
F_4	Single seed descent	8
F_3	Single seed descent	4
F_3	Bulk selection	4

Our group members from Thailand BIOTEC, Drs. Theerayut Toojinda and Jonaliza Lanceras-Siangliw, have also genotyped most of the drought-tolerant materials obtained in the region, as well as parents used to develop populations, with 90 molecular markers. Some markers appear useful for identifying lines that possess favorable QTL regions on chromosomes 1, 3, 4, 8, and 9 that have been identified to contain drought-tolerance characters. For example, the Lao tolerant line TDK-21-B-24-19-1-B, selected from a cross between IR57514-PMI-5-B-1-2 (tolerant line) and popular variety TDK1, had the same alleles for all 25 markers in the five chromosome regions except for two markers in the chromosome 3 region. The drought-tolerant

IR57514 parent had the same expression as TDK-21-B-24-19-1-B for the two markers, indicating a possibility that the region may be responsible for drought tolerance in this case. Several markers in chromosome regions 1, 3, 4, and 8 showed different allele patterns between drought-tolerant and recipient genotypes from all three countries. We need to examine further to see whether they could be used to select progenies for drought tolerance in a marker-assisted selection program.

We are also continuing research on water environment characterization to identify drought-prone areas in Laos. The use of drought-resistant varieties will be most effective in these areas.

References

Basnayake J, Fukai S, Ouk M. 2006. Contribution of potential yield, drought response index and days to flower for drought adaptation in rainfed lowland rice in Cambodia. Proceedings of the 13th Agronomy Conference in Australia, Perth, Australia, 9-14 September 2006. www.cropscience.org.au.

Bidinger FR, Mahalakshmei V, Rao GDP. 1987. Assessment of drought resistance in pearl millet [*Pennisetum americanum* (L.) Leeke]. I. Factors affecting yield under stress. Aust. J. Agric. Res. 38:37-48.

Cooper M, Rajatasereekul S, Immark S, Fukai S, Basnayake J. 1999. Rainfed lowland rice breeding strategies for northeast Thailand. 1. Genotypic variation and genotype × environment interactions for grain yield. Field Crops Res. 64:131-152.

Fukai S. 2004. Stress physiology for crop improvement: emphasis on drought resistance. In: Vanavichit A, editor. Proceedings of the 1st International Conference on Rice for the Future. Kasetsart University, Bangkok, 31 August-3 September 2004. p 187-194.

Fukai S, Basnayake J, Cooper M. 2001. Modeling water availability, crop growth, and yield of rainfed lowland rice genotypes in northeast Thailand. In: Tuong TP, Kam SP, Wade L, Pandey S, Bouman BAM, Hardy B, editors. Characterizing and understanding rainfed lowland environments. Proceedings of the International Workshop on Characterizing and Understanding Rainfed Environments, 5-9 December 1999, Bali, Indonesia. Manila (Philippines): International Rice Research Institute. p 111-130.

Fukai S, Cooper M. 1995. Development of drought-resistant cultivars using physio-morphological traits in rice. Field Crops Res. 40:67-86.

Fukai S, Kamoshita A. 2005. Ecological, morphological and physiological aspects of drought adaptation of rice in upland and rainfed lowland systems. In: Toriyama K, Heong KL, Hardy B, editors. Rice is life: scientific perspectives for the 21st century. Proceedings of World Rice Research Conference, Tokyo and Tsukuba, Japan, 5-7 November, 2004. IRRI, Los Baños, Philippines, and Japan International Research Center for Agricultural Sciences, Tsukuba, Japan. p 448-451.

Fukai S, Pantuwan G, Jongdee B, Cooper M. 1999. Screening for drought resistance in rainfed lowland rice. Field Crops Res. 64:61-74.

Jongdee B, Fukai S, Cooper M. 2002. Leaf water potential and osmotic adjustment as physiological traits to improve drought tolerance in rice. Field Crops Res. 76:153-163.

Jongdee B, Pantuwan G, Fukai S, Fischer K. 2006. Improving drought tolerance in rainfed lowland rice: an example from Thailand. Agric. Water Manage. 80:225-240.

Lilley JM, Fukai S. 1994. Effect of timing and severity of water deficit on four diverse rice cultivars. I. Rooting pattern and soil water extraction. Field Crops Res. 37:205-213.

Mitchell JH, Siamhan D, Wamala MH, Risimeri JB, Chinyamakobvu E, Henderson SA, Fukai S. 1998. The use of seedling leaf death score for evaluation of drought resistance of rice. Field Crops Res. 55:129-139.
Ouk M, Basnayake J, Tsubo M, Fukai S, Fischer KS, Cooper M, Nesbitt H. 2006. Use of drought response index for identification of drought resistant genotypes in rainfed lowland rice. Field Crops Res. 99:48-58.
Pantuwan G, Fukai S, Cooper M, Rajatasereekul S, O'Toole JC. 2002a. Yield response of rice (*Oryza sativa* L.) genotypes to different types of drought under rainfed lowlands. Part 1. Grain yield and yield components. Field Crops Res. 73:153-168.
Pantuwan G, Fukai S, Cooper M, Rajatasereekul S, O'Toole JC. 2002b. Yield response of rice (*Oryza sativa* L.) genotypes to drought under rainfed lowlands. 2. Selection of drought resistant genotypes. Field Crops Res. 73:169-180.
Pantuwan G, Fukai S, Cooper M, Rajatasereekul S, O'Toole JC. 2002c.Yield response of rice (*Oryza sativa* L.) genotypes to drought under rainfed lowland. 3. Plant factors contributing to drought resistance. Field Crops Res. 73:181-200.
Pantuwan G, Fukai S, Cooper M, Rajatasereekul S, O'Toole JC, Basnayake J. 2004. Drought resistance among diverse rainfed lowland rice (*Oryza sativa* L.) genotypes screened at the vegetative stage in dry season and its association with grain yield obtained under drought conditions in wet season. Field Crops Res. 89:281-297.
Rajatasereekul S, Sriwisut S, Porn-uraisanit P, Ruangsook S, Mitchell JH, Fukai S. 1997. Phenology requirement for rainfed lowland rice in Thailand and Lao PDR. Proceedings of an International Workshop, Ubon Ratchathani, Thailand, 1996, Australian Centre for International Agricultural Research, Canberra, ACT. p 97-103.
Sibounheuang V, Basnayake J, Fukai S. 2006. Genotypic consistency in the expression of leaf water potential in rice (*Oryza sativa* L.). Field Crops Res. 97:142-154.
Tsubo M, Basnayake J, Fukai S, Sihathep V, Siyavong P, Sipaseuth, Chanphengsay M. 2006. Toposequential effects on water balance and productivity in rainfed lowland rice ecosystem in Southern Laos. Field Crops Res. 97:209-220.
Tsubo M, Fukai S, Basnayake J, Tuong TP, Bouman B, Harnpichitvitaya D. 2007. Effects of soil clay content on water balance and productivity in rainfed lowland rice ecosystem in northeast Thailand. Plant Prod. Sci. 10(2):232-241.

Notes

Authors' addresses: S. Fukai and J. Basnayake, The University of Queensland, Brisbane, Australia; O. Makara, Cambodia Agricultural Research and Development Institute, Phnom Penh, Cambodia.
Acknowledgments: Many scientists from Thailand, Laos, Cambodia, and Australia contributed to the projects, which were mostly funded by the Australian Centre for International Agricultural Research and the Rockefeller Foundation.

Molecular breeding for drought-tolerant rice (*Oryza sativa* L.): progress and perspectives

Zhi-Kang Li and Yong-Ming Gao

Rice production consumes disproportionally large amounts of water worldwide and there is an urgent need to develop drought-tolerant or water-saving rice cultivars because of the increasing threats of water shortage for arable lands worldwide. Few drought-tolerant rice varieties have been commercially released in the past decades, which could largely be attributed to the lack of breeding efforts specifically targeting improving drought tolerance (DT) and water-use efficiency (WUE), and partially to the complexity of genetics and physiology associated with DT/WUE in rice. Efforts have been limited in applying marker-assisted selection to improve DT in rice despite the numerous studies in genetically dissecting DT in rice using the QTL mapping approach. Progress was made recently in developing drought-tolerant rice cultivars at IRRI, which indicates that the conventional breeding approach is effective for breeding DT for higher-elevation rainfed ecosystems of Asia. However, developing hybrid rice cultivars should be an effective strategy to combine high yield potential with a good level of WUE/DT for most shallow rainfed lowlands of Asia. Backcross breeding and designed QTL pyramiding appear to be a new and promising breeding strategy for the development of large numbers of DT introgression lines in elite genetic backgrounds by large-scale backcross breeding; deep exploitation of useful genetic diversity for DT from the primary gene pool of rice; effective selection, discovery, allelic mining, and characterization of QTL networks for DT; and directed trait improvement by designed QTL pyramiding based on accurate genetic information of QTL networks. Many promising drought-tolerant lines have been developed in the program, even though the theoretical aspects underlying the genetic networks of the target traits and QTL pyramiding by design remain to be fully established.

Rice is the staple food for more than 3 billion people in Asia, where more than 90% of the world's rice is produced and consumed. As a semiaquatic plant species originated from tropical swamps, rice loves water. Rice production in Asia has more than tripled in the past three decades, resulting primarily from the Green Revolution, which dramatically increased rice productivity in high-input irrigated systems (Khush 1999, 2001).

However, rice production requires the use of large amounts of water. Although current high-yielding semidwarf rice cultivars have similar water productivity with respect to transpiration efficiency as other C_3 cereals such as wheat, at about 2 kg grain m^{-3} water transpired (Bouman and Tuong 2001), the total seasonal water input to rice fields is 2–3 times more than for other cereals because of the additional water required for land preparation and higher evaporation rates from the water layer in rice fields (Tuong et al 2005). Thus, water deficiency or drought has been the principal factor limiting rice yields on approximately 46 million ha of rainfed rice fields of Asia (Pandey et al 2000). Most modern semidwarf cultivars are not adapted to rainfed systems, which are characterized by many abiotic stresses. As a result, low-yielding traditional varieties are still grown on about 50% of the rainfed area of Asia, with an average yield of 1–2 t ha^{-1} largely because of their better adaptation to different stresses and favored grain quality (Mackill et al 1996). Also, a significant portion of traditionally irrigated rice-growing areas in Asia has become rainfed fields since water resources are rapidly diminishing due to rapid population growth, water pollution, industrialization, and urbanization. High investment is required to achieve high yields in these areas, thus reducing rice farmers' already very low income from rice production.

Drought stress is a complex phenomenon. It can occur at any time during a cropping season and it fluctuates considerably across years and locations in the rainfed areas of Asia. Drought at the early (germination, seedling, and tillering) stages causes delayed transplanting (in rainfed lowlands) or delayed germination (in uplands) and slowed growth, resulting in poor crop establishment. This reduces the number of panicles per unit area and panicle size. Drought at the reproductive stage (panicle initiation, flowering, and grain filling) causes varied degrees of spikelet sterility and poor grain filling. This latter type of terminal stress tends to cause more severe yield loss because rice is extremely sensitive to drought at the reproductive stage (Cruz and O'Toole 1984). In rainfed systems, rice crops may encounter either or both types of drought in a single season, but the frequency of occurrence of each drought type tends to show specific patterns in different geographic ecosystems, providing robust target environments for breeders.

Mechanisms of drought tolerance in rice

To most plant breeders and physiologists, the final and meaningful definition of drought tolerance (DT) is yield loss under stress and this definition will be used throughout this chapter. Rice plants may achieve their adaptation to either type of drought by complex mechanisms in both physiology and phenology, which are important for identifying specific traits related to DT and for developing appropriate screening techniques in breeding programs. For DT, these systems include a better water uptake system such as deep and thicker roots (Lafitte et al 2002), traits that reduce transpiration or nonproductive water loss from the shoots such as cuticular resistance to water vapor/leaf surface wax (Haque et al 1992, O'Toole and Cruz 1983), high water-use efficiency/rapid stomatal closure and leaf rolling (Dingkuhn et al 1989), and rapid osmotic adjustment and dehydration tolerance (Lilley and Ludlow 1996).

Drought escape by accelerated or delayed flowering under stress may also contribute significantly to rice adaptation to drought depending on specific situations (Lafitte et al 2004, 2006, Xu et al 2005).

The above-mentioned morphological and physiological mechanisms underlying DT are important for understanding how rice plants adapt to stressful conditions of drought. Nonetheless, important questions remain regarding how to determine correct target traits and develop appropriate screening techniques in breeding programs to improve DT because there is no convincing evidence that any DT mechanisms (traits) alone would be sufficient to give rice plants ability to adapt well to specific stress conditions in terms of grain yield.

Genetic basis of DT in rice

It is well known that significant genotypic variation exists among different rice germplasm accessions for DT and its components (Babu et al 1999, 2001, Price et al 1997, Lafitte et al 2006). The complex physiology and phenology involved in DT already imply a complicated genetic basis for this variation in DT. The complex and quantitative nature of DT explains, at least partially, the frustration of breeders in breeding drought-adapted cultivars resulting from the substantial genotype × environment interaction (Fukai and Cooper 1995, Pantuwan et al 2002, Lafitte and Courtois 2002, Kamoshita et al 2002b). Over the past decade, great efforts have been made to genetically dissect DT and its component traits in rice using the QTL mapping approach. Table 1 shows the reported results in mapping QTLs affecting DT and its components in rice from 31 independent studies on 12 different rice populations, which could be summarized in the following four points. First, the number of loci affecting DT and each of its components is very large and these loci are widely distributed across the rice genome, but only a few QTLs are detectable in any specific population/environment. Second, QTLs with large and consistent effects on DT and related traits are few. Third, each individual component trait contributes little to DT and this is true for most QTLs affecting DT component traits. Fourth, epistasis, or interactions among QTLs affecting DT and its components, has not been examined adequately in most studies. Thus, it remains unclear how to apply QTL information from mapping populations to improve DT in breeding populations unrelated to the reference mapping populations because of possible epistasis and QTL × environment interaction, uncertain relationships between secondary traits and grain yield under drought, and unknown allelic diversity at identified DT QTLs in breeding materials (Li et al 2000).

Breeding for improved DT in rice

Improving DT by the conventional breeding approach

Developing drought-tolerant rice varieties has long been recognized as the most efficient way to overcome the problem of drought. However, progress in developing drought-tolerant rice cultivars has been slow. For example, most rice cultivars grown in the rainfed areas of Asia today remain traditional landraces (Pandey, personal

Table 1. Summarized results in mapping QTLs affecting DT and its component traits in rice.

Trait	Pop.[a]	No.[b]	Env.[c]	Type[d]	QTL no.	Reference
Tiller and root traits	1	4	2	RIL	18	Champoux et al (1995)
Tiller and root traits	1	4	1	RIL	29	Ray et al (1996)
Osmotic adjustment and DT	1	1	1	RIL	7	Lilley et al (1996)
Root traits	2	10	1	DH	39	Yadav et al (1997)
Root traits	2	4	1	DH	12	Zheng et al 2000
Drought score	2	1	2	DH	2	Hemamalini et al (2000)
Shoot traits	2	3	2	DH	16	Hemamalini et al (2000)
Root traits	2	5	2	DH	23	Hemamalini et al (2000)
Shoot traits	2	4	3	DH	42	Courtois et al (2000)
Yield and root traits	2	–	2	DH	?	Venuprasad et al (2002)
Root traits	2	4	2	DH/NIL	9	Shen et al (2001)
Plant height and tillering	2	2	2	DH/NIL	3	Shen et al (2001)
Leaf rolling and stomatal conductance	3	2	1	F_2	8	Price et al (1997)
Root traits	3	8	1	F_2	24	Price and Tomos (1997)
Root traits and tillering	3	4	1	RIL	18	Price et al (2000)
Root traits	3	8	2	RIL	24	Price et al (2002)
Dehydration avoidance traits	3	–	2	RIL	17	Price et al (2002)
Grain yield	3	1	2	RIL	3	Lafitte et al (2004)
Yield components	3	6	2	RIL	48	Lafitte et al (2004)
Heading date, plant height	3	2	2	RIL	15	Lafitte et al (2004)
Root thickness	3	1	2	RIL	2	Lafitte et al (2004)
Biomass	3	1	2	RIL	4	Lafitte et al (2004)
Harvest index	3	1	2	RIL	5	Lafitte et al (2004)
Root traits	4	5	2	RIL	28	Ali et al (2000)
Root traits and shoot biomass	4	7	1	RIL	22	Kamoshita et al (2002a)
Root traits	5	7	1	DH	35	Zhang et al (2001)
Osmotic adjustment	5	1	1	DH	5	Zhang et al (2001)
Cellular membrane stability	5	1	1	DH	9	Tripathy et al (2000)
Shoot biomass and root traits	5	7	4	DH	15	Kamoshita et al (2002b)
Root traits	5	7	1	DH	37	Nguyen et al (2004)
Osmotic adjustment	5	1	1	DH	5	Nguyen et al (2004)
Grain yield	5	1	1	DH	5	Babu et al (2003)
Relative yield	5	1	1	DH	2	Babu et al (2003)
Yield components	5	3	1	DH	12	Babu et al (2003)
Heading date and plant height	5	2	1	DH	14	Babu et al (2003)
Shoot traits	5	4	1	DH	8	Babu et al (2003)
Grain yield (DT)	5	1	5	DH	7	Lanceras et al (2004)
Biomass	5	1	5	DH	8	Lanceras et al (2004)
Harvest index	5	1	5	DH	6	Lanceras et al (2004)
Yield components	5	3	5	DH	40	Lanceras et al (2004)
Heading date, plant height	5	2	5	DH	16	Lanceras et al (2004)
Relative yield	7	1	2	RIL	4	Yue et al (2005)
Relative spikelet fertility	7	1	2	RIL	5	Yue et al (2005)

Continued on next page

Table 1 continued.

Trait	Pop.[a]	No.[b]	Env.[c]	Type[d]	QTL no.	Reference
Drought response index	7	1	2	RIL	7	Yue et al (2005)
Leaf traits	7	3	2	RIL	16	Yue et al (2005)
Heading date	7	1	2	RIL	7	Yue et al (2005)
Yield components	7	6	2	RIL	27	Yue et al (2006)
Grain yield	7	1	2	RIL	5	Zou et al (2005)
Yield components	7	4	2	RIL	27	Zou et al (2005)
Relative yield	7	1	2	RIL	3	Yue et al (2006)
Relative yield components	7	6	2	RIL	15	Yue et al (2006)
Root traits	7	11	2	RIL	38	Yue et al (2006)
Leaf drying and rolling	7	2	2	RIL	10	Yue et al (2006)
Osmotic adjustment	8	1	1	BC_3F_3	14	Robin et al (2003)
Heading date and plant height	9	2	2	NIL	26	Xu et al (2005)
Grain yield	9	1	2	NIL	10	Xu et al (2005)
Yield per plant	10	1	2	BC_2F_2	2	Moncada et al (2001)
Yield components	10	4	2	BC_2F_2	13	Moncada et al (2001)
Heading date and plant height	10	2	2	BC_2F_2	10	Moncada et al (2001)
Leaf size/ABA accumulation	11	3	1	F_2	17	Quarrie et al (1997)
Root traits	12	7	2	RIL	40	Li et al (2005)

[a]Populations: 1 = CO39 × Moroberekan, 2 = IR64 × Azucena, 3 = Azucena × Bala, 4 = IR58821 × IR52561, 5 = CT9993 × IR62266, 6 = Kalinga III × Azucena, 7 = Zhenshan97 × IRAT109, 8 = IR62266 × IR60080-46A, 9 = Teqing × Lemont, 10 = Caiapo × *O. rufipogon* L., 11 = IR20 × 63-83, 12 = Yuefeng × IRAT109, 13 = IR1552 × Azucena. [b]No. = the number of component traits studied. [c]The number of environments in which the studies were conducted. [d]DHL, RIL, and NIL represent doubled haploid lines, recombinant inbred lines, and near-isogenic lines.

communication). There are two major reasons for this. First, rice cultivation has been historically accompanied by steadily improved irrigation and yield potential. Thus, past rice breeding efforts worldwide were largely devoted to increasing yield potential under high-input conditions. In other words, breeding for improved DT was largely neglected in most Asian breeding programs in the past. As a result, most modern high-yielding semidwarf rice varieties were poorly adapted to the water-limited conditions of rainfed systems where low-yielding traditional landraces are still widely grown because of their better DT. This situation is changing as breeding for improved DT has recently become the research priority of the International Rice Research Institute (IRRI) to reduce poverty in fragile rainfed systems (see IRRI Medium-Term Plan of 2007) and of many national breeding programs of Asian countries. Second, the complex physiology, phenology, and genetics as well as large environmental effects and genotype × environment interaction involved in DT make it difficult to combine high yield potential of modern rice cultivars with a desirable level of DT through the conventional breeding strategy. For example, different levels of DT are required to achieve yield stability for different target environments. For most shallow rainfed lowlands of Asia that are characterized by high (or potentially high) productivity but become increasingly water-deficit or drought-prone, a new variety should have a good level of water-use efficiency (WUE)/DT at the reproductive stage combined with high yield potential and some other desirable properties such as good grain quality and biotic stress resistance in order to be beneficial to farmers in these areas. On the other hand, a good level of tolerance of delayed transplanting plus a high level of DT at the reproductive stage are required to adapt well to higher-elevation rainfed areas of South/Southeast Asia where drought occurs more frequently. For the upland ecosystem with frequent and severe drought, a new variety should have a high level of DT during its whole life cycle plus excellent resistance to rice blast.

The conventional breeding approach based on line crossing and mass selection remains the predominant method in all rice breeding programs worldwide. The two key elements for success in improving rice DT using the conventional breeding method are to generate sufficient genetic variation for DT in breeding populations and to develop a reliable and feasible screening protocol to identify individuals with target traits from large segregating populations. Although yield under stress is used as a target trait for DT by most rice breeders today, this trait itself can result from a wide range of adaptive strategies in breeding populations segregating for flowering time, including different types of DT, drought avoidance, drought escape, and general adaptability to specific environments. Lafitte et al (2006) tested 166 rice germplasm accessions from around the world under mild terminal drought in lowland conditions that reduced, on average, grain yield to 84% of the control value. The tested varieties showed a wide range of yield responses to the stress—some lines produced up to 150% as much grain yield under stress as in the control, whereas others suffered a yield reduction of more than 90% (Table 2). They found that yield under full irrigation was positively correlated with yield under stress ($r = 0.55$, $P < 0.001$), even though cultivars with greater yield potential tended to be affected more by stress than low-potential or poorly adapted cultivars.

In practical breeding, breeders tend to use upland rice landraces, the only ecotype that adapts well to the more extreme drought in the rainfed uplands of Asia, as donor parents for improving DT in their breeding programs. Line crossing between drought-tolerant upland ecotypes and high-yielding lowland varieties creates tremendous segregation for DT and related traits in breeding populations, as seen in most QTL mapping populations (Table 1). But it is also difficult to break undesirable linkage between DT and poor yield potential associated with most upland landraces, particularly when breeders are targeting developing high-yielding and drought-tolerant varieties for shallow rainfed lowlands.

Marker-assisted selection (MAS) for improving DT in rice

As mentioned above, past efforts in identifying QTLs affecting DT and its components were primarily aiming at improving DT by MAS if QTLs affecting secondary traits of DT could be accurately mapped and characterized (Lafitte and Courtois 2000). To date, no drought-tolerant rice varieties have been developed and released to farmers by MAS even though a few attempts have been made to apply MAS to improving DT in rice. Shen et al (2001) reported an effort at IRRI to introgress a large segment of rice chromosome 1 containing a putative QTL for deep and thick roots from an upland

Table 2. Summarized statistics of the performance of 166 parental lines under continuously flooded lowland conditions (L irr), lowland conditions but with stress imposed near heading (L stress), upland (aerobic soil) conditions with frequent irrigation (U irr), or upland conditions with restricted irrigation to impose stress (U stress). Grain yield in the upland experiment is the average measured across the two irrigation regimes.

Water level	Grain yield (g m^{-2})		Plant height (cm)		Flowering date (d)	
	Mean ± SD	Range	Mean ± SD	Range	Mean ± SD	Range
L irr	304 ± 135	42–667	103 ± 22	40–157	82 ± 12	48–106
L stress	236 ± 114	3–547	99 ± 20	45–145	83 ± 13	50–106
U irr	50 ± 30[a]	0–141[b]	75 ± 15	38–135	86 ± 12	52–116
U stress			66 ± 14	45–123	88 ± 12	54–112
% change L	–16 ± 42	–94–152	–3 ± 10	–28–37	1 ± 15	–18– 18
% change U**	–83 ± 21	–100– 58	–11 ± 10	–30–30	4 ± 7	–15– 27

[a]Grain yield reported for upland experiments is the average of both irrigation levels. [b]For yield, this is % change relative to irrigated lowland. For other traits, this is the % change from the upland irrigated treatment to the upland stress treatment.
Source: Lafitte et al (2006).

cultivar, Azucena, into IR64 using MAS and found that the majority of BC progeny carrying the desired introgressions failed to show expected deeper roots than IR64. In a six-year effort, Steele et al (2006) were able to put four QTLs for deep roots from Azucena into an elite cultivar, Kalinga III, by MAS, but only one of the four target QTLs expressed the expected effect for increased root length.

Today, most rice breeders are still reluctant to apply MAS to improving complex traits such as DT in their breeding programs. This is not surprising because, in addition to relatively high costs, most information is missing for breeders to choose appropriate target QTLs, which includes the magnitudes and consistency of identified QTLs in the target genetic backgrounds and environments and the possible genetic drag associated with the transfer of target QTLs.

Improving rice DT by backcross breeding and designed QTL pyramiding

Recently, a new strategy, "trait improvement by designed QTL pyramiding," has been successfully applied to combine high yield potential with significantly improved DT in rice as part of the International Rice Molecular Breeding Network coordinated at IRRI (Yu et al 2003, Lafitte et al 2006, Li et al 2005, Li 2006). Technically, this strategy has three steps: (1) developing introgression lines (ILs) for DT by backcross breeding, (2) identifying genes/QTLs and genetic networks for DT by using ILs and DNA markers, and (3) developing drought-tolerant rice cultivars by designed QTL pyramiding (Fig. 1). These steps are described separately as follows.

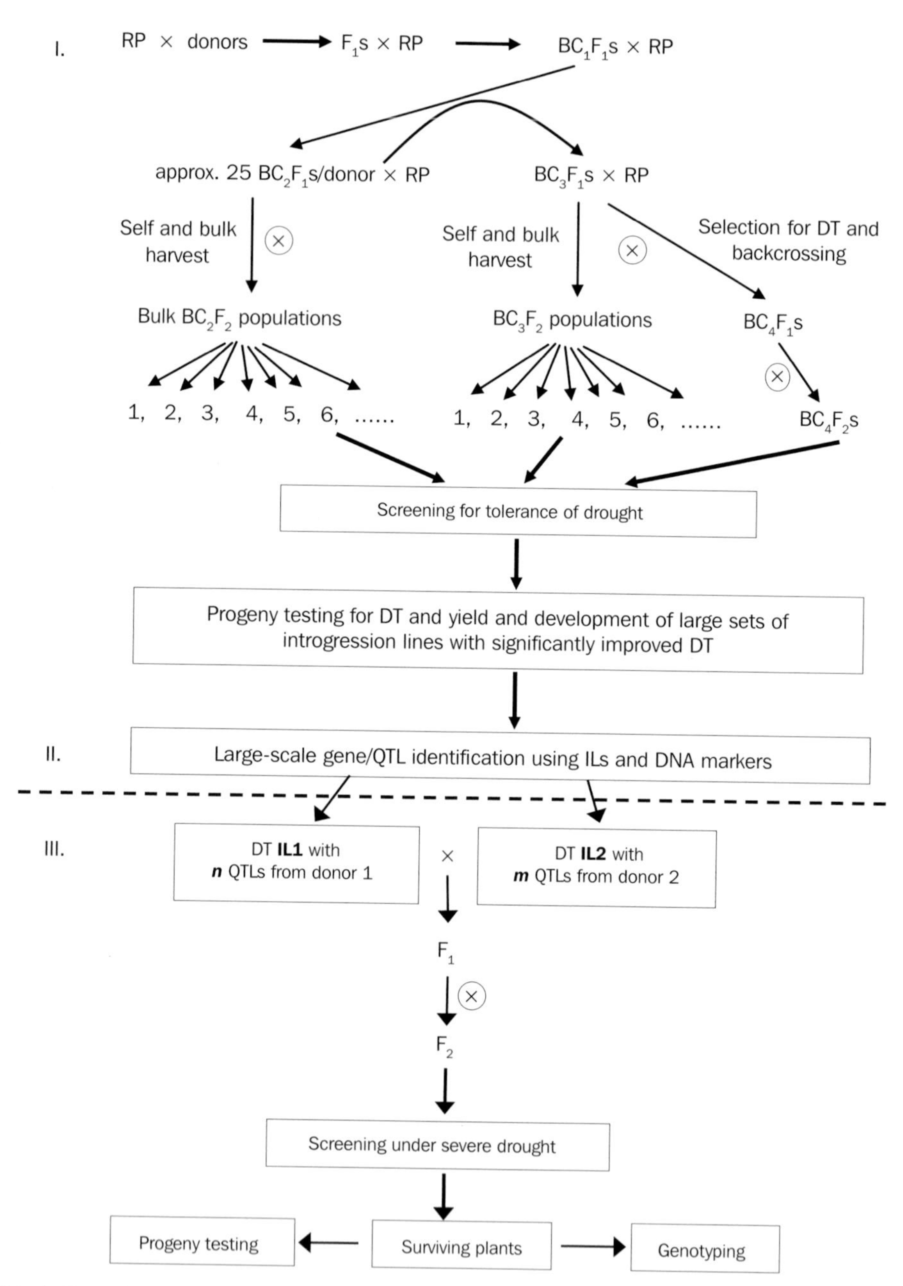

Fig. 1. Procedures of backcross breeding (I) for developing DT introgression lines (ILs), identifying QTLs/networks (II), and developing DT rice cultivars by designed QTL pyramiding (III); RP = recurrent parent.

Table 3. Summary of the selection experiments of the BC_2F_2 progeny for drought tolerance in two dry seasons at IRRI.[a]

Year	Stress	RP	N	No. of surviving (selected) plants per population			
				n	Range	Total	SI (%)
2000	LL	IR64	64	64	2– 23	583	4.55
2000	UL	IR64	25	25	5– 29	326	7.41
2001	LL	IR64	60	59	0–110	2,192	18.26
2000	LL	Teqing	31	31	3– 15	279	4.50
2000	UL	Teqing	22	22	2– 22	182	4.70
2001	LL	Teqing	47	28	0– 30	210	2.24
2000	LL	NPT	62	32	0– 9	62	0.50
2001	LL	NPT	51	42	0– 66	835	8.18
		Total	362	303		4,669	6.8

[a]RP, N, n, and SI are the recurrent parents, the total number of BC_2F_2 bulk populations screened, the number of BC_2F_2 populations with surviving plants, and the selection intensity; LL and UL are the lowland and upland stress conditions. In lowland conditions, the stress was applied at the reproductive stage, and, in both seasons, the recurrent parents did not survive. In the upland conditions of 2000, IR64 did not survive but Teqing did (those plants that performed better than Teqing were selected).
Source: Lafitte et al (2006).

Developing introgression lines for DT by backcross breeding

In the first step, a large-scale backcross breeding program was used to introgress useful genes/QTLs from the primary gene pool of rice into elite genetic backgrounds and develop large numbers of ILs with significantly improved DT. At IRRI, three elite rice lines, IR64 and Teqing (high-yielding and widely adaptable indica varieties), and a new plant type (NPT, a high-yielding tropical japonica line), were used as the recurrent parents (RPs) and crossed with 195 diverse donors, and backcrossed twice to the RPs to create many BC_2F_2 bulk populations (Fig. 1, Ali et al 2006). The parental lines of the backcross breeding program originated from 34 countries worldwide and represent a significant portion of the genetic diversity in the primary gene pool of rice according to a survey with 101 well-distributed SSR markers (Yu et al 2003).

For DT, 362 BC_2F_2 bulk populations were screened under two types (lowland and upland stress) of severe drought that killed the RPs (Lafitte et al 2006), resulting in 4,669 selected BC_2F_2 plants that showed better DT than the RPs (Table 3). Progeny testing indicated that most selected BC progeny indeed had improved DT when compared with the respective RPs (Lafitte et al 2007). Interestingly, transgressive drought-tolerant plants were identified from 83.7% of the screened BC populations, including 99.3% of the IR64 populations, 81.0% of the Teqing populations, and 65.5% of the NPT populations. The number of survival plants selected from each bulk ranged from 0 to 110, with an average selection intensity of 6.8% (10.6% for the IR64 populations, 3.5% for the Teqing populations, and 4.0% for the NPT populations).

Table 4. Genetic background effects on the selection of drought-tolerant BC_2F_2 plants under the lowland water stress during the 2000-01 dry season.

Donor	Performance (%)[a]	Recurrent parent			Donor	Performance (%)	Recurrent parent		
		IR64	Teqing	NPT			IR64	Teqing	NPT
B4122	–47.6	37	2	1	Hei Mi Chan	46.6	52	2	
Shwewartun	–41.5	5	0	6	Tek Si Chut	–50.0	13	0	
Pokhreli	152.4	119	7	30	Sadajira 19	–100.0	55	0	
Khole marshi	–39.2	84	30	14	Dacca 6	–58.0	20	19	
UPR191-66	–40.6	68	0	24	Zale	–9.8	2	0	
ASD18	20.3	59	0	54	Giza 14	–	29	4	
IRBB60	–36.0	110	0	66	M202	86.3	46		45
SML242	–81.6	6	0	5	Jumli Marshi	82.6	72		40
Rusty Late	–31.5	38	2	4	Rasi	–	63		9
CHIPDA	–19.1	47	0	85	Moroberekan	71.8	13		11
Ziri	6.4	10	5	0	TGMS29	33.5	22		3
Vary Lava 16	58.1	24	0	5	Palung 2	–51.5	33		36
LA 110	–13.3	25	5	47	SLG-1	–100.0	26		0
Khumal 4	21.2	0	0	0	Dhan4	–51.5		1	0
Pusa	–33.8	15	4		ASD 16	–9.8		10	0
Guang122	13.7	52	0		Jalmagna	–		0	28
Minghui63	–27.5	23	0		TKM 6	63.8		11	5
MR 77	0.5	31	0		UP 15	–41.7		22	24
Budda	–18.9	75	11		UZ-Ros 275	50.0		6	37
Doddi	–19.3	81	2		Chorofa	–48.7		1	20
Gajale	–	61	22		Mean		41.6	4.9	21.4

[a]Population size was 250 plants per population and donor performance was the percentage of yield reduction under the lowland stress condition.
Source: Lafitte et al (2006).

The unique feature of the backcross breeding program is that many donors of divergent origins were used and all backcross populations were screened for DT regardless of donor performance for the target traits. Progeny testing indicated that most selected backcross progeny indeed had improved DT, even though individual selected BC progeny did vary for the amount of tolerance. Several important results were obtained. First, there are tremendous amounts of "hidden" diversity in the primary gene pool of rice for DT, reflected by the fact that BC progeny showed transgressive performance of DT over the parental lines in most backcross populations regardless of the performance of their donors. In other words, DT genes appear to be widely and randomly distributed in the primary gene pool of rice. Thus, the common practice of selecting donor parents based on phenotype practiced by most breeders is a poor way to exploit this hidden diversity. Second, it was common to identify BC progeny with extreme phenotypes. Third, selection efficiency, defined as the number of superior progeny identified per backcross population, is highly dependent upon (1) the recipient genetic background (Table 4), (2) the recipient by donor combinations, and

(3) the levels of stress applied. More severe stresses could significantly increase the accuracy of selection and reduce the number of total selected plants to a manageable size. Third, the first round of selection (screening) should be done in the BC_2 instead of BC_3 generation because much reduced selection efficiency was observed in the latter. Fourth, backcross breeding combined with direct selection for yield under severe stress is a highly effective way to improve DT in rice because most individuals in a BC population have the same genetic background, and are less affected by the genetic "noise" from co-segregating nontarget traits, such as heading date and plant size. It is also easier to apply a uniform severe stress at the critical developmental stage(s) and to identify superior BC progeny based on direct comparison with RPs.

One potential limitation of this approach is that the level of stress needed to reveal genetic variation in single-plant screens may be unrealistically severe. Further evaluation of the selected lines has to be conducted to establish gains resulting from the first round of screening. In addition, applying an appropriate level of stress for DT remains a major challenge under field conditions. Nevertheless, the high probability of being able to identify large numbers of drought-tolerant progeny in our advanced backcross populations demonstrated that, despite the complex genetics and diverse physiological mechanisms, introgression of genes from a diverse source of donors into elite genetic backgrounds through backcross breeding and efficient selection is a powerful way to exploit hidden diversity for the genetic improvement of DT in rice.

Discovering and mining QTL alleles for DT using ILs and DNA markers

As mentioned above, three large sets of ILs with significantly improved DT have been developed at IRRI, which are unique in two aspects. First, all sister lines within a single set of ILs are in the same elite genetic background but each has a few introgressed genomic segments associated with DT from a known donor. These ILs are valuable genetic materials for characterizing the effect of specific introgressions on DT and related traits. Second, the three sets of ILs together contain a wide range of DT types and QTLs from many donors of diverse origin, providing a unique set of genome-wide genetic stocks for large-scale QTL/allele discovery and functional genomics research on DT in rice (Li et al 2005). Third, further improvement of DT can be achieved by designed trait/QTL pyramiding using populations derived from crosses between promising sister ILs carrying different sets of target QTLs, which will be described in the next section.

The second step is to characterize the genome-wide introgression patterns in the ILs and identify DT QTLs (donor segments that are responsive to selection for DT) using DNA markers (Fig. 1). The principle of using selected ILs and DNA markers to identify and map QTLs affecting DT is straightforward and takes advantage of both linkage mapping and linkage disequilibrium (LD) mapping (Li et al 2005). Figure 2 shows the introgression pattern and the X^2 profiles along chromosomes 1 and 2, based on SSR marker genotypes, of 38 DT ILs selected under severe lowland (at the reproductive stage) and upland drought from the IR64/Type3 BC_2F_2 population to demonstrate the methodology of QTL identification using ILs. A total of 46 DT QTLs were detected in which the Type3 (donor) allele and genotypic frequencies deviated

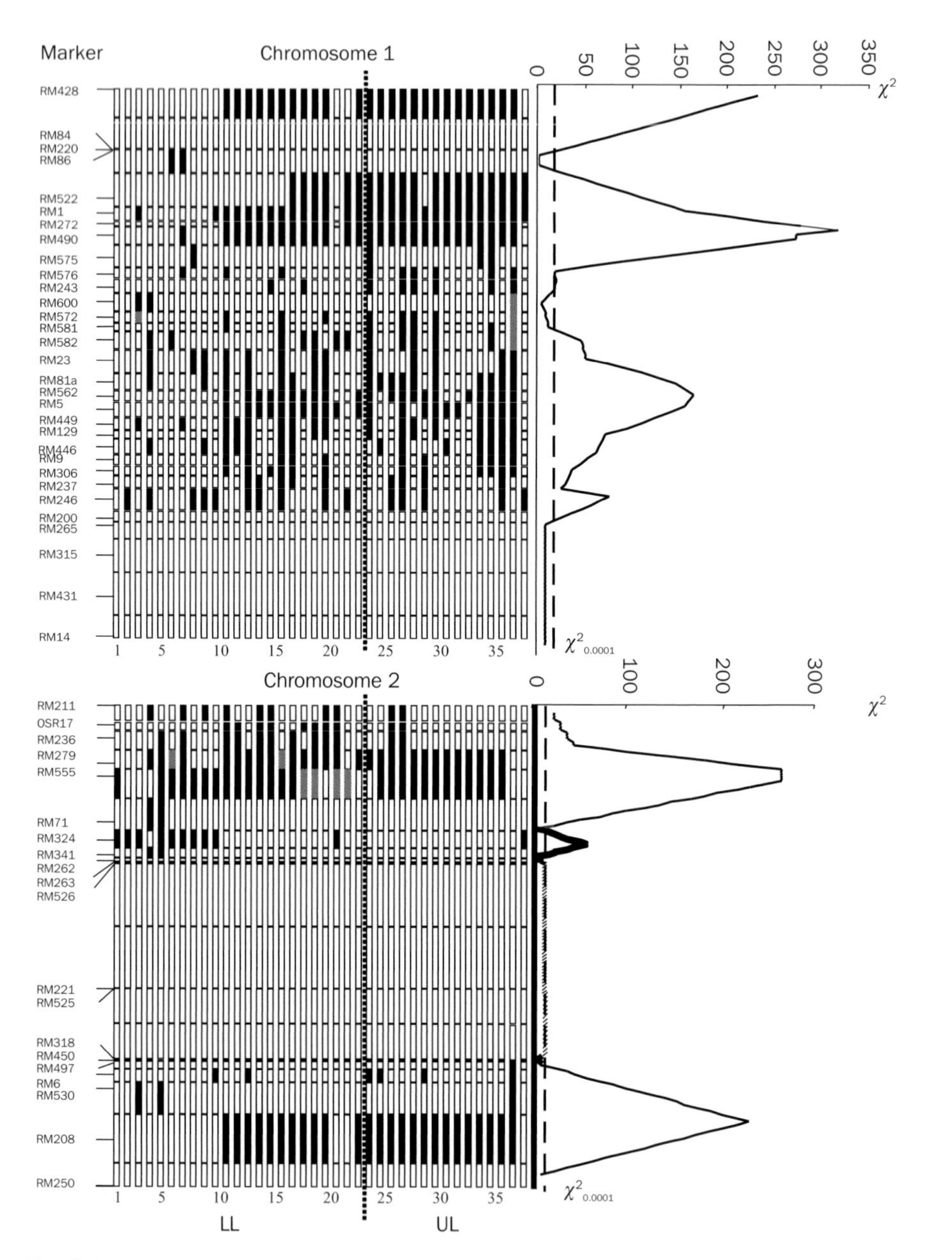

Fig. 2. Introgression patterns and chromosomal locations of QTLs for drought tolerance detected by linkage and linkage disequilibrium of SSR markers in 38 introgression lines selected from the IR64 × Type3 BC_2F_2 population, which survived very severe lowland drought (LL) and upland drought (UL). χ^2 values are obtained from the deviations of the observed genotypic frequencies from the expectations (see Table 5).

Table 5. Forty-six QTLs for drought tolerance detected by linkage or linkage disequilibrium in 38 introgression lines selected from the IR64/Type3 BC_2F_2 population under lowland and upland drought.

QTL[a]			Lowland (n = 23)[b]		Upland (n = 15)		Total (n = 38)		
Marker	Bins	AL	Frequency	X^2_2	Frequency	X^2	Frequency	X^2_2	GA[c]
RM428	1.2	AL1	0.565	70.2	0.933	194.1	0.711	231.1	A
RM490	1.4	AL1	0.652	117.4	1.000	225.0	0.789	318.5	A
RM562	1.6	AL1	0.523	49.6	0.833	118.9	0.649	152.5	A
RM5	1.7	AL1	0.500	68.0	0.667	93.7	0.566	156.2	A
RM246	1.8	–	0.455	45.1	0.533	32.2	0.486	69.1	A
OSR17	2.1	AL1	0.326	23.5	0.300	7.9	0.316	20.8	A
RM555	2.3	AL1	0.826	204.3	0.867	165.6	0.842	368.3	A
RM324	2.5	AL2	0.500	55.0	0.067	–	0.500	55.0	A
RM6	2.9	–	–	–	0.333	10.8	0.333	10.8	A
RM208	2.11	AL1	0.522	67.9	0.933	194.1	0.684	229.9	A
RM81B	3.1	–	0.391	43.8	0.400	30.0	0.395	73.9	A
RM231	3.2	AL1	0.630	100.3	0.933	194.1	0.750	272.9	A
RM411	3.8	–	0.348	23.4	0.500	27.1	0.408	43.6	CD
RM504	3.9	–	0.182	7.8	0.643	80.8	0.361	57.2	A
RM293	3.10	AL1	0.565	99.9	0.867	165.6	0.684	251.3	A
RM85	3.12	–	0.674	118.4	0.500	27.1	0.605	131.4	A
RM307	4.1	AL1	0.435	55.6	0.667	93.7	0.526	140.9	A
RM471	4.4	I	0.565	82.9	0.933	194.1	0.711	250.7	A
RM241	4.5	–	–	–	0.500	29.5	0.500	29.5	CD
RM349	4.8	AL1	0.674	118.4	0.933	194.1	0.776	295.6	A
RM122	5.1	AL1	0.413	26.3	0.767	96.8	0.553	102.6	A
RM13	5.2	AL3	0.043	–	–	–	–	–	A
RM509	5.4	AL1	0.870	229.7	0.967	194.6	0.908	422.6	A
RM161	5.5	AL3	0.043	–	–	–	–	–	A
RM276	6.3	AL1	0.696	136.8	0.933	194.1	0.789	318.5	A
RM527	6.5	AL1	0.261	9.5	0.567	56.8	0.382	50.7	A
RM528	6.7	–	0.391	22.0	0.500	31.2	0.434	46.7	PD
RM340	6.8	AL1	0.326	12.1	0.643	51.9	0.446	51.3	A
RM494	6.9	AL1	0.283	7.3	0.467	23.2	0.355	26.1	PD
RM432	7.3	AL1	0.391	43.8	0.867	165.6	0.579	174.1	A
RM346	7.4	AL2	–	–	0.067	–	–	–	A
RM172	7.7	AL1	0.630	117.1	0.933	194.1	0.750	294.9	A
RM408	8.1	AL2	0.130	–	0.067	–	–	–	A
RM38	8.2	–	0.262	6.1	0.700	93.5	0.433	58.4	A
RM126	8.3	AL1	0.761	159.1	1.000	225.0	0.855	369.0	A
RM331	8.4	–	0.283	16.1	0.433	18.4	0.342	27.4	A
RM223	8.5	AL1	0.478	44.3	0.333	10.8	0.421	52.4	A
RM264	8.8	AL2	0.478	68.9	0.067	–	0.478	68.9	A
RM321	9.3	AL2	–	–	0.067	–	–	–	A
RM242	9.6	AL3	0.043	–	–	–	–	–	A
RM271	10.4	AL1	0.370	23.3	0.567	44.1	0.447	62.6	A

Continued on next page

Table 5 continued.

QTL[a]			Lowland (n = 23)[b]		Upland (n = 15)		Total (n = 38)		
Marker	Bins	AL	Frequency	X^2_2	Frequency	X^2	Frequency	X^2_2	GA[c]
RM171	10.5	AL3	0.043	–	–	–	–	–	A
RM228	10.6	AL2	–	–	0.067	–	–	–	A
RM202	11.3	AL2	0.130	–	0.067	–	–	–	A
RM206	11.6	AL1	0.500	42.0	0.700	93.5	0.579	116.6	A
RM19	12.2	AL2	0.130	–	0.067	–	–	–	A

[a]Markers located at the peaks of the X^2 statistics (Fig. 2) and the bins each represent a genomic region of approximately 20 cM, in which the number before the dot indicating the chromosome and the number after the dot indicating the position of the bin, starting from the top of the chromosome. Grouping of the QTLs was based on the results of LD analyses and underlined markers are those involved in association loops or ALs (Fig. 3A and 3B). [b]The frequency indicates the frequency of introgression at each marker and X^2 statistics were obtained based on the deviations of the observed genotypic frequencies in the selected ILs from the expectations in a BC_2F_2 population. X^2_2 values at significance levels of P = 0.05, 0.01, 0.001, and 0.0001 are 6.0, 9.2, 13.8, and 18.4. [c]Gene action is inferred based on the observed genotypic frequencies of the selected ILs, in which additivity (A) is suggested by excess donor homozygote, complete or partial dominance (CD or PD) by excess of both donor homozygote and heterozygote, and overdominance (OD) by excess heterozygote. The underlined markers are those detected by LD analyses at a threshold of $P < 0.0000001$ (Fig. 3).
Source: Li et al (2005).

significantly from expectation (Table 5). These included 34 QTLs with excessive introgression in ILs selected under both lowland and upland conditions and 2 QTLs on chromosomes 2 and 4 (bins 2.9 and 4.5) specifically detected in the 15 upland-selected ILs. Most DT QTLs appeared to be additive because the donor homozygote at these loci was apparently favored by selection, and only four loci (bins 3.8, 4.5, 6.7, and 6.9) appeared to be dominant or partially dominant with both the donor homozygote and heterozygote favored at these loci. Nineteen of the QTLs appeared to have large effects on DT with introgression frequencies > 0.55 (Table 5). In addition, eight QTLs (underlined) were detected in two association loops from LD analyses (Fig. 3).

Gametic LD analyses revealed large numbers of nonrandom associations between or among the introgressed donor loci in the 38 selected DT ILs from the IR64/Type3 BC_2F_2 population and most of these nonrandom associations occurred between unlinked loci (Li et al 2005). Figure 3 shows two high-confidence genetic networks constructed based on the principle of hierarchy and complete genetic overlap between loci. Figure 3A is the genetic network constructed based on 244 nonredundant significant *Ds* between 36 loci detected in the 23 lowland-selected ILs and Figure 3B is the one built upon 270 nonredundant significant LDs between 33 DT loci detected in the 15 upland-selected ILs from the same population. The genetic overlap between the two networks from the independent selection experiments of the same BC population in the lowland and upland drought, measured as the percentage of the same loci in both networks, was 85.1%. These results indicate that large numbers of QTLs are acting in a hierarchical manner in response to the strong directional selection for DT. The strong and positive associations between unlinked DT loci within each of the QTL

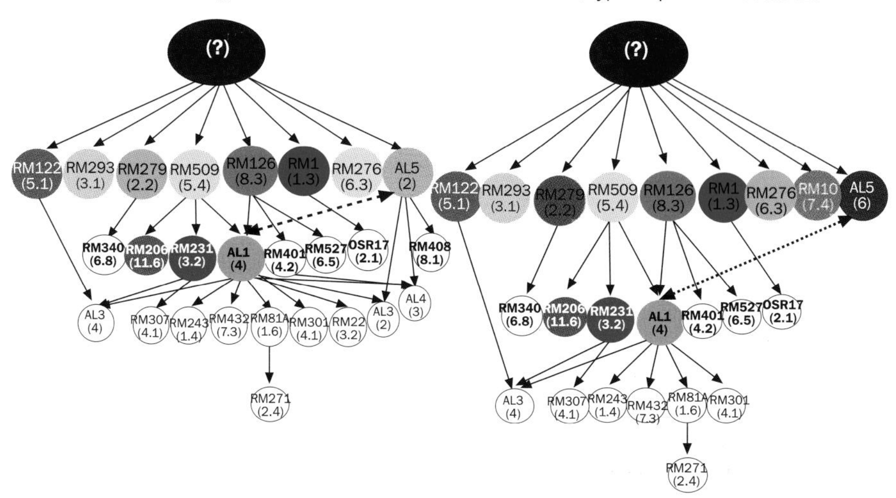

Fig. 3. Genetic networks constructed based on the principle of hierarchy and the level of introgression of the DT QTLs and association loops (ALs) detected in 23 lowland-selected DT ILs (A) and 15 upland-selected ILs (B) from the IR64/Type3 BC_2F_2 population. Here, an AL is defined as a group (2–6) of unlinked but perfectly and positively associated loci in the selected ILs from the BC population (Li et al 2005). The hierarchy of each network was determined based on the level of introgression in which a hypothetical locus/AL with 100% introgression was put on the top (presumably as a gene for signal transduction), followed by loci or ALs with successively decreasing introgression. The arrows in the networks each have three meanings: (1) the level of introgression of the locus or AL it points to, (2) the level of the absolute overlap (D′ = 1.00) between the two loci/ALs it is linked to, and (3) the inclusive relationship of the locus/AL at both the lower and upper levels. The number under each marker represents the chromosome and the bin it is in and the number under an AL indicates the number of loci it contains (Gao et al 2008).

association loops (ALs) indicate that QTLs are acting in groups and loci within a QTL group were possibly co-regulated in response to selection. This type of multilocus structure and similar genetic networks were detected in the 793 DT ILs selected from 67 BC_2 populations. Further data analyses from progeny testing have detected large effects on multiple phenotypes associated with individual ALs, indicating that these ALs were indeed the targets of selection (data not shown). Thus, identification and characterization of the genetic networks associated with DT and salt tolerance should be an important task in future QTL mapping studies.

Developing drought-tolerant rice cultivars by designed QTL pyramiding

In the third step (Fig. 1), promising ILs that have the same or better yield potential and unrelated DT QTLs from different donors are identified based on results from steps I and II and used as parents for QTL pyramiding. Crosses are designed and made between promising sister ILs to produce segregating F_2 populations, which are then screened for DT under severe stress to identify superior individuals that have

Table 6. Ten pyramiding F_2 populations from crosses between 14 promising DT IR64 introgression lines from which 560 surviving F_2 plants carrying DT QTLs from two different donors were selected under severe lowland drought during the 2002-03 dry season.

Cross	Female introgression line			Male introgression line			F_2 population	
	Code	Donor name[a]	Origin	Code	Donor	Origin	N	n
1	1	STY (I)	Myanmar	5	BR24 (I)	Bangladesh	237	25
2	1	STY (I)	Myanmar	6	BR24 (I)	Bangladesh	190	55
3	2	BR24 (I)	Bangladesh	10	Zihui100 (I)	China	299	30
4	3	BR24 (I)	Bangladesh	11	Binam (J)	Iran	318	90
5	3	BR24 (I)	Bangladesh	12	OM1723 (I)	Vietnam	305	105
6	4	BR24 (I)	Bangladesh	12	OM1723 (I)	Vietnam	248	55
7	4	BR24 (I)	Bangladesh	11	Binam (J)	Iran	154	30
8	7	Type3 (I)	India	13	Haoannong (J)	China	255	70
9	8	Type3 (I)	India	10	Zihui100 (I)	China	235	70
10	9	Zihui100 (I)	China	14	Haoannong (J)	China	219	30

[a]STY = Shwe-Thwe-Yin; I is indica and J is japonica.

desirable QTLs for DT from two different donors and good yield potential. This third step can be repeated to pyramid multiple QTLs from 4 and 8 different donors in the second- and third-round QTL pyramiding to develop superior drought-tolerant rice cultivars. At IRRI, 10 F_2 populations developed this way were screened under very severe lowland drought, resulting in 560 drought-tolerant F_3 progeny (Table 6). The selection efficiency of 22.8% in the designed QTL pyramiding of F_2 populations was much higher than the 6.8% in the first-round screened BC_2 populations under less severe stress (see Table 3).

Figure 4 shows the genotypic frequencies of 25 DT QTLs segregating in the 25 surviving plants from the first F_2 population in Table 6, in which the female IL has 16 DT QTLs from a Bangladeshi upland variety, BR24, and the male IL has 9 DT QTLs from a commercial variety, Shwe-Thwe-Yin, from Myanmar. At all QTLs, the observed allelic and genotypic frequencies deviated significantly from the expectations as one allele was significantly more frequent than the others, including six loci (QTLs 4, 7, 9, 11, 13, and 14) at which one of the alleles was fixed (Fig. 4). All QTLs, except QTL 20 on chromosome 10, showed excess homozygosity and the heterozygote was virtually eliminated at most loci. On average, each of the selected F_2 lines had 22.5 DT QTLs, ranging from 12 to 25. LD analyses again detected many highly significant ($P < 0.001$) nonrandom associations between the segregating DT QTLs, which resulted in a genetic network consisting of two ALs of 11 largely unlinked loci, indicating the presence of complex epistatic relationships among many of the DT QTLs (Fig. 5). This type of non-Mendelian segregation in large numbers of segregating DT QTLs across the genome in this population was observed in all 10 pyramiding F_2 populations, and was later found to be under apparently epigenetic control (data not shown).

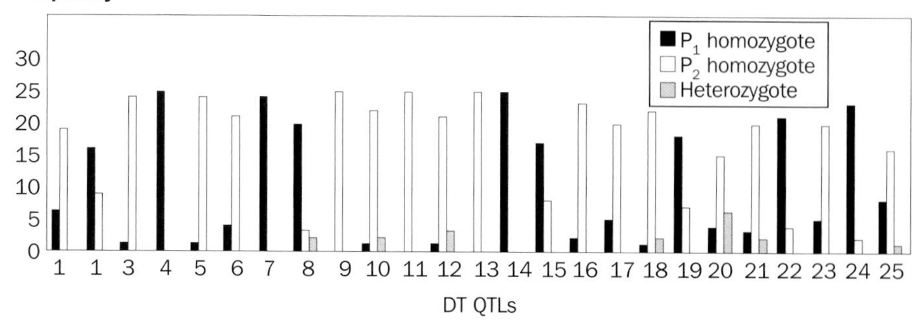

Fig. 4. Genotypic frequencies of 25 segregating DT QTLs in 25 selected drought-tolerant F_2 plants.

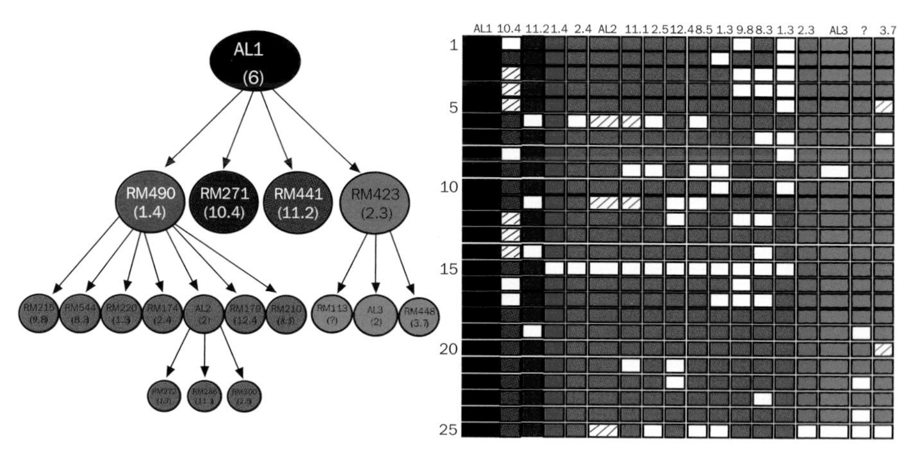

Fig. 5. Graphical genotypes of 25 segregating DT loci detected in 25 progeny selected under severe drought from an F_2 population derived from a cross between two ILs, DGI 21 (IR64/STY) and DGI 60 (IR64/BR24), and the genetic network consisting of all 25 loci identified from the population.

Progeny testing in replicated experiments in 2004 indicated that, on average, the 25 pyramiding F_4 lines were flowering 3 days earlier and yielded 2.84 times as much as IR64 under severe terminal stress. Under nonstress conditions, these lines were 2 days earlier in heading and 2.5 cm taller than IR64, and yielded the same as IR64. However, there was considerable variation among the 25 lines under both stress and nonstress conditions, which led to the identification of four promising lines that, on average, outyielded IR64 by 37.8% under nonstress conditions and by 238% under stress (Tables 7 and 8). In fact, many promising lines have been developed from the 10 pyramiding populations in Table 6, which had significantly improved DT and yield potential when compared with the recurrent parent, IR64, the most widely grown variety in South and Southeast Asia (data not shown).

Table 7. Performance of 25 selected drought-tolerant F_4 plants with pyramided QTLs in replicated experiments under terminal stress and nonstress conditions during the 2004 dry season.

Traits		QTL pyramiding F_4 lines				IR64	
		Mean	SD	Min	Max	Mean	SD
Days to heading (no.)	(S)	87.2	2.9	79.3	91.0	89.7	1.8
	(N)	80.1	2.1	75.0	83.5	82.0	2.3
S – N		7.1**		4.3	7.5	7.7**	
Plant height (in cm)	(S)	60.7	3.7	50.6	69.1	58.4	3.3
	(N)	79.9	5.6	67.8	88.4	77.4	3.2
S – N		–19.1**		–17.2	–19.4	–19.0**	
Tillers/plant	(S)	22.6	1.5	20.0	26.0	23.3	2.5
	(N)	23.3	2.3	19.3	28.3	23.5	3.2
S – N		–0.7		0.7	–2.3	–0.2	
Panicles/plant	(S)	14.4	2.2	9.0	18.7	14.6	3.5
	(N)	22.3	2.0	18.6	26.3	21.9	3.0
S – N		–7.9**		–9.6	–7.7	–7.4**	
Fertility (%)	(S)	58.0	11.3	29.7	76.9	33.8	16.0
	(N)	87.5	5.3	73.5	96.6	89.0	7.6
S – N		–29.5**		–43.8	–19.7	–55.3**	
Grain yield/plant (g)	(S)	16.5	9.8	3.4	35.9	7.2	4.8
	(N)	179.4	38.0	109.4	246.9	167.7	46.4
S – N		–162.9**		–106.1	–210.9	–160.5**	

*Significant difference with $P<0.05$, ** significant difference with $P<0.01$.

Table 8. Performance of top five pyramided F_4 lines in replicated experiments under terminal stress and nonstress conditions during the 2004 dry season.[a]

Lines (no.)	HD (d) (S)	PH (cm) (S)	TN (S)	PN (S)	FT% (S)	GY (g) (S)	HD (d) (N)	PH (cm) (N)	TN (N)	PN (N)	FT% (N)	GY (g) (N)
6	87.0	58.8	21.0	14.7	60.8	21.4	80.0	86.9	19.3	18.6	94.6	246.9
8	87.7	61.5	25.0	17.0	58.7	18.1	80.0	77.6	24.2	23.8	91.3	220.2
17	84.7	61.3	23.2	18.1	71.4	25.4	77.3	75.3	22.6	22.1	85.5	215.9
23	85.3	65.4	24.1	18.7	72.0	32.3	79.7	86.2	23.2	22.1	91.6	227.9
Mean	86.2	61.8	23.3	17.1	65.7	24.3	79.3	81.5	22.3	21.7	90.8	227.7
SD	1.2	2.4	1.5	1.5	6.0	5.3	1.1	5.1	1.8	1.9	3.3	11.9
IR64	89.7	58.4	23.3	14.6	33.8	7.2	82.0	77.4	23.5	21.9	89.0	167.7
SD	1.8	3.3	2.5	3.5	16.0	4.8	2.3	3.2	3.2	3.0	7.6	46.4

[a]HD = heading days, PH = plant height, TN = tiller number, PN = panicle number, FT = fertility rate, GY = grain yield, S = stress, N = nonstress.

References

Ali AJ, Xu JL, Ismail AM, Fu BY, Vijaykumar CHM, Gao YM, Domingo J, Maghirang R, Yu SB, Gregorio G, Yanaghihara S, Cohen M, Carmen B, Mackill D, Li ZK. 2006. Hidden diversity for abiotic and biotic stress tolerances in the primary gene pool of rice revealed by a large backcross breeding program. Field Crops Res. 97:66-76.

Ali ML, Pathan MS, Zhang J, Bai G, Sarkarung S, Nguyen HT. 2000. Mapping QTLs for root traits in a recombinant inbred population from two indica ecotypes in rice. Theor. Appl. Genet. 101:756-766.

Babu RC, Nguyen BD, Chamarerk V, Shanmugasundaram P, Chezhian P, Jeyaprakash P, Ganesh SK, Palchamy A, Sadasivam S, Sarkarung S, Wade LJ, Nguyen HT. 2003. Genetic analysis of drought resistance in rice by molecular markers: association between secondary traits and field performance. Crop Sci. 43:1457-1469.

Babu RC, Pathan MS, Blum A, Nguyen HT. 1999. Comparison of measurement methods of osmotic adjustment in rice cultivars. Crop Sci. 39:150-158.

Babu RC, Shashidhar HE, Lilley JM, Thanh ND, Ray JD, Sadasivam S, Sarkarung S, O'Toole JC, Nguyen HT. 2001. Variation in root penetration ability, osmotic adjustment and dehydration tolerance among accessions of rice adapted to rainfed lowland and upland ecosystems. Plant Breed. 120:233-238.

Bouman BAM, Tuong TP. 2001. Field water management to save water and increase its productivity in irrigated rice. Agric. Water Manage. 49(1):11-30.

Champoux MC, Wang G, Sarkarung S, Mackill DJ, O'Toole JC, Huang N, McCouch SR. 1995. Locating genes associated with root morphology and drought avoidance in rice via linkage to molecular markers. Theor. Appl. Genet. 90:969-981.

Courtois B, McLaren G, Sinha PK, Prasad K, Yadav R, Shen L. 2000. Mapping QTLs associated with drought avoidance in upland rice. Mol. Breed. 6:55-66.

Cruz RT, O'Toole JC. 1984. Dry land rice response to an irrigation gradient at flowering stage. Agron. J. 76:178-183.

Dingkuhn M, Cruz RT, O'Toole JC, Doerffling K. 1989. Net photosynthesis, water use efficiency, leaf water potential, and leaf rolling as affected by water stress in tropical upland rice. Aust. J. Agric. Res. 40:1171-1181.

Fukai S, Cooper M. 1995. Development of drought-resistant cultivars using physio-morphological traits in rice. Field Crops Res. 40:67-86.

Haque MM, Mackill DJ, Ingram T. 1992. Inheritance of leaf epicuticular wax content in rice. Crop Sci. 32:865-868.

Hemamalini GS, Shashidhar HE, Hittalmani S. 2000. Molecular marker assisted tagging of root morphological traits under two contrasting moisture regimes at peak vegetative stage in rice (*Oryza sativa* L.). Euphytica 112:69-78.

Kamoshita A, Wade J, Ali L, Pathan S, Zhang J, Sarkarung S, Nguyen HT. 2002a. Mapping QTLs for root morphology of a rice population adapted to rainfed lowland conditions. Theor. Appl. Genet. 104(5):880-893.

Kamoshita A, Zhang J, Siopongco J, Sarkarung S, Nguyen HT, LJ Wade. 2002b. Effects of phenotyping environment on identification of quantitative trait loci for rice root morphology under anaerobic conditions. Crop Sci. 42(1):255-265.

Khush GS. 1999. Green revolution: preparing for the 21st century. Genome 42:646-655.

Khush GS. 2001. Green revolution: the way forward. Nat. Rev. Genet. 2:815-822.

Lafitte HR, Courtois B. 2000. Genetic variation in performance under reproductive stage water deficit in a doubled-haploid rice population in upland fields. In: Ribaut JM, Poland D, editors. Molecular approaches for the genetic improvement of cereal production in water-limited environments. A strategic planning workshop held on 21-25 June 1999. El Batán (Mexico): International Maize and Wheat Improvement Center. p 97-102.
Lafitte HR, Courtois B. 2002. Interpreting cultivar-environment interactions for yield in upland rice: assigning value to drought-adaptive traits. Crop Sci. 42:1409-1420.
Lafitte HR, Courtois B, Arraudeau M. 2002. Genetic improvement of rice in aerobic systems: progress from yield to genes. Field Crops Res. 75:171-190.
Lafitte HR, Guan Yongsheng, Shi Yan, Li Z-K. 2007. Whole plant responses, key processes, and adaptation to drought stress: the case of rice. J. Exp. Bot. 58:169-175.
Lafitte HR, Price AH, Courtois B. 2004. Yield response to water deficit in an upland rice mapping population: associations among traits and genetic markers. Theor. Appl. Genet. 109(6):1237-1246.
Lafitte HR, Vijayakumar CHM, Gao YM, Shi Y, Xu JL, Fu BY, Yu SB, Ali AJ, Domingo J, Maghirang R, Torres R, Mackill D, ZK Li. 2006. Improvement of rice drought tolerance through backcross breeding: evaluation of donors and results from drought nurseries. Field Crops Res. 97:77-86.
Lanceras JC, Pantuwan G, Jongdee B, Toojinda T. 2004. Quantitative trait loci associated with drought tolerance at reproductive stage in rice. Plant Physiol. 135:384-399.
Li ZK, Fu BY, Gao YM, Xu JL, Ali J, Lafitte HR, Jiang YZ, Rey JD, Vijayakumar CHM, Maghirang R, Zheng TQ, Zhu LH. 2005. Genome-wide introgression lines and a forward genetics strategy for functional genomic research of complex phenotypes in rice. Plant Mol. Biol. 59:33-52.
Li Z, Shen LS, Courtois B, Lafitte R. 2000. Development of near isogenic introgression line (NIIL) sets for QTLs associated with drought tolerance in rice. In: Ribaut JM, Poland D, editors. Molecular approaches for the genetic improvement of cereal production in water-limited environments. A strategic planning workshop held on 21-25 June 1999. El Batán (Mexico): International Maize and Wheat Improvement Center. p 103-107.
Lilley JM, Ludlow MM. 1996. Expression of osmotic adjustment and dehydration tolerance in diverse rice lines. Field Crops Res. 48:185-197.
Lilley JM, Ludlow MM, McCouch SR, Champoux MC, O'Toole JC. 1996. Locating QTL for osmotic adjustment and dehydration tolerance in rice. J. Exp. Bot. 47:1427-1436.
Mackill D, Coffman W, Garrity D. 1996. Rainfed lowland rice improvement. Manila (Philippines): International Rice Research Institute.
Moncada MP, Martínez CP, Tohme J, Guimarães E, Chatel M, Borrero J, Gauch H, McCouch SR. 2001. Quantitative trait loci for yield and yield components in an *Oryza sativa* × *Oryza rufipogon* BC_2F_2 population evaluated in an upland environment. Theor. Appl. Genet. 102:41-52.
Nguyen TT, Klueva N, Chamareck V, Aarti A, Magpantay G, Millena ACM, Pathan MS, Nguyen HT. 2004. Saturation mapping of QTL regions and identification of putative candidate genes for drought tolerance in rice. Mol. Gen. Genomics 272:35-46.
O'Toole JC, Cruz RT. 1983. Genotypic variation in epicuticular wax of rice. Crop Sci. 23:392-394.
Pandey S, Behura D, Villano R, Naik D. 2000. Economic cost of drought and farmers' coping mechanisms: a study of rainfed rice in eastern India. IRRI Discussion Paper Series. Los Baños (Philippines): International Rice Research Institute. p 1-35.

Pantuwan G, Fukai S, Cooper M, Rajatasereekul S, O'Toole JC. 2002. Yield response of rice (*Oryza sativa* L.) genotypes to different types of drought under rainfed lowlands. Part 3. Plant factors contributing to drought resistance. Field Crops Res. 73:181-200.

Price AH, Tomos AD. 1997. Genetic dissection of root growth in rice (*Oryza sativa* L.). II. Mapping quantitative trait loci using molecular markers. Theor. Appl. Genet. 95:143-152.

Price AH, Tomos AD, Virk DS. 1997. Genetic dissection of root growth in rice (*Oryza sativa* L.). I. A hydroponic screen. Theor. Appl. Genet. 95:132-142.

Price AH, Townend J, Jones MP, Audebert A, Courtois B. 2002. Mapping QTLs associated with drought avoidance in upland rice grown in the Philippines and West Africa. Plant Mol. Biol. 48(5-6):683-695.

Price AH, Steele KA, Moore BJ, Barraclough PB, Clark LJ. 2000. A combined RFLP and AFLP linkage map of upland rice (*Oryza sativa* L.) used to identify QTLs for root-penetration ability. Theor. Appl. Genet. 100:49-56.

Quarrie SA, Laurie DA, Zhu J, Lebreton C, Semikhodskii A, Steed A, Witsenboer H, Calestani C. 1997. QTL analysis to study the association between leaf size and abscisic acid accumulation in droughted rice leaves and comparisons across cereals. Plant Mol. Biol. 35(1-2):155-165.

Ray JD, Yu L, McCouch SR, Champoux MC, Wang G, Nguyen HT. 1996. Mapping quantitative trait loci associated with root penetration ability in rice (*Oryza sativa* L.). Theor. Appl. Genet. 92:627-636.

Robin S, Pathan MS, Courtois B, Lafitte R, Carandang S, Lanceras S, Amante M, Nguyen HT, Li Z. 2003. Mapping osmotic adjustment in an advanced back-cross inbred population of rice. Theor. Appl. Genet. 107(7):1288-1296.

Shen L, Courtois B, McNally K, Robin S, Li ZK. 2001. Evaluation of near-isogenic lines of rice introgressed with QTLs for root traits through marker-aided selection. Theor. Appl. Genet. 103:70-83.

Steele KA, Price AH, Shashidhar HE, Witcombe JR. 2006. Marker-assisted selection to introgress into an Indian upland rice variety. Theor. Appl. Genet. 112:208-221.

Tripathy JN, Zhang J, Robin S, Nguyen HT. 2000. QTLs for cell-membrane stability mapped in rice (*Oryza sativa* L.) under drought stress. Theor. Appl. Genet. 100:1197-1202.

Tuong TP, Bouman BAM, Mortimer M. 2005. More rice, less water – integrated approaches for increasing water productivity in irrigated rice-based systems in Asia. Plant Prod. Sci. 8(3):231-241.

Venuprasad R, Shashidhar HE, Hittalmani S, Hemamalini GS. 2002. Tagging quantitative trait loci associated with grain yield and root morphological traits in rice (*Oryza sativa* L.) under contrasting moisture regimes. Euphytica 128:293-300.

Xu JL, Lafitte HR, Gao YM, Fu BY, Torres R, Li ZK. 2005. QTLs for drought avoidance and tolerance identified in a set of random introgression lines of rice. Theor. Appl. Genet. 111:1642-1650.

Yadav R, Courtois B, Huang N, McLaren G. 1997. Mapping genes controlling root morphology and root distribution in a doubled-haploid population of rice. Theor. Appl. Genet. 94:619-632.

Yu SB, WJ Xu, Vijayakumar CHM, Ali J, Fu BY, Xu JL, Marghirang R, Domingo J, Jiang YZ, Aquino C, Virmani SS, Li ZK. 2003. Molecular diversity and multilocus organization of the parental lines used in the International Rice Molecular Breeding Program. Theor. Appl. Genet. 108:131-140.

Yue Bing, Lizhong Xiong, Weiya Xue, Yongzhong Xing, Lijun Luo, Caiguo Xu. 2005. Genetic analysis for drought resistance in field with different types of soil. Theor. Appl. Genet. 111:1127-1136.

Yue Bing, Weiya Xue, Lizhong Xiong, Xinqiao Yu, Lijun Luo, Kehui Cui, Deming Jin, Yongzhong Xing, Qifa Zhang. 2006. Genetic basis of drought resistance at reproductive stage in rice: separation of drought tolerance from drought avoidance. Genetics 172:1213-1228.

Zhang J, Zheng HG, Aarti A, Pantuwan G, Nguyen TT, Tripathy JN, Sarial AK, Robin S, Babu RC, Nguyen BD, Sarkarung S, Blum A, Nguyen HT. 2001. Locating genomic regions associated with components of drought resistance in rice: comparative mapping within and across species. Theor. Appl. Genet. 103:19-29.

Zheng HG, Babu RC, Sathan PM, Ali L, Huang N, Courtois B, Nguyen HT. 2000. Quantitative trait loci for root-penetration ability and root thickness in rice: comparison of genetic backgrounds. Genome 43(1):53-61.

Zou GH, Mei HW, Liu HY, Liu GL, Hu SP, Yu XQ, Li MS, Wu JH, Luo LJ. 2005. Grain yield responses to moisture regimes in a rice population: association among traits and genetic markers. Theor. Appl. Genet. 112(1):106-113.

Notes

Authors' addresses: Z.-K. Li, International Rice Research Institute, DAPO Box 7777, Metro Manila, Philippines; Z.-K. Li and Y.-M. Gao, Institute of Crop Sciences/National Key Facility for Crop Gene Resources and Genetic Improvement, Chinese Academy of Agricultural Sciences, e-mail: lizhk@caas.net.cn or z.li@cgiar.org.

Recent efforts to improve drought resistance of rice in Brazil

Flavio Breseghello, Cleber Moraes Guimarães, and Beatriz da Silveira Pinheiro

Rainfed upland rice is a major crop in Brazil, contributing approximately 30% of the total rice production. Although upland rice is planted in areas with favorable rainfall distribution, much could be gained by increasing its resistance to drought. The total amount of rain in the season is normally sufficient; however, rainfall distribution is irregular in some years and the soil has low water retention capacity, and thus is unable to buffer those irregularities. Consequently, upland rice is regarded as a high-risk crop, which receives low inputs, resulting in low mean yield. Research on drought resistance in rice at Embrapa had three periods: more intense in the 1980s, with the release of varieties with reasonably high drought resistance; low emphasis in the 1990s, with a change in focus to plant type and grain quality; a rebirth in the 2000s, with projects focusing especially on the root system. Roots are being evaluated in soil columns or in the field during the dry season. Research projects are being implemented and breeding populations are being prepared, although no resistant varieties have been released yet as a result from those recent projects. Embrapa recognizes drought in rice as an important research theme, which will require medium-term multidisciplinary projects to deliver a significant impact on upland rice production.

Upland rice is a major crop in Brazil. Total rice harvest in the country is approximately 12 million t year^{-1}, 30% of which comes from the aerobic, well-drained areas. In 2002-04, an average of 2 million ha of upland rice was grown yearly, with a mean yield of 1.9 t ha^{-1} (IBGE 2006). Although irrigated rice is concentrated in the south, especially in the state of Rio Grande do Sul, upland rice is widespread, constituting a traditional staple crop for small farmers, a pioneer crop in newly deforested areas, a grain crop planted in the periodical renewal of pastures, as well as a rotation crop for high-input soybean production.

Although the mean yield of upland rice in Brazil is still low, its potential is above 6 t ha^{-1}, as frequently seen in farmers' fields. Those high yields are more frequent under equatorial climate, in the Amazon region, where annual rainfall exceeds 2,000 mm. In the Cerrado, where annual rainfall is around 1,500 mm, mean yield is lower. This is due to occasional crop failures as a consequence of drought stress, but also because farmers tend to be more parsimonious in the inputs applied, which in turn is related to the perceived risk of the upland rice crop in that region.

Early studies on drought resistance at Embrapa

During the 1970s and 1980s, when the deforestation frontier was in the Cerrado, upland rice was used as the pioneer crop in virtually all areas, in the first 1–3 years after clearing natural vegetation. Rice's resistance to soil acidity, relatively low fertility requirements, and the existence of a local market for the grain made rice almost the only option as a crop for those areas, although mean yield was low and crop failure due to drought was relatively frequent. In that period, drought received the highest priority in the research agenda of the National Rice Research Program, led by the National Center of Rice and Beans Research (currently Embrapa Rice and Beans). The breeding program focused on drought resistance during the reproductive period, based on the evaluation of potential parents as well as elite lines prior to release (Pinheiro 2003). Crosses have been mainly confined within the japonica group, using Brazilian and African lines as sources of drought resistance (e.g., Brazilian varieties IAC 25 and IAC 47 and African variety 63-83). Cultivars released in that period (e.g., Rio Paranaíba, Guarani) showed a level of drought resistance comparable with that of the parents.

Carbon isotope discrimination (Δ) has been considered a potential parameter for indirect selection of drought-resistant lines. The efficiency of Δ as an indicator of drought resistance in the reproductive stage was tested in Brazilian germplasm by Pinheiro et al (2000). That study indicated that drought altered Δ, especially in the last stem internode; however, its relationship with spikelet fertility and grain yield was weak. Those results indicated that Δ had limited value for selection for drought resistance in the breeding program. Studies involving drought stress in the reproductive stage indicated that the density of the root system at 60–80 cm of depth presents higher correlation with yield (Pinheiro et al 1985). For this reason, deep rooting has been considered a primary target for upland rice breeding in drought-prone areas.

In the late 1980s, the deforestation frontier advanced northward, to more rainy regions, on the outskirts of the Amazon forest. Upland rice, as a pioneer crop, moved along with it. Consequently, losses due to drought became less frequent, average yield increased, and lodging became a more limiting factor for yield gain. Studies on crop physiology (Pinheiro et al 1985) and agrometeorology (Steinmetz et al 1985) demonstrated that the breeding program should change its focus to a modern plant type, with high yield potential and lodging resistance, for cropping under favorable rainfall distribution or supplementary irrigation. In order to promote those changes, Embrapa and partners, especially the Centro Internacional de Agricultura Tropical (CIAT), introduced indica germplasm into the upland rice breeding program. Derived cultivars released in the 1990s contributed to an increase in the mean yield of upland rice. The potential yield of Brazilian traditional varieties is approximately 4.5 t ha^{-1}, with a leaf area index (LAI) of 3.5–4.0 (Pinheiro and Guimarães 1990), whereas new aerobic releases (e.g., Maravilha, Canastra, BRS Bonança) can attain experimental yields around 7 t ha^{-1}, with LAI of 6–8 and no significant lodging.

Although the contribution of the modern plant type has been very important, this change of focus caused the evaluation of drought tolerance to be discontinued, and drought physiology studies to receive low priority during that period. Consequently, modern cultivars showed a decrease in drought resistance compared with traditional upland varieties (Pinheiro 2003).

Defining the target environment

The target environment for drought resistance of rice in Brazil is the Cerrado, an extensive region in central Brazil, characterized by savanna-like vegetation and alternation of rainy summers and dry, warm winters. The soils are predominantly Latosols, deep, well drained, and distrofic. Annual rainfall is 1,400–1,700 mm, distributed mostly from October to March. Upland rice is planted from October to December. The total amount of rain during the crop cycle is normally sufficient. Nevertheless, dry periods with a duration of 1–2 weeks during the wet season, known as "*veranicos*," are relatively frequent, and represent a risk of crop failure. High temperature, low air moisture, little or no cloudiness, combined with soils with low water retention capacity, result in quick soil dehydration, from top to bottom. In this situation, deep-rooting plants can withstand longer dry periods, tapping water from deeper soil layers. Nevertheless, it is assumed that rice will remain in the more favorable regions within the Cerrado macro-region, since agrometeorology studies demonstrated that the drier parts of the Cerrado present a high risk of crop failure.

Potential benefits of increased drought resistance

Raising mean yield of upland rice. National mean yield would increase if upland rice cultivars were able to better extract and use the water available in the soil during periods of regular rainfall, and withstand periods of water shortage with less yield reduction. Although rice is unlikely to reach high levels of drought resistance, even a moderate gain would have a tangible effect on the country's mean yield. Considering that the incidence of blast disease (caused by *Magnaporthe grisea*) and root-feeding termites (*Procornitermes* spp., *Syntermes* sp.) tends to increase in water-stressed fields, the benefits of drought resistance could be amplified by reducing opportunities for those highly destructive biotic stresses in the tropics. Field observations indicate that drought-resistant varieties retard the spread of blast symptoms. Equally, more developed root systems appear to withstand some loss due to termite feeding, without clear effects on the shoot.

Increasing food security for small farmers. Rice is a staple food in Brazil; therefore, small landholders normally grow rice for consumption. Although this system does not contribute a large portion of Brazilian rice production, it has significant social relevance. In those cases, drought resistance is important because crop failure has a direct impact on the quality of life throughout the year. The development of rice cultivars with good yield stability and grain quality and suitable for hand-harvest is a valid objective for breeding, especially for public institutions.

Reversing the northward migration of the crop. The deforestation front is currently in the plain Amazon forest, and upland rice plays an important role as a pioneer crop in the newly deforested areas. However, the rhythm of deforestation must be restrained, and deforested areas in the Amazon should be occupied preferentially by sustainable agroforestry crops. Under this scenario, upland rice will have to find new niches. The most promising opportunities are as a rotation crop under intensive agriculture and as a cash crop in pasture renovation (Pinheiro et al 2006). Soybean occupies approximately 24 million ha each year in Brazil, and the area occupied by cultivated pastures is approximately 100 million ha (www.sidra.ibge.gov.br). Even a small portion of those areas planted with upland rice each year would result in very significant rice production.

Improving the adaptation of rice to the no-tillage system. With soaring oil prices and growing concerns about soil erosion, reduced-tillage or no-tillage systems are becoming predominant in the Cerrado. Embrapa supports the no-tillage system, and the current view is that rice must adapt to the system, rather than the opposite. The rice root system is sensitive to soil compaction, which may be associated with the no-tillage system (Guimarães et al 2001). Under this situation, the root system remains confined near the soil surface; thus, even a few days without rain may result in drought stress. Embrapa is testing the hypothesis that cultivars with deep and thick root systems are better adapted to the no-tillage system.

Establishing a robust cropping system for upland rice

The definition of crop management recommendations depends on the amount of drought resistance of the cultivar since drought interacts with canopy establishment and planting time. Reduced seeding rate normally results in reduced leaf area index and consequently lower transpiration rate at the crop canopy level. Under drought, a less dense crop can result in higher spikelet fertility and better grain filling, partially preventing yield loss and quality degradation. However, in the absence of drought, densely planted fields tend to give higher yields because of better radiation-use efficiency. Additionally, a lower plant population aggravates weed competition, a major problem in fields under intensive agriculture in the tropics.

The time of occurrence of *veranicos* over a number of years and locations allowed Embrapa to elaborate a climatic risk zoning for rice, indicating the planting times expected to have low, intermediate, and high risk of crop failure (Steinmetz et al 1988). Although those guidelines are expected to reduce the failure rate on average, the climatic statistics have high variances associated with them. Furthermore, with a global climate change scenario, projections made on the basis of past events may have little predictive power. Finally, since *veranicos* can occur at any time, staggered planting could at least prevent complete crop failure, as part of the field would escape drought. This measure carries the risk, however, of building up blast inoculum over time, thus damaging areas planted late in the season.

Rice breeding objectives for the Cerrado

For a rice cultivar to thrive in the Cerrado, it must have relatively high yield potential and superior grain quality, be lodging resistant, and yet be stable in the face of three major constraints: blast disease, weed competition, and *veranicos*. On the one hand, grain quality is a determinant of cultivar adoption because the Brazilian market strictly demands long-and-slender nonsticky rice. On the other hand, agronomic performance is essential for rice to be competitive as a field crop. For example, soybean growers require, for a rotation crop, a comparable level of yield stability and well-defined technology as there is for soybean, including the option for a no-tillage system. Unfortunately, upland rice is still regarded as a high-risk crop in the Cerrado.

Drought research projects at Embrapa

Drought resistance returned to the rice research agenda recently at Embrapa (Pinheiro 2003). Because of the high expectations around the new science of genomics, the focus of current research projects remains on genetics and breeding. The project "*Orygens*," led by Embrapa Genetic Resources and Biotechnology, aims at developing tools for molecular breeding in rice, especially for drought resistance. The approach for genetic analysis in this project includes QTL mapping, association analysis, and allele mining. The project "*Drought phenotyping network*," led by Embrapa Maize and Sorghum and supported by the Generation Challenge Programme, has a main objective of establishing drought phenotyping sites and methodology for rice, maize, sorghum, and wheat. The environmental conditions in the Cerrado are similar to those of the African savanna; thus, results and protocols are likely to be transferable between those environments, and this transferability establishes a connection between Brazilian rice breeding and the priorities of international agricultural research. The rice breeding program of Embrapa Rice and Beans is branching off a subprogram for drought-prone areas, with the objective of developing breeding populations and selection methods. This project is expected to develop into an elite breeding program with capacity to release competitive cultivars for the more favorable areas of the Cerrado. The material sought must have a good balance of drought resistance, blast resistance, and early vigor. As a way of increasing selection efficiency, the project will initially rely only on early-flowering germplasm. Reduction of variation in flowering time is convenient to avoid confounding factors in drought resistance phenotyping. This project is a potential client for results from upstream research projects, especially for selectable markers with a proven effect on drought or blast resistance.

Phenotyping methods for drought resistance

The projects described above are integrated and use similar evaluation protocols. They focus on root architecture as a key trait for drought resistance in the conditions prevalent in the Cerrado.

Root architecture in the greenhouse. Plants are evaluated in cylindrical soil columns, 25 cm wide and 80 cm tall. From planting to heading, pots are well watered, keeping water potential above –35 kPa, monitored by tensiometers at 15 cm of depth. From this point on, two water treatments are applied: well irrigated and water stressed. Columns are weighed daily, and water is added according to the reduction in weight, being 100% for the nonstressed treatment and 50% for the stressed treatment. Density of the root system is evaluated at harvesting, in four column segments of 20 cm. Other variables measured are yield, harvest index, shoot dry mass, tiller fertility, 100-seed weight, panicle size, plant height, spikelet sterility, and leaf temperature.

Drought resistance in the field. Evaluation of drought resistance in the field is done in the dry season in Porangatu, a location in the Cerrado region with a warm and reliable dry season from April to September. Uniform irrigation is achieved with a line sprinkler system. The nonstressed treatment is irrigated as needed to keep water potential at 15 cm in the soil above –35 kPa, whereas the stressed treatment is irrigated at about half that frequency. The stress treatment starts at 25–30 days after emergence. The root system is evaluated by augering at 20–40 and 40–60 cm of depth with two samples per plot, at the flowering stage. Yield reduction between water treatments is approximately 50–70%. However, because of the extremely low air moisture in the dry season, the "nonstressed" treatment often does not achieve full yield as compared to the normal growing season. Lines are selected based on a drought susceptibility index (Fisher and Maurer 1978) and yield potential. Other evaluations include canopy temperature, leaf-rolling score, tiller number, tiller fertility, flowering delay, spikelets per panicle, spikelet sterility, 100-grain weight, and harvest index.

Preliminary results

The first results from the projects cited are confirming that there is significant variability for drought resistance among tropical japonica germplasm in Brazil.

Greenhouse evaluations. Drought × genotype interaction in greenhouse stress trials in soil columns was significant for yield but not for yield components (Table 1A). Evaluation of root density in those experiments revealed interesting correlations (Table 2). Pot grain yield was inversely correlated with shoot dry biomass, indicating that large plants demanded more water, and thus were more stressed under limited water supply in the pots. Accordingly, a dense root system at depth (>40 cm) was not favorable for yield in pots, probably because those plants quickly depleted the water supplied to the bottom of the column, running into stress earlier than genotypes with lower density of deep roots. Leaf temperature tended to be higher in larger plants, confirming the stressed status of the plants. A change in root architecture due to drought was genotype-dependent. Some cultivars presented similar root densities in both treatments, whereas others changed root architecture, producing more deep roots when water stressed. Despite the inverse correlation found in pots, deep roots in the greenhouse evaluation appear to correlate positively with drought resistance in the field (results not shown). This probably can be explained by the unconstrained volume of soil in the field as opposed to the constrained volume of the pots.

Field evaluations. The effect of drought × genotype interaction was significant for yield and spikelet sterility, indicating that varieties reacted differently to drought stress in the field. No significant interaction was observed for panicle size and tiller

Table 1. F values of the analysis of variance of agronomic traits for drought stress treatments and genotypes of the Brazilian rice collection in a field (Porangatu 2004) and greenhouse (2005).[a]

Source of variation	Yield	Spikelet sterility	Spikelets per panicle	Tiller fertility	Plant height
(A) Greenhouse					
Drought	79.29**	12.38*	1.66ns	0.64 n.s.	58.09**
Genotype	3.89**	2.15**	4.51**	2.30**	8.32**
Drought × genotype	1.86**	1.14 n.s.	0.52 n.s.	1.24 n.s.	0.70 n.s.
(B) Field					
Drought	58.65**	60.97**	23.17*	24.24*	18.68*
Genotype	2.00**	1.71**	2.27**	1.57**	6.73**
Drought × genotype	1.50**	1.54**	1.08 n.s.	1.15 n.s.	1.34*

[a]n.s. = nonsignificant, * = significant at 5%, ** = significant at 1%.

Table 2. Correlation coefficients and *P* values between grain yield, shoot dry biomass, root density at four column sections (0–20, 20–40, 40–60, and 60–80 cm), and leaf temperature in the greenhouse.

	Grain yield	Shoot biomass	Root $_{0–20}$	Root $_{20–40}$	Root $_{40–60}$	Root $_{60–80}$
Shoot biomass	–0.501 0.002					
Root $_{0–20}$	n.s.	n.s.				
Root $_{20–40}$	n.s.	0.659 <0.001	0.477 0.003			
Root $_{40–60}$	–0.401 0.014	0.704 <0.001	0.470 0.003	0.698 <0.001		
Root $_{60–80}$	–0.556 <0.001	0.604 <0.001	n.s.	0.418 0.010	0.591 <0.001	
Leaf temperature	–0.327 0.048	0.476 0.005	n.s.	0.433 0.007	0.498 0.002	n.s.

fertility, indicating that those yield components may be more difficult to manipulate for improving drought resistance (Table 1B). Since drought is not ubiquitous in the Cerrado, high yield capacity in the absence of stress is also important. For this reason, lines are being selected based on a two-way dispersion plot of drought susceptibility index × yield without stress. It was also observed that susceptible lines had higher canopy temperature and lower flowering delay than resistant lines. Elite lines are being evaluated for drought resistance in the field, for detection of potential parents for a breeding program focused on drought-prone areas. In the first evaluations, stress was relatively mild; however, some interesting differences were detected (Table 3).

The elite lines BRA 01596, BRA 01506, and BRA 01600 presented high yield under nonstress conditions, and relatively mild yield reduction under stress (Guimarães et al 2007). Those materials result from crosses between varieties from Brazil and the United States.

Development of new populations

Embrapa is developing new genetic stocks for breeding and research, based on japonica germplasm, both traditional and improved. Those populations are described below.

A new recurrent selection population. Three early-flowering populations of upland rice, CNA6, CG1, and CG3, will be evaluated for drought resistance in the dry seasons of 2007, 2008, and 2009, respectively. Those populations have been developed under a recurrent selection scheme since the mid-1990s, with a 3-year recombination cycle. CNA6 is derived from the population CNA-IRAT5, which was synthesized from 26 drought- and blast-resistant lines. CNA-IRAT5 was introgressed with 25 new parents, mostly Brazilian landraces, to form CNA6 (Morais et al 1997). The population CG1 was synthesized by intercrossing 19 pure lines and cultivars selected for yield, grain quality, and plant type (Castro et al 2000). The population CG3 was synthesized from a selection of 61 families from the early-flowering elite breeding program at Embrapa Rice and Beans, based on yield, phenotypic acceptance, disease resistance, and grain quality. In 2010, the lines selected from the three populations will be intercrossed to form a new population. That population will combine favorable alleles for drought resistance from a broad genetic base, and as such will represent a valuable resource for breeding and genetic analysis of drought resistance mechanisms.

A panel for association mapping. A panel of approximately 200 lines derived from the population CNA6 will be evaluated for agronomic performance in the wet seasons of 2006-07 and 2007-08, and for drought resistance in the dry seasons of 2007 and 2008. In parallel with field trials, families will be inbred to near fixation by the end of 2009. The population structure of this panel is being characterized with unlinked SSR markers for future association studies. The identification of significant associations between molecular markers and stress resistance in this panel will allow marker-assisted selection in the CNA6 population, according to the guidelines described by Breseghello and Sorrells (2006).

QTL mapping populations. Three mapping populations are being developed for quantitative trait loci (QTL) analysis of drought resistance, agronomic traits, and grain quality. Those populations derive from the crosses BRS Curinga/BRS Soberana, BRS Primavera/BRS Curinga, and BRS Primavera/BRS Douradão. BRS Primavera and BRS Soberana are considered more drought sensitive than BRS Curinga and BRS Douradão; therefore, segregation for drought resistance is expected in all three crosses. Approximately 200 recombinant inbred lines will be available for each population early in 2009. Those populations will be phenotyped for drought resistance in 2009 and 2010 and for agronomic traits in the wet season of 2009-10. Complete genetic maps are being planned for those populations; however, they will depend on future projects.

Table 3. Means of yield, spikelet sterility, and tiller sterility for Brazilian varieties and breeding lines under contrasting water-stress treatments (Porangatu 2006).

Variety or inbred line	Grain yield ($kg\ ha^{-1}$)		Relative yield	Spikelet sterility (%)		Tiller sterility (%)	
	Nonstress	Stress	Stress/ nonstress	Nonstress	Stress	Nonstress	Stress
BRA 01596	5,117	2,720	0.532	10.3	34.0	8.3	21.1
BRA 01506	4,679	2,571	0.550	24.9	30.5	8.5	19.3
BRA 01600	4,534	2,535	0.559	12.8	31.6	9.0	25.7
CNAs9019	3,222	2,467	0.766	21.4	42.0	9.2	14.8
CNAs9025	4,001	1,961	0.490	15.0	29.9	16.6	15.0
Bonança	3,513	2,234	0.636	23.6	33.9	13.5	17.6
Curinga	3,471	2,160	0.622	22.3	48.7	18.1	12.6
BRA 02601	2,949	1,969	0.668	24.7	29.5	14.9	17.3
BRA 02598	2,420	2,219	0.917	29.2	26.8	37.4	26.0
Vencedora	2,994	1,768	0.590	27.3	25.5	9.6	19.2
CNAs9045	3,755	1,391	0.371	23.1	35.2	9.1	25.9
Soberana	2,519	1,281	0.508	35.1	49.6	18.6	31.3
Primavera	4,407	533	0.121	24.5	64.6	13.4	17.9
Mean	3,660	1,985	0.563	22.6	37.1	14.3	20.3
CV[a]	21.4%			34.6%		8.9%	

[a]Coefficient of variation of the joint analysis of the stress and nonstress experiments.

Conclusions

After a decade when drought resistance had low priority as a breeding objective, it has recently reemerged as a valuable and achievable goal. In this new phase, plant breeders, crop physiologists, and molecular biologists are working together as a team. There is increasing awareness that none of those disciplines alone will deliver the needed breakthroughs. Proper phenotyping recovered its status as a fundamental component of genetic analysis. Breeders are developing populations that will bridge the gap between molecular mapping and marker-assisted selection. Molecular biologists are increasingly interested in traits with real value in field conditions. With global warming and water scarcity looming, the stage is set for this team to make a huge contribution to food security and preservation of natural resources.

References

Breseghello F, Sorrells ME. 2006. Association analysis as a strategy for improvement of quantitative traits in plants. Crop Sci. 46:1323-1330.

Castro EM, Morais OP, Sant'Ana EP, Breseghello F, Moura Neto F. 2000. Mejoramiento poblacional de arroz de tierras altas en Brasil. In: Guimarães EP, editor. Avances en el mejoramiento poblacional en arroz. Santo Antonio de Goiás (Brazil): Embrapa. p 221-240.

Fisher RA, Maurer R. 1978. Drought resistance in spring wheat cultivars. I. Grain yield responses. Aust. J. Agric. Res. 29:897-912.

Guimarães CM, Prabhu AS, Castro EM, Ferreira E, Cobucci T. 2001. Cultivo do arroz em rotação com soja. Circular Técnica 41. Santo Antônio de Goiás (Brazil): Embrapa Arroz e Feijão. 8 p.

Guimarães CM, Stone LF, Morais OP. 2007. Resposta de cultivares e linhagens elites de arroz de terras altas ao deficit hídrico. Santo Antônio de Goiás: Embrapa Arroz e Feijão. 4 p.

IBGE. 2006. Produção agrícola municipal: culturas temporárias e permanentes 1990-2004. www.sidra.ibge.gov.br.

Morais OP, Castro EM, Sant'Ana EP. 1997. Selección recurrente en arroz de secano en Brasil. In: Guimarães EP, editor. Selección recurrente en arroz. Cali (Colombia): International Center for Tropical Agriculture. p 99-115.

Pinheiro BS. 2003. Integrating selection for drought tolerance into a breeding program: the Brazilian experience. In: Fischer KS, Lafitte R, Fukai S, Atlin G, Hardy B, editors. Breeding rice for drought-prone environments. Los Baños (Philippines): International Rice Research Institute. p 75-83.

Pinheiro BS, Austin RB, Carmo M, Hall MA. 2000. Carbon isotope discrimination and yield of upland rice as affected by drought at flowering. Pesq. Agropecu. Bras. 35:1939-1947.

Pinheiro BD, Castro EDM, Guimarães CM. 2006. Sustainability and profitability of aerobic rice production in Brazil. Field Crops Res. 97:34-42.

Pinheiro BS, Guimarães EP. 1990. Índice de área foliar e produtividade do arroz de sequeiro. I. Níveis limitantes. Pesq. Agropecu. Bras. 25:863-872.

Pinheiro BS, Steinmetz S, Stone LF, Guimarães EP. 1985. Tipo de planta, regime hídrico e produtividade do arroz de sequeiro. Pesq. Agropecu. Bras. 20:87-95.

Steinmetz S, Reyniers FN, Forest F. 1985. Evaluation of the climatic risk on upland rice in Brazil. In: Colloquium "Resistance en la recherche en milieu intertropical: quelles recherches pour le moyen terme? 1984, Dakar. Proceedings. Paris (France): CIRAD. p 43-54.

Steinmetz S, Reyniers FN, Forest F. 1988. Caracterização do regime pluviométrico e do balanço hídrico do arroz de sequeiro em distintas regiões produtoras do Brasil: síntese e interpretação dos resultados. EMBRAPA-CNPAF Documentos 23. Goiânia (Brazil): EMBRAPA-CNPAF. v. 1. 66 p.

Notes

Authors' addresses: F. Breseghello, Embrapa Rice and Beans, Plant Breeding and Genetics, C.P. 179, 75375-000 Santo Antônio de Goiás, GO, Brazil, flavio@cnpaf.embrapa.br; C.M. Guimarães and B. da Silveira Pinheiro, Embrapa Rice and Beans, Crop Physiology.

Harnessing quantitative genetics and genomics for understanding and improving complex traits in crops

James B. Holland and Andrea J. Cardinal

Classical quantitative genetics aids crop improvement by providing the means to estimate heritability, genetic correlations, and predicted responses to various selection schemes. Genomics has the potential to aid quantitative genetics and applied crop improvement programs via large-scale high-throughput DNA sequencing and fingerprinting, gene expression analyses, and reverse genetics methods. To date, these techniques have mainly been useful in the identification of genes with discrete or at least moderate effects on high-value traits. A practical result of this research is the development of allele-specific markers that tend to be useful across many breeding populations. For example, knowledge of the fatty acid biosynthesis pathway in plants and the sequencing of genes in that pathway is being exploited to produce DNA markers to aid selection for specific modified fatty acid traits in soybean. The application of large-scale gene mapping techniques to the improvement of highly quantitative traits (controlled by many genes of small effects) is not yet proven, however, and, even if useful, may not be highly cost-effective unless large-scale genomics infrastructure is already in place to aid breeding programs. An example of the application of genomics to large-scale genetic diversity studies is the large-scale maize QTL mapping study under way based on the development of 26 related RIL populations that capture a large portion of the genetic variation available worldwide among public lines. For genomics to be useful for the improvement of drought resistance in rice, the identification of component traits with relatively simpler architecture and/or a very large-scale investment in genomics-assisted breeding may be required.

Historically, quantitative genetics has aided crop improvement programs by providing means to estimate population-based parameters such as heritability and genetic correlations. Such estimates can be applied with selection response theory to make predictions about genetic gain, compare different selection strategies and evaluation designs, develop optimal multitrait selection indices, and predict correlated changes in unselected traits due to selection on target traits (Falconer and Mackay 1996, Hallauer and Miranda 1988). These approaches have been remarkably successful, particularly given that much of quantitative genetic theory is based on assumptions that do not seem to be true (Holland 2006). By grounding quantitative genetics in biological reality, it should be possible to improve breeding methods and make greater improvements in highly complex traits that have been recalcitrant to breeding, such as drought resistance. Modern genomics technologies and approaches combined with large-scale phenotypic evaluation (which together might be termed "phenomics") may provide the improved understanding needed to increase our ability to manipulate complex traits in crops.

The promise of quantitative trait locus mapping and marker-assisted selection techniques has been apparent for years now. However, despite suggestions that breeders should be able to "control all allelic variation for all genes of agronomic importance (...) through a combination of precise genetic mapping, high-resolution chromosome haplotyping, and extensive phenotyping" (Peleman and van der Voort 2003), markers now are not generally useful for manipulating complex traits such as drought resistance. Marker assay costs continue to decrease, which will increase their utility to breeding programs. Nevertheless, breeders will still face some fundamental obstacles to applying markers in breeding programs. Such obstacles include the limited precision of QTL mapping experiments, genetic heterogeneity of complex traits, and poor integration between mapping experiments and conventional breeding methods. Therefore, researchers will need to design strategies to overcome these obstacles for marker-assisted selection to be of general use for the improvement of drought resistance or other complex traits.

Marker-assisted selection today

To date, DNA markers have been most widely used in practical plant breeding to assist in the backcrossing of major genes into elite cultivars that were previously developed through conventional plant breeding. Markers aid backcross selection by (1) aiding selection on target alleles whose effects are difficult to observe phenotypically, (2) selecting recombination events near target genes to minimize linkage drag, and (3) selecting progeny with higher proportions of the recurrent parent genetic background to reduce the number of generations needed to recover a line that is nearly isogenic to the recurrent parent (Holland 2004). Numerous examples exist to demonstrate that marker-assisted backcrossing is being used fairly widespread across crops, particularly in commercial companies and the larger public breeding programs (Cahill and Schmidt 2004, Chen et al 2000, Dubcovsky 2004, Eagles et al 2001). By design, backcrossing is a conservative breeding method; however, modified backcrossing methods have been used to introgress exotic and wild species germplasm into adapted backgrounds, particularly in self-pollinated crops (Tanksley and McCouch 1997). These methods have resulted in the identification of exotic-derived alleles with favorable effects that can enhance the breeding gene pools of these species, thus facilitating future genetic gains (Li et al 2005, Zamir 2001).

Markers are also used currently to select alleles with major effects on high-value traits. For the most part, these are disease resistance genes, such as cereal cyst nematode resistance genes in wheat (Eagles et al 2001) and soybean cyst nematode resistance genes in soybean (Young 1999). It is not a coincidence that markers have been most useful for selecting resistance to root diseases, as these are some of the most difficult resistances to accurately phenotype, and marker assays are likely to be more accurate than phenotypic scores, and may well be less expensive. In addition, markers are being used to select for grain quality traits, and even some abiotic stress resistances (Eagles et al 2001). A valuable marker for abiotic stress resistance in rice is the *Sub1* gene, which confers submergence tolerance. Markers linked to this gene have been shown to be diagnostic across numerous rice lines (Xu et al 2004), and identification of the specific nucleotide sequence changes that differentiate tolerant from nontolerant alleles (Xu et al 2006) allows the development of perfect markers for tracking this gene in breeding programs.

The examples where marker-assisted selection (MAS) has been, or is expected to soon be, an important part of mainstream breeding programs have in common two important factors. First, the markers are tightly linked to (or target directly) a few loci with relatively large effects on traits that are difficult or costly to accurately phenotype. For example, the *Sub1* gene in rice is the primary genetic determinant of submergence tolerance (Xu et al 2004). Second, specific marker alleles are associated with desired alleles at target loci consistently across multiple breeding populations. This second point is key because it eliminates the need to establish the linkage phase between markers and their target alleles in every population. Markers must be consistently diagnostic for target alleles to implement MAS in breeding programs in which many crosses are made annually between constantly changing sets of breeding parents. Holland (2004) reviewed the factors that affect the utility of markers across breeding populations. Sequencing the variant alleles at the target gene may be required to create optimally useful markers.

Once consistently useful marker loci are identified, MAS can be more effective than phenotypic selection because MAS can be implemented on a single-plant basis (such as in early breeding generations) when conventional selection would not even be attempted because of the extremely low heritability of the trait on a single-plant basis. Bonnet et al (2004) demonstrated that selection for F_2 plants that are either homozygous or heterozygous for the desired alleles at several marker loci ("F_2 enrichment") is an efficient method to reduce the number of lines that require extensive phenotypic evaluation in later generations.

Marker-assisted selection for polygenic traits

Many important agricultural traits are controlled by many genes, are highly influenced by the environment, and exhibit substantial genotype-by-environment interactions. Traits conferring productivity in drought environments, even traits that represent components of yield or relatively simple traits contributing to yield, are by and large quantitative traits (Richards 2006). For example, in one rice population, 27 quantitative trait loci (QTLs) were detected for traits related to yield potential or yield under drought and 38 QTLs were detected for root traits under drought (Yue et al 2008).

Worse, there was little repeatability of QTLs across the two years in this study, even though it was a partly controlled environment (drought was induced specifically at flowering time using a combination of rainout shelters and PVC pipes).

Polygenic traits are the most difficult to breed for, typically requiring large-scale multienvironment testing in order to make progress from selection. For that reason, DNA markers could have a great impact on plant breeding if they could be used to aid in selection for quantitative traits. The numerous success stories of the use of MAS for polygenic traits in plants are, unfortunately, countered by at least as many examples where MAS was not sufficiently better than conventional selection to justify its cost (Holland 2004).

The reasons why MAS has not been more generally effective for polygenic traits include genetic heterogeneity and inaccurate estimation of the positions and effects of the QTLs using typical mapping schemes and population sizes. Genetic heterogeneity refers to the situation in which one trait is affected by different genes in different populations (Holland 2007). This is a common problem with polygenic traits—with so many genes involved, only some subset of them is likely to be involved in one particular mapping population. Furthermore, recalling that one of the conditions under which MAS is most effective is when a trait is controlled by a few genes with moderate to large effects, we can see that a major hindrance to MAS for polygenic traits is the small effects of the large majority of QTLs on such genes. Unfortunately, in many cases, this is a biological reality and there may be no technological approach that can avoid this problem.

Inaccurate estimates of QTL positions and effects are caused primarily by the limited genotypic sample sizes used in most studies. Typical approximate confidence intervals for QTL positions are on the order of 20 cM (Dekkers and Hospital 2002, Kearsey and Farquhar 1998, Lee 1995). The size of QTL confidence intervals cannot be reduced simply by increasing marker density beyond about 10-cM spacing. Instead, improved resolution of QTL positions requires increased mapping population sizes or the creation of near-isogenic lines with overlapping introgressions near the QTL (Stuber 1998, Zamir 2001).

Beyond the inaccuracies in estimating QTL positions, the magnitudes of their effects are typically overestimated, often leading to unrealistic expectations for gain from MAS (Bohn et al 2001, Dekkers and Hospital 2002). In any one sample of progeny from a mapping population, too few QTLs are identified as significant, and the effects of those QTLs that are identified are overestimated (Beavis 1998, Melchinger et al 1998, Utz et al 2000). The problem gets worse as heritability and population size decrease and the true number of QTLs increases (Beavis 1998). Even if the actual genes are known, estimation of the effects of many genes simultaneously is so difficult that MAS can offer little advantage over phenotypic selection (Bernardo 2001).

Thus, attempting to directly use markers to select for QTLs with small effects on complex traits exhibiting significant genetic heterogeneity has not been generally effective. Recent approaches to the use of molecular markers for improving quantitative traits involve "genomewide selection" originally proposed for animal breeding populations by Muewissen et al (2001). This method avoids the difficult issue of significance testing required to declare which genome regions contain QTLs by predicting the effects of *all* markers on the target trait using mixed models analysis. Then, rather than selecting plants or lines carrying desired alleles at a subset of loci considered significant, selections are made on the basis of the net value of alleles carried by a line at all marker loci. The objective of this model is not to identify genome regions carrying QTLs, so accuracy of any individual locus prediction is not of concern; rather, accuracy of the net genotypic value prediction across all loci of each line is the goal.

Genomewide selection methods become most effective when coupled with one or more generations of selection in off-season nurseries purely on the basis of marker-based predictions, when phenotypic selection is either not possible or not correlated to expression of the target traits in the target production environments. Hospital et al (1997) demonstrated that the application of marker-assisted selection in off-season nurseries where phenotypic selection is not applicable can increase gain from selection per year. Eathington et al (2007) reported that Monsanto Company's plant breeding programs have increased gains from selection for multiple quantitative traits by applying marker selection based on a proprietary method in off-season nurseries. Bernardo and Yu (2007) simulated a genomewide selection method for maize involving phenotypic evaluation and marker model development in testcrosses of doubled haploid lines from a breeding cross. This is followed by two generations of off-season selection based solely on marker-based genotype predictions using marker allele effect predictions from the initial testcross phenotype evaluations. Genomewide selection in off-season nurseries resulted in significant improvement in predicted gains per year beyond the initial gain from direct phenotypic selection, and genomewide selection was superior to estimating allele effects only at a subset of loci declared significant for QTL effects (Bernardo and Yu 2007).

Genomewide selection holds promise for improving gains from selection for quantitative traits, but broad application of this approach requires a massive marker data collection and analysis infrastructure, as is in place in several industrial-scale commercial programs. The method still relies upon accurate phenotypic evaluations to obtain good marker allele effect estimates, and essentially requires re-estimating allele effects in every breeding population. No estimates of the economic efficiency of the huge investments required to conduct this large-scale approach to marker-assisted breeding have been reported.

Bernardo and Yu (2007) referred to genomewide selection for quantitative traits as a "brute-force and black-box procedure" to maximize genetic gain without necessarily illuminating the biological underpinnings of the target traits. An alternative approach is to attempt to understand target complex traits at the molecular, biochemical, and physiological levels. This may involve the study of component traits or the study of biochemical pathways that lead to definable phenotypes of interest. If the genetic controls of the trait can be sufficiently well defined, then markers designed to tag genes in the biochemical or physiological pathway of the trait can be used, with a reasonable probability that some of them will define major-effect genes with relatively consistent effects across segregating populations.

From basic biology to developing markers useful for practical breeding

As an example of the application of biochemical and genomic information to develop useful DNA markers to aid selection, several research groups are attempting to identify diagnostic markers for fatty acid content in soybean oil. Fatty acid contents in soybean are not highly polygenic traits, but do share some aspects in common with complex traits. Genetic studies have concluded that the low palmitic, low linolenic, and increased stearic acid traits are controlled by a few genes with major effects and by some modifier genes with small effects (Cardinal 2008). However, although these traits were treated as qualitatively inherited in most studies, they are actually measured quantitatively. Furthermore, the elevated oleic acid trait is inherited as a quantitative trait and it is highly influenced by environmental conditions (Burton et al 1983, Oliva et al 2006), although at least one major induced mutant gene has been described (Takagi and Rahman 1996).

Traditional breeding for these traits has resulted in the successful development of cultivars with altered fatty acid profiles via recurrent selection, mutagenesis, or screening for natural mutations. In parallel, the understanding of the biosynthesis of fatty acids in plants has improved tremendously over the last 10 years and most of the genes involved in the fatty acid synthesis pathway in plants have been cloned and sequenced in several plant species, including soybean. Importantly, each of five major fatty acids present in soybean triacylglycerols (palmitic, stearic, oleic, linoleic, and linolenic acids) is derived from a common biosynthetic pathway (Ohlrogge and Browse 1995, Kinney 1997). These fatty acids differ in the number of C atoms in their chain (16 or 18) and the number of unsaturated hydrocarbon bonds they possess (zero to three). For the purposes of identifying the candidate genes for mutations that affect different fatty acid contents, the enzymes that cause these specific differences are of greatest interest. These include plastid enzymes that synthesize palmitoyl-ACP and stearoyl-ACP, acyl-ACP delta-9 desaturase (which catalyzes the synthesis of oleolyl-ACP from stearoyl-ACP), and the acyl-ACP thioesterases that release fatty acids into the cytoplasm so that they can be sterified to form the acyl-CoA pool in the cytoplasm. Acyl chains are attached to each carbon of the glycerol molecule by the action of three microsomal acyl transferases. In addition, microsomal desaturases that catalyze the synthesis of linoleyl-PC and linolenoyl-PC are of great importance.

Sequence information of genes involved in the lipid biosynthetic pathway in soybeans provides a tool to develop allele-specific markers to investigate the relationship between QTLs or mutations for seed fatty acid components and specific genes of the pathway and their interactions. However, the success of this approach depends on (1) the stability of a specific allele variant effect across different genetic backgrounds and environments, (2) the knowledge of a particular biochemical pathway and availability of sequence information for several candidate genes in the pathway, (3) the level of genome duplication of a crop, and (4) the amount of genetic diversity in a crop and consequently the amount of variation at the DNA level.

In soybeans, the effects of most of the fatty acid mutant alleles that breeders have developed are very stable across genetic backgrounds and environments except for some high oleic acid alleles, the biochemical pathway of fatty acid biosynthesis is well understood, and genes involved in this pathway have been cloned and sequenced in several plant species, including soybeans; so the candidate gene approach should be feasible. However, the soybean genome has gone through two rounds of duplications and therefore the sequencing of candidate genes has been complicated by this phenomenon. When several copies of a candidate gene exist that share a very high level of sequence similarity, specific PCR amplification of particular genes or alleles can be extremely difficult, hindering both sequence and genetic analysis of the orthologous genes. For example, four isoforms each of candidate genes responsible for low linolenic and low palmitic acid mutations have been observed and some isoforms share more than 96% sequence identity, complicating the sequence effort tremendously (Anai et al 2005, Bilyeu et al 2003, Cardinal et al 2007). Furthermore, once all isoforms of a candidate gene are sequenced, cosegregation analysis between the inheritance of a specific allele of a particular isoform and the trait of interest has to be performed in order to determine which specific loci affect the trait.

Soybeans have a very narrow genetic base and the chances of finding a random single nucleotide polymorphism (SNP) at the cDNA level for any given gene isoform are very small, requiring the sequencing of the introns and 5′ and 3′ untranslated regions of the transcribed gene to increase the chances of finding polymorphisms. It is important to detect polymorphisms for each isoform of a candidate gene to perform cosegregation studies with the trait of interest in several populations segregating for different mutant alleles of a trait. Ideally, the allele-specific marker should be defined by the SNP or INDEL (insertion/deletion) causing the mutation; however, this is not a requisite for the methodology to work. The allele-specific marker of a candidate gene will work in a MAS strategy for most populations as long as the SNP or INDEL that is defining the allele specificity is very rare in the genetic base of a particular crop.

Allele-specific markers have been developed for one low palmitic acid mutant, several low linolenic acid mutants, and one mid oleic acid mutant (Alt et al 2005, Anai et al 2005, Bilyeu et al 2003, 2005, Cardinal et al 2007). These markers should be useful for the selection of desired traits in any population in which those specific fatty acid mutations or alleles are segregating, as demonstrated by the specificity of the allele-specific markers across a broad sampling of elite soybean germplasm. However, these markers have not been used in MAS schemes because of intellectual property concerns with some of the germplasm sources of altered fatty acid contents. In addition, allele-specific markers have not been developed for all the low palmitic, low linolenic, and mid oleic acid mutations that breeders have developed. Therefore, until such markers are developed for all the mutations available and being used in combination by breeders, phenotypic selection will be more readily employed.

A top-down approach to understanding the genomics of complex traits

Another approach to tackling the genetics of complex traits with genomics is to attempt large-scale integrating mapping studies that capture a substantial proportion of the genetic variation in a breeding pool, so that QTL effects can be defined in relation to the larger gene pool, rather than in relation to only a two-parent mapping population. In maize, we are taking this approach in the U.S. National Science Foundation–funded project "Molecular and Functional Diversity of the Maize Genome."

A major aspect of this program is a large-scale QTL mapping study that will attempt to define the effects of much of the genetic variation that is available among public maize inbred lines worldwide. Based on previous SSR surveys of a large number of public maize inbreds (Liu et al 2003), a core set of 27 inbred lines was selected to capture a large proportion (approx. 85%) of the observed molecular variation in the larger set. This set includes B73, one of the most important public maize inbreds, which is represented in the pedigrees of virtually all commercial maize hybrids in the U.S. (Troyer 1999). Because of the importance of B73 commercially, its excellent wide adaptation, and its use as the reference genotype for public-sector maize genome sequencing projects, it was chosen as the reference line in our QTL mapping experiment. The crossing design was to cross each of the other 26 inbred lines as males to B73, and to create 200 F_5-derived lines without conscious selection from each cross. We included the previously developed intermated B73 × Mo17 recombinant inbred population (Lee et al 2002) as one of the 26 cross populations. The 5,000 new mapping lines have been released for public distribution (http://maizecoop.cropsci.uiuc.edu/nam-rils.php). Each line was also genotyped at 1,105 biallelic single nucleotide polymorphism (SNP) markers. SNP markers were chosen on the basis of reliability, genome distribution, and informativeness across all 26 populations (www.panzea.org).

We evaluated the overall set of 5,200 mapping lines plus 280 diverse inbreds representing the maize association mapping panel (Flint-Garcia et al 2005) in six environments in 2006 and five environments in 2007. The very large number of entries in the experiment necessitated a large physical size of the evaluation fields, leading to concerns about within-replication error variation. Therefore, a field experimental design was created to help adjust for spatial environmental trends within fields of this very large experiment. One replication of the experiment was planted at each location. Each population was considered a set in the field design, and the order of sets is randomized across environments. Across populations, each set was arranged as a 10 × 20 alpha lattice design and two check inbreds (B73 and the other parent of the particular population) were included within each incomplete block. These repeated check plots provide information on the within-location spatial environmental trends at the cost of an additional 10% plots. Phenotypic data were analyzed with mixed models analysis using ASReml software (Gilmour et al 2006) to appropriately model heterogeneous error variances and field spatial effects within each environment, to partition genetic variance among and within populations, permitting heterogeneous within-population variances, and to obtain best linear unbiased predictors for each genotype.

QTL analysis was implemented based on the SNP data and the best linear unbiased predictors for each trait, following the nested association mapping procedure recently proposed by Yu et al (2008). To date, we have implemented QTL analysis based on fitting unique SNP allele effects within populations, which is expected to detect effects of QTLs linked to the SNPs because, although linkage disequilibrium (LD) across populations is low, LD within populations will follow the typical pattern for a two-parent RIL population (Burr et al 1988). Once the critical QTL positions are identified in this way, more precise mapping can be achieved by projecting progenitor sequence variants onto the mapping panel, based on flanking SNP markers. Simulation analysis of this approach suggests that the NAM analysis will provide 80% power to detect QTLs with effects as small as 1.2% of the phenotypic variation (Yu et al 2008). Further, since LD in maize rapidly approaches zero over several kb in most gene regions (Remington et al 2001), SNPs will be detected in this analysis only if they are actually in the causative gene itself. If a SNP is linked to, but in linkage equilibrium with, a QTL, the main effect of the SNP across populations is expected to be unaffected by the linked QTL. Thus, the precision of the NAM procedure is a function of linkage disequilibrium among the founder lines of the population, which in this set of maize lines should be very low. Thus, we expect that QTLs will be reduced to quantitative trait nucleotides (QTN) in this analysis. Haplotype-based modification of NAM may be possible in regions where LD extends over longer distances, leading to reduced precision in the mapping, but higher power to detect genome regions affecting the complex traits.

What can we expect from this experiment? If the results of other large-scale maize mapping studies are any guide, the genetic architecture of complex traits in maize appears to be, well, complex. Laurie et al (2004) developed a high-resolution mapping population by crossing lines from opposite populations resulting from divergent selection for high or low oil content for 70 generations, intermating the progeny for 10 generations, and then selfing to produce 500 mapping lines. Lines were genotyped at 488 markers and more than 50 QTLs were detected for oil content in this population. All of the QTLs had small effects, and, even with this large number of QTLs detected, they explained only about half of the genetic variance. This suggests that many more QTLs of even smaller effect explain most of the remaining genetic variation (Laurie et al 2004). A separate study conducted on 976 maize inbred testcrosses evaluated in 19 environments found 15 QTLs for yield and more than 25 QTLs each for grain moisture and plant height (Schon et al 2004). Although the line mean heritability values were high, the QTL models for none of the traits explained much more than half of the genetic variance, again suggesting that many QTLs with even smaller effects remained undetected.

In fact, our results demonstrate many of these same features of complex traits. For flowering time, which had an entry mean heritability of almost 95% in this experiment, we have detected about 50 QTLs explaining 85% of the genetic variation even at a stringent significance threshold of $P < 0.0001$. None of the QTLs have large effects; the largest allele effect confers less than a two-day difference in flowering time compared to the B73 allele, despite the fact that some of the founders flower about a month later than B73 in some environments. At each of the QTLs, it is clear that functionally different alleles exist, suggesting that that even when finite numbers of genes are involved, many more functionally distinct alleles may be segregating in diverse populations.

Let us assume that our 5,000-plus line mapping experiment is successful for other important agronomic traits, and that we detect a large number of QTLs cumulatively affecting a large proportion of the genetic variation in the traits studied (although probably individual QTL effects will be very small). How will this information be useful for breeding? The most important practical result of this experiment may be the first clear evaluation of the distribution of allelic effects across diverse maize germplasm. We expect to find at least some alleles in the diverse lines that are more favorable than the corresponding B73 allele. But the question remains: Will there be any QTLs of sizable effect where an exotic allele is favorable? And, what proportion of diverse line alleles will be more favorable than B73 alleles? The results of our experiment will allow us to predict specific optimal combinations of alleles derived from 26 different maize lines. These predictions are testable, and a clear line of investigation to follow will be to undertake a genotype-building selection program to assemble sets of favorable alleles from different lines into a single line. Understanding which QTL alleles are carried in key ancestral lines provides a means to positively exploit the genetic heterogeneity of complex traits (Holland 2007). However, if the QTL effects are all very small, the gain from this breeding program may not be very large. The question then may well be whether the huge investment in the genomics agenda will have been worthwhile relative to the phenotype-driven exotic maize incorporation programs that are already under way (Goodman et al 2000).

Conclusions

What are the implications of this discussion for improving drought resistance in rice? Because of the highly complex nature of drought responses in rice, the challenge of improving drought resistance is daunting (Lafitte et al 2006, Yue et al 2006). The development of DNA markers to aid selection for drought resistance may require the identification of component traits that can be understood at the gene level, so that specific gene sequence variants can be related to phenotypic changes of interest. Alternatively, a large-scale top-down approach of diversity-based QTL mapping could lead to more reliable identification of QTLs with small effects, accompanied by an understanding of their distribution across populations. This would help to bring genetic heterogeneity problems under control, but, if the QTL effects are very small, will the effort be worth it? Finally, industrial-scale methods such as genomewide selection that essentially predict marker allele effects in every breeding cross and exploit population-specific linkage disequilibrium between markers and QTLs in off-season nurseries are another option to enhance gain from selection for complex traits, but also require tremendous investments. The economic efficiency of these strategies relative to phenotype-based selection is not well understood at this time.

References

Alt JL, Fehr WR, Welke GA, Sandhu D. 2005. Phenotypic and molecular analysis of oleate content in the mutant soybean line M23. Crop Sci. 45:1997-2000.

Anai T, Yamada T, Kinoshita T, Rahman SM, Takagi Y. 2005. Identification of corresponding genes for three low-α-linolenic acid mutants and elucidation of their contribution to fatty acid biosynthesis in soybean seed. Plant Sci. 168:1615-1623.

Beavis WD. 1998. QTL analyses: power, precision, and accuracy. In: Paterson AH, editor. Molecular dissection of complex traits. Boca Raton, Fla. (USA): CRC Press. p 145-162.

Bernardo R. 2001. What if we knew all the genes for a quantitative trait in hybrid crops? Crop Sci. 41:1-4.

Bernardo R, Yu J. 2007. Prospects for genomewide selection for quantitative traits in maize. Crop Sci. 47:1082-1090.

Bilyeu K, Palavalli L, Sleper D, Beuselinck P. 2005. Mutations in soybean microsomal omega-3 fatty acid desaturase genes reduce linolenic acid concentration in soybean seeds. Crop Sci. 45:1830-1836.

Bilyeu KD, Palavalli L, Sleper DA, Beuselinck PR. 2003. Three microsomal omega-3 fatty-acid desaturase genes contribute to soybean linolenic acid levels. Crop Sci. 43:1833-1838.

Bohn M, Groh S, Khairallah MM, Hoisington DA, Utz HF, Melchinger AE. 2001. Re-evaluation of the prospects of marker-assisted selection for improving insect resistance against *Diatraea* spp. in tropical maize by cross validation and independent validation. Theor. Appl. Genet. 103:1059-1067.

Bonnet DG, Rebetzke GJ, Spielmeyer W. 2004. Strategies for efficient implementation of molecular markers in wheat breeding programs. Mol. Breed. 15:75-85.

Burr B, Burr FA, Thompson KH, Albertson MC, Stuber CW. 1988. Gene mapping with recombinant inbreds of maize. Genetics 118:519-526.

Burton JW, Wilson RF, Brim CA. 1983. Recurrent selection in soybean. IV. Selection for increased oleic acid percentage in seed oil. Crop Sci. 23:744-747.

Cahill DJ, Schmidt DH. 2004. Use of marker assisted selection in a product development breeding program. In: Fischer T, Turner N, Angus J, McIntyre L, Robertson M, Borrell A, Lloyd D, editors. New directions for a diverse planet: Proceedings for the 4th International Crop Science Congress. Brisbane, Australia.

Camacho-Roger AM. 2006. Molecular markers and genes associated with low palmitic and low linolenic acid content in N97-3681-11 and N97-3708-13 soybean lines. M.S. thesis, North Carolina State University.

Cardinal AJ. 2008. Molecular genetics and breeding for fatty acid manipulation in soybean. Plant Breed. Rev. 30:259-294.

Cardinal, AJ, Burton JW, Camacho-Roger AM, Yang JH, Wilson RF, Dewey RE. 2007. Molecular analysis of soybean lines with low palmitic acid content in the seed oil. Crop Sci. 47:304-310.

Chen S, Lin XH, Xu CG, Zhang Q. 2000. Improvement of bacterial blight resistance in 'Minghui 63', an elite restorer line of hybrid rice, by molecular marker-assisted selection. Crop Sci. 40:239-244.

Dekkers JCM, Hospital F. 2002. The use of molecular genetics in the improvement of agricultural populations. Nature Rev. 3:22-32.

Dubcovsky J. 2004. Marker-assisted selection in public breeding programs: the wheat experience. Crop Sci 44:1895-1898.

Eagles HA, Bariana HS, Ogbonnaya FC, Rebetzke GJ, Hollamby GJ, Henry RJ, Henschke PH, Carter M. 2001. Implementation of markers in Australian wheat breeding. Aust. J. Agric. Res. 52:1349-1356.

Eathington, SR, Crosbie TM, Edwards MD, Reiter RS, Bull JK. 2007. Molecular markers in a commercial breeding program. Crop Sci. 47:S-154-163.

Falconer DS, Mackay TFC. 1996. Introduction to quantitative genetics. 4th ed. Longman Technical, Essex, UK.

Flint-Garcia SA, Thuillet AC, Yu J, Pressoir G, Romero SM, Mitchell SE, Doebley J, Kresovich S, Goodman MM, Buckler ES.2005. Maize association population: a high-resolution platform for quantitative trait locus dissection. Plant J. 44:1054-1064.

Gilmour AR, Gogel BJ, Cullis BR, Thompson R. 2006. ASReml User Guide Release 2.0. VSN International Ltd., Hemel, Hempstead, UK.

Goodman MM, Moreno J, Castillo F, Holley RN, Carson ML. 2000. Using tropical maize germplasm for temperate breeding. Maydica 45:221-234.

Hallauer AR, Miranda JB. 1988. Quantitative genetics in maize breeding. 2nd edition. Iowa State Univ. Press, Ames, Iowa, USA.

Holland JB. 2004. Implementation of molecular markers for quantitative traits in breeding programs: challenges and opportunities. In: Fischer T, Turner N, Angus J, McIntyre L, Robertson M, Borrell A, Lloyd D, editors. New directions for a diverse planet: Proceedings for the 4th International Crop Science Congress. Brisbane, Australia.

Holland JB. 2006. Theoretical and biological foundations of plant breeding. In: Lamkey KR, Lee M, editors. Plant breeding: The Arnel R. Hallauer International Symposium. Ames, Iowa (USA): Blackwell. p 127-140.

Holland, J.B. 2007. Genetic architecture of complex traits in plants. Curr. Opin. Plant Biol. 10:156-161.

Hospital F, Moreau L, Lacoudre F, Charcosset A, Gallais A. 1997. More on the efficiency of marker-assisted selection. Theor. Appl. Genet. 95:1181-1189.

Kearsey MJ, Farquhar AGL. 1998. QTL analysis in plants: Where are we now? Heredity 80:137-142.

Kinney AJ. 1997. Genetic engineering of oilseeds for desired traits. In: Setlow JK, editor. Genetic engineering, principles and methods. New York: Plenum Press.

Lafitte HR, Li ZK, et al. 2006. Improvement of rice drought tolerance through backcross breeding: evaluation of donors and selection in drought nurseries. Field Crops Res. 97:77-86.

Laurie CC, Chasalow SD, et al. 2004. The genetic architecture of response to long-term artificial selection for oil concentration in the maize kernel. Genetics 168:2141-2155.

Lee M. 1995. DNA markers and plant breeding programs. Adv. Agron. 55:265-344.

Lee M, Sharopova N, Beavis WD, Grant D, Katt M, Blair D, Hallauer A. 2002. Expanding the genetic map of maize with the intermated B73 × Mo17 (IBM) population. Plant Mol. Biol. 48:453-461.

Li Z-K, Fu B-Y, et al. 2005. Genome-wide introgression lines and their use in genetic and molecular dissection of complex phenotypes in rice (*Oryza sativa* L.). Plant Mol. Biol. 59:33-52.

Liu K, Goodman M, Muse S, Smith JS, Buckler E, Doebley J. 2003. Genetic structure and diversity among maize inbred lines as inferred from DNA microsatellites. Genetics 165:2117-2128.

Melchinger AE, Utz HF, Schon CC. 1998. Quantitative trait locus (QTL) mapping using different testers and independent population samples in maize reveal low power of QTL detection and large bias in estimates of QTL effects. Genetics 149:383-403.
Muewisse THE, Hayes BJ, Goddard ME. 2001. Prediction of total genetic value using genome-wide dense marker maps. Genetics 157:1819-1829.
Ohlrogge J, Browse J. 1995. Lipid biosynthesis. Plant Cell 7:957-970.
Oliva ML, Shannon JG, Sleper DA, Ellersieck MR, Cardinal AJ, Paris RL, Lee JD. 2006. Stability of fatty acid profile in soybean genotypes with modified seed oil composition. Crop Sci. 46:2069-2075.
Peleman JD, van der Voort JR. 2003. Breeding by design. Trends Plant Sci. 8:330-334.
Remington DL, Thornsberry JM, Matsuoka Y, Wilson LM, Whitt SR, Doebley J, Kresovich S, Goodman MM, Buckler ES, IV. 2001. Structure of linkage disequilibrium and phenotypic associations in the maize genome. Proc. Natl. Acad. Sci. USA 98:11479-11484.
Richards RA. 2006. Physiological traits used in the breeding of new cultivars for water scarce environments. Agric. Water Manage. 80:197-211.
Schon CC, Utz HF, Groh S, Truberg B, Openshaw S, Melchinger AE. 2004. Quantitative trait locus mapping based on resampling in a vast maize testcross experiment and its relevance to quantitative genetics for complex traits. Genetics 167:485-498.
Stuber CW. 1998. Case history in crop improvement: yield heterosis in maize. In: Paterson AH, editor. Molecular dissection of complex traits. Boca Raton, Fla. (USA): CRC Press. p 197-206.
Takagi Y, Rahman SM. 1996. Inheritance of high oleic acid content in the seed oil of soybean mutant M23. Theor. Appl. Genet. 92:179-182.
Tanksley SD, McCouch SR. 1997. Seed banks and molecular maps: unlocking genetic potential from the wild. Science 277:1063-1066.
Troyer AF. 1999. Background of U.S. hybrid corn. Crop Sci. 39:601-626.
Utz HF, Melchinger AE, Schon CC. 2000. Bias and sampling error of the estimated proportion of genotypic variance explained by quantitative trait loci determined from experimental data in maize using cross validation and validation with independent samples. Genetics 154:1839-1849.
Wilson RF, Marquardt TC, Novitsky WP, Burton JW, Wilcox JR, Kinney AJ, Dewey RE. 2001. Metabolic mechanisms associated with alleles governing the 16:0 concentration of soybean oil. J. Am. Oil Chem. Soc. 78:335-340.
Xu K, Deb R, Mackill DJ. 2004. A microsatellite marker and a codominant PCR-based marker for marker-assisted selection of submergence tolerance in rice. Crop Sci. 44:248-253.
Xu K, Xu X, et al. 2006. *Sub1A* is an ethylene-response-factor-like gene that confers submergence tolerance to rice. Nature 442:705-708.
Young ND. 1999. A cautiously optimistic vision for marker-assisted breeding. Mol. Breed. 5:505-510.
Yu J, Holland JB, McMullen MD, Buckler ES. 2008. Genetic design and statistical power of nested association mapping in maize. Genetics 178:539-551.
Yue B, Xue W, Xiong L, Yu X, Luo L, Cui K, Jin D, Xing Y, Zhang Q. 2006. Genetic basis of drought resistance at reproductive stage in rice: separation of drought tolerance from drought avoidance. Genetics 172:1213-1228.
Zamir D. 2001. Improving plant breeding with exotic genetic libraries. Nature Rev. Genet. 2:983-989.

Notes

Authors' addresses: J.B. Holland, USDA-ARS Plant Science Research Unit, Raleigh, NC 27695, USA; J.B. Holland and A.J. Cardinal, Department of Crop Science, North Carolina State University, Raleigh, NC 27695, USA. Some of the research described here was supported by grants from the U.S. National Science Foundation (DBI-0321467) to JBH and from U.S. Department of Agriculture National Research Initiative (Competitive Grants Program award numbers 2001-35301-10601 to JBH and 2003-00691 to AJC).

Physiological and molecular mechanisms of drought resistance

Drought-resistant rice: physiological framework for an integrated research strategy

R. Serraj, G. Dimayuga, V. Gowda, Y. Guan, Hong He, S. Impa, D.C. Liu, R.C. Mabesa, R. Sellamuthu, and R. Torres

The unpredictability of drought patterns and the complexity of the physiological responses involved have made it difficult to characterize component traits required for improved performance, thus limiting crop improvement to enhance drought resistance. The various stress response mechanisms and options to enhance plant survival under severe stress do not usually translate into yield stability under water deficit. Increased crop yield and water productivity require the optimization of the physiological processes involved in the initial critical stages of plant response to soil drying, water-use efficiency, and dehydration avoidance mechanisms. New high-throughput phenotyping methodologies allow fast and detailed evaluation of potential drought-resistant donors and the large number of lines developed by drought breeding programs. Similarly, large collections of rice germplasm, including minicore sets, wild relatives, and mutant lines, are screened for drought-resistance traits. Genetic sources of drought resistance have now been identified for all major rice agroecosystems and some of the associated traits have been characterized. The identification and genetic mapping of QTLs for yield and related physiological traits under drought stress across environments are currently a major focus. This approach provides a powerful tool to dissect the genetic and physiological bases of drought resistance. If validated with accurate phenotyping and properly integrated in marker-assisted breeding programs, this will accelerate the development of drought-resistant genotypes. This paper reviews IRRI's recent progress and achievements in understanding the physiology of drought resistance in rice and presents future perspectives on the genetic enhancement of drought adaptation.

Drought is the most important source of climate-related risk for rice production in rainfed areas (Pandey et al 2007). Rice cultivars combining improved drought resistance with yield potential under favorable conditions are the most promising and deliverable technology for increasing productivity in drought-prone areas. To take advantage of the recent advances in genomics and biotechnology, it is well accepted that the complexity of drought resistance can only be tackled with a holistic approach integrating plant breeding with physiological dissection of the resistance traits and molecular genetic tools together with agronomic practices that lead to better conservation and use of soil moisture and matching crop genotypes with the environment (Crouch and Serraj 2002).

Crop physiology has made a significant contribution to understanding the mechanisms underlying crop growth and development, and bridging the "phenotype gap" generated by the recent progress in genomics (Miflin 2000, Boote and Sinclair 2006). However, despite the various efforts deployed over the past decades for dissecting drought resistance, the identification and characterization of component traits, which can be transferred through plant breeding into cultivars with high-yielding genetic backgrounds, have generally had very limited success. Several reasons have been proposed to explain this lack of success and to document the few cases in which trait-based selection for drought resistance resulted in actual yield improvement (Sinclair et al 2004, Richards 2006). A common feature of these success stories is that the timing and intensity of drought are critical for these traits to be effective, in addition to the time scale and crop phenological stage in which they operate. This also indicates that any putative drought-resistance trait is unlikely to be relevant under all water-deficit scenarios due to the high levels of G × E interactions generally observed in the phenotypic expression of component traits, and their impacts on crop productivity (Hammer and Jordan, this volume).

While trait-based selection for improved drought resistance has progressed very slowly over the past decades (Fukai and Cooper 1995, Bernier et al 2008a), direct empirical selection for grain yield under managed drought stress has been more successful recently in both rice (Venuprasad et al 2007a) and other crops (Edmeades et al 1999, Nigam et al 2005, Bänziger et al 2006). However, this approach also has to face the challenges of applying large-scale and reliable protocols for managed-stress screening in rice breeding programs, and resolving G × E interactions in the various drought-prone target environments. Genotype-by-environment interaction for drought-resistance component traits results to a large extent from variation in the weather scenarios among growing environments. The impact of any given drought trait is usually highly sensitive to the rate at which the drought develops, the timing when it occurs in the season, and the severity of the stress. A drought trait that might offer substantial benefit in one weather scenario of developing drought might well result in a negative response in another scenario (Sinclair and Muchow 2001).

One way to overcome the large G × E limitation is to understand the basic processes accounting for the drought-resistance trait and how the mechanism reacts under a range of weather scenarios. Simulation models can also help overcome G × E by combining a mechanistic understanding of a drought trait with a range of weather scenarios (Chapman 2008). Breeding for specific drought-resistance characters can thus be targeted to those geographical regions that would have the highest probability of frequent yield increases. Based on the well-known equation of Passioura (1977), modeling grain yield as a function of water transpired, transpiration efficiency, and harvest index, three major breeding goals have been designed for increasing crop yield and water productivity of rainfed grain crops (Condon et al 2004): (1) increase plant water uptake while minimizing water losses through soil evaporation, percolation beyond the root zone, and residual water left behind in the root zone after harvest; (2) acquire more photosynthate in exchange for each unit of water transpired during CO_2 fixation by the crop, generally referred to as water-use efficiency; (3) partition more of the acquired assimilates into harvestable product. This physiological framework has provided simple yet clear guidelines for crop improvement, which has led to successful breeding applications, all related to one or the other of the three strategies of Passioura's yield model (Richards 2006). This has provided the proof of concept that well-designed and careful trait-based dissection of yield components coupled with in-depth understanding of the underlying physiological mechanisms of those traits can lead to tangible and reproducible success in breeding for drought adaptation.

Recent progress in molecular genetics and genome sequencing technologies is now enabling high-throughput whole-genome genotyping platforms, and the cost will decline further (Leung 2008). By comparison, the phenotyping of large germplasm collections and populations for drought-related traits in field trials is still laborious, imprecise, and costly. This has made phenotyping the current bottleneck of crop improvement and molecular mapping. Precision phenotyping requires a physiological understanding of the underlying mechanisms of the traits and dynamics involved in crop responses to water deficits. Most phenotypic traits have been generally assumed to be fixed. However, crop growth and development traits are dynamic processes controlled by a complex network of genes. Analyzing phenotypic data measured at a single point in time in mapping studies would not be suitable to capture the genetic control of developmental changes in response to stress.

Crop response to water deficit is a continuous process that involves distinct phases of water use and stress profiles. Although substantial research efforts have been devoted to investigating the mechanisms of "drought tolerance," focusing on survival stage under severe stress, this has led to little progress in crop improvement. Dehydration avoidance is more relevant as a strategy for relieving agricultural drought and maintaining crop performance, before survival drought develops. In rice, the conclusion emerging from long-term multilocation drought studies was that rainfed lowland rice is mostly a drought avoider, with those genotypes that produce higher grain yield under drought being those able to maintain better plant water status around flowering and grain setting (Fukai et al, this volume).

The objectives of this paper are to describe the recent progress in drought physiology and phenotyping at IRRI and to review the current knowledge of the key traits and physiological processes involved in dehydration avoidance, growth regulation, and reproductive-stage processes under drought stress, toward an integrated strategy for improving drought resistance in rice.

Drought or droughts?

Water deficit is a major challenge for all agricultural crops, but for rice it is even more so, because of its semiaquatic phylogenetic origins and the diversity of rice ecosystems and growing conditions (O'Toole 2004). Drought is conceptually defined in terms of rainfall shortage compared with a normal average value in the target region.

However, drought occurrence and effects on rice productivity depend more on rainfall distribution than on the total seasonal rainfall. A typical example was given in a recent screening experiment at IRRI during the wet season of 2006, when seasonal rainfall exceeded 1,200 mm, including a major typhoon (Milenyo) with around 320 mm of rainfall in a single day. Yet, a short dry spell that coincided with the flowering stage resulted in a dramatic decrease in grain yield and harvest index, compared to the irrigated control (Serraj et al, unpublished). Beyond the search for global solutions to generic "drought," the precise characterization of droughts in the target population of environments (TPE) is a prerequisite for better understanding their consequences for crop production (Heinemann et al 2008).

Chronic, catastrophic, and inherent droughts

Drought definitions depend on the disciplinary outlook, including meteorological, hydrological, and agricultural perspectives. Agricultural drought occurs when soil moisture is insufficient to meet crop water requirements, resulting in yield losses. Depending on timing, duration, and severity, this can result in catastrophic, chronic, or inherent drought stress, which would require different coping mechanisms, adaptation strategies, and breeding objectives.

The 2002 drought in India could be described typically as a catastrophic event, as it affected 55% of the country's area and 300 million people. Rice production declined by 20% from the trend values (Pandey et al 2007). Similarly, the 2004 drought in Thailand affected more than 8 million people in almost all provinces. Severe droughts generally result in starvation and impoverishment of the affected population, resulting in production losses during years of complete crop failure, with dramatic socioeconomic consequences for human populations (Pandey and Bhandari, this volume). Production losses to drought of milder intensity, although not so alarming, can be substantial. The average rice yield in rainfed eastern India during "normal" years still varies between 2.0 and 2.5 t ha^{-1}, far below yield potential. Chronic dry spells of relatively short duration can often result in substantial yield losses, especially if they occur around flowering stage. In addition, drought risk reduces productivity even during favorable years in drought-prone areas because farmers avoid investing in inputs when they fear crop loss (Pandey et al 2007). Inherent drought is associated with the increasing problem of water scarcity, even in traditionally irrigated areas, due to rising demand and competition for water uses. This is, for instance, the case in China, where the increasing shortage of water for rice production is a major concern, although rice production is mostly irrigated (Ding et al 2005).

Increasing rice productivity in drought-prone rainfed areas requires adapted solutions and strategies in response to the different types of drought, based on precise characterization of the TPEs. With milder chronic droughts being generally more frequent than catastrophic ones, overall crop productivity in rainfed areas would probably benefit more from breeding for enhanced water productivity and resistance to chronic water deficits.

Drought-prone TPEs

A new rice field classification system, proposed by IRRI (Bouman, pers. comm.), has defined four major classes of drought-prone rainfed environments, with more than 50% probability that one of the following may occur:

1. Early-season drought risk in lowland (DEL), with nonflooded soils and root zone below saturation for at least 10 consecutive days before flowering.
2. Flowering-stage drought risk in lowland (DFL), with nonflooded soils and root zone below saturation for at least 7 days, around anthesis.
3. Late-season drought risk in lowland (DLL), with nonflooded soils and root zone below saturation for at least 10 consecutive days after flowering.
4. Flowering-stage drought risk in upland (DFU), with fields without rainfall or irrigation for at least 7 days around anthesis and groundwater table deeper than 100 cm.

A detailed characterization of the drought-prone rainfed environments, according to the new field classification system, is still to be done at a fine level in the major target environments across Asia and sub-Saharan Africa. Drought-prone rainfed rice ecosystems were previously classified based on toposequence position and defined by the water regime encountered (Garrity et al 1986). The upland ecosystem occupies more than 10 million ha in Asia and was previously structured based on production systems and agroecological characterization (Courtois and Lafitte 1999). Several studies have also previously discussed the biophysical characteristics of the rainfed lowland ecosystem and their implications for breeding (Mackill et al 1996, Fukai et al 2001, Wade et al 1999a). The highly unstable dynamics of hydrology, with frequent shifting between flooded and aerobic conditions within a paddy, impose a large amount of environmental variability and result in strong impact of spatial variation in the toposequence on crop growth and performance parameters (Cooper et al 1999a).

There are 23 million ha of drought-prone rice area in Asia alone, with more than half across the uplands and rainfed lowlands of India. Recent analysis of rainfall distribution data in eastern India, northeast Thailand, and southern China for the period 1970-2003 (Pandey et al 2007) indicated that drought is a recurrent phenomenon in all three regions, with probabilities of drought occurrence variable between 0.1 and 0.4 and highest in eastern India. The probability of late-season drought is also found to be spatially more covariate than early-season drought, which highlights the critical importance of rice sensitivity to drought during the reproductive stage (Pandey et al 2007).

In northwest Bangladesh, the average annual rainfall varies between 1,500 and 2,000 mm, with more than 200 mm of rainfall per month during the monsoon period (June to September), where transplanted aman rice (T. aman) is grown mostly under rainfed conditions. However, the erratic rainfall distribution causes drought frequently in this region, and results in yield losses that are generally higher than the damage caused by flooding and submergence (Towfiqul Islam 2008). A recent characterization and modeling study showed that the recurrence interval of drought is around 2–3 years, especially during the latest part of T. aman, generally recognized as terminal drought (Towfiqul Islam 2008). Short-duration varieties such as BRRI dhan 39 are generally used to escape terminal drought in this region. However, the risk of early droughts is also very serious, with a return period of 10 mm rainfall deficit up to 1.3 years in some districts, which requires a new set of drought-adapted T. aman rice varieties.

Plant water use and responses to water deficits

It is well known that plants can sense water availability around the roots and respond by sending hydraulic and/or chemical signals to the shoot to elicit several adaptive responses, including stomatal closure, decrease in leaf expansion, and gas exchange (Tardieu and Davies 1993). The typical response curve of a particular physiological process to plant-available soil water can be described with two straight lines that intersect at the threshold value for which the rate of the process in stressed plants starts to diverge from a reference value (Sadras and Milroy 1996, Ray and Sinclair 1997). Based on this relationship, plant responses to soil water deficits can be typically described as a sequence of three successive stages of soil dehydration. Stage I occurs before reaching the threshold value, when water is still freely available from the soil and transpiration is not limited by soil water availability. Stage II starts when the plant reaches the threshold value of available water and the rate of water uptake cannot match the potential transpiration rate. Stomatal conductance declines, limiting the transpiration rate to a level similar to that of soil water uptake, and resulting in the maintenance of plant water balance. Stage III is reached when the plants are no longer able to limit transpiration through stomatal conductance; they must then resort to other mechanisms of drought adaptation for survival.

Virtually all major processes contributing to crop yield, including leaf expansion, photosynthetic rate, and growth, start to be down-regulated late in stage I or early in stage II of soil drying (Serraj et al 1999). At the end of stage II, these growth-supporting processes have effectively reached zero and no further net growth occurs in the plants. The focus of stage III is mostly on survival, which generally involves mechanisms such as osmotic adjustment (Serraj and Sinclair 2002) and that can be critical in natural dry-land ecosystems, but have little relevance to increasing/stabilizing crop yield in most agricultural situations. Thus, increased crop yields and water productivity require the optimization of the physiological processes involved in the critical stages (mainly stage II) of plant response to soil drying.

Extensive experimental evidence has established a general response of plant gas exchange to soil drying when expressed as a function of the fraction of extractable volumetric soil water content (Sadras and Milroy 1996). Plant gas exchange is generally constant until about FTSW (fraction of transpirable soil water) 0.4 to 0.3; then, soil drying results in a linear decrease in leaf gas exchange until available soil water is almost nil. This response pattern has been observed over a broad range of environments, species, and soils (review by Sadras and Milroy 1996). Transpiration rate on drying soil relative to plants on well-watered soil (relative transpiration rate, RT) was found to be described as a function of the fraction of transpirable soil water, using the following equation for soybean (Serraj and Sinclair 1997):

$$RT = 1 / [1 + 5.25*\exp(-11.23*FTSW)] \quad [1]$$

This pattern has been confirmed in rice by describing the response using a two-segment model with a linear plateau segment for the initial phase and a linear decreasing segment for the second phase of soil drying (Fig. 1). We evaluated the genotypic variation of transpiration response to progressive soil water deficit in rice genotypes under controlled conditions. The relationships of normalized transpiration rate (NTR) to soil drying, as measured by FTSW, were measured according to the dry-down protocol described previously (Sinclair and Ludlow 1986, Serraj et al 1999). The NTR response curves were well described by linear-plateau functions that allowed the calculation of the soil-water thresholds at which transpiration of drought-stressed plants began to decrease compared with the well-watered treatment (Ray and Sinclair 1997). The FTSW soil-water thresholds varied significantly among rice genotypes, varying between 0.347 in upland-adapted cultivar Apo and 0.735 in lowland-adapted cultivar IR72 (Fig. 1). Based on this analysis, cultivars with lower threshold values (e.g., Apo) are able to maintain higher NTR during the drying cycle than those with higher thresholds (e.g., IR72), which would be associated with genetic differences in the control of transpiration and drought avoidance mechanisms. Those with lower thresholds would allow a greater amount of dry matter accumulation during drought, which might result in higher transpiration efficiencies, as has been recently shown in groundnut cultivars transformed with DREB1A (Bhatnagar-Mathur et al 2007). Similar data of genotypic variation in NTR response curves were also observed with other species, including maize (Ray and Sinclair 1997). However, as vapor pressure deficit (VPD) also greatly influences transpiration rates, with possible genotypic variation in the magnitude of the response (Sinclair et al 2008), this would affect NTR response to soil dry-down. Experimental work is now under way to investigate the interaction of FTSW and VPD on leaf gas exchange during progressive soil drying.

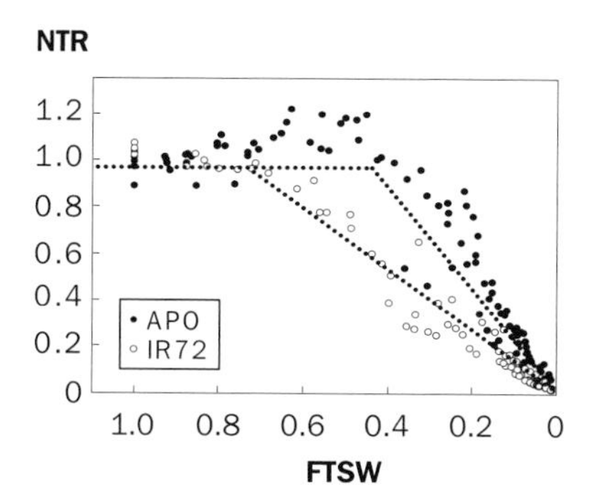

Fig. 1. Plot of normalized leaf transpiration rate (NTR) against the fraction of transpirable soil water (FTSW) in Apo and IR72 plants grown in the greenhouse and exposed to gradual soil drying.

Given the little scope for improving rice performance under drought based on survival mechanisms of stage III, dehydration avoidance mechanisms offer more promising avenues by improving plant characteristics to prolong phases I and II during soil drying. Some of these characteristics are deeper rooting and ability to overcome soil physical barriers and hardpans, and water conservation by controlling transpiration response to soil drying and decreasing stomatal conductance under high VPD conditions.

Concepts and tools for precise high-throughput phenotyping

Improving the precision and throughput of phenotyping is now often highlighted as the bottleneck and the main priority of drought-resistance studies, but a relevant measure of plant sensitivity to water deficit dynamics is still a challenging question. Time (days after stress application) is often used in drought experiments, but time is obviously an inaccurate co-variable as plant physiological responses depend mainly on environmental conditions and stress occurrence parameters. Similarly, the use of plant water status parameters such as leaf water potential or relative water content as stress co-variables is also often biased by variability, G × E interaction, and mostly by impracticality for high-throughput field applications (Lafitte 2002).

Beyond the search for generic drought phenotyping recipes that would fit all situations and environments, our current approach for developing a relevant phenotyping methodology starts by answering the question: What is the independent variable in the soil-plant-atmosphere continuum system that can be quantitatively and reproducibly related to crop response to water deficits?

It has long been found that genetic differences in the physiological responses to water deficits are mainly related to differences in soil water extraction. In rice, Lilley and Fukai (1994a) demonstrated that cultivar differences in the rates of stress development were strongly associated with the variation in extractable soil water and water extraction rate. After accounting for differences in water extraction ability, cultivar differences in sensitivity of physiological processes to water deficit were small (Lilley and Fukai 1994b). It is, hence, crucial for phenotyping crop responses to drought to characterize precisely the dynamics of soil water extraction.

FTSW dry-down approach

The total amount of soil water available to support plant water uptake was defined by Sinclair and Ludlow (1986) as the "transpirable soil water" and the relative dryness of the soil between upper and lower limits as the fraction of transpirable soil water (FTSW). By definition, FTSW has a value of 1.0 at the upper limit and 0 at the lower limit. An FTSW decrease results in progressive water deficit that influences many physiological processes such as transpiration, photosynthesis, or leaf expansion (Serraj et al 1999). These processes are generally inhibited when FTSW decreases to values in the range of 0.4–0.5, with a consistent trend across a wide range of environments and plant genotypes (Sadras and Milroy 1996).

A simple derivation model was proposed by Sinclair (2005), defining plant water flux in drying soil relative to that in well-watered soil and by examining the response in a range of soil volumetric water contents. This derivation resulted in a relatively simple expression that predicted daily transpiration rate response to drying soil, which was consistently independent of the absolute value of transpiration rate, root length density, and soil depth:

$$RT = (\Psi_{soil} - \Psi_{leaf}) / [\alpha E_w/(1{,}000\ dK) - \Psi_{leaf}]$$

where RT is relative daily transpiration rate, Ψ_{soil} and Ψ_{leaf} are, respectively, the hydrostatic pressure in the soil and leaves during the period of active transpiration (MPa), α is a variable (cm^2) for geometry of soil water extraction around roots, E_w is the upper limit for daily water loss rate (equal to 0.8 cm d^{-1}), d is depth of soil water extraction (cm), and K is the soil hydraulic conductivity (cm d^{-1}). The expression of relative transpiration as a function of volumetric soil water content available to support transpiration minimized the influence of soil texture on the overall response of water flux to progressive soil drying. The derivation offers a theoretical basis to explain the stability in daily transpiration response to drying soil that has been observed over a wide range of crops and conditions. This also supports the robustness of using FTSW as a stress co-variable in drought studies.

We developed a drought phenotyping platform using FTSW as a soil moisture co-variable for monitoring and comparing the dynamic responses of rice genotypes to progressive soil drying under controlled environmental conditions. The relationships of relative transpiration to FTSW were well described by linear-plateau functions that allowed the determination of the soil-water thresholds at which transpiration of drought-stressed plants began to decrease compared with the well-watered treatment (Fig. 1). Comparative FTSW response curves were also established for key physiological processes such as photosynthesis, stomatal conductance, transpiration, and tissue expansion in rice accessions commonly used as parents of mapping populations. This technique is now being tested for a field-based system (Fig. 2), under both upland and lowland conditions, where soil moisture profiles are automatically monitored in parallel to plant water status and leaf gas exchange measurements to analyze the dynamics of rice response to water deficits. The use of FTSW as a stress co-variable makes possible the integration of on-station field experiments with multilocation trials and controlled-environment experiments that analyze specific trait responses to stress and related gene expression.

Building on the previous progress in dry-season field screening at IRRI, to reliably select drought-resistant germplasm and backcross families (Lafitte and Courtois 2002, Atlin et al, this volume), the focus is now on further enhancing the precision and throughput of drought phenotyping protocols. Our general approach to drought phenotyping aims at capturing the dynamics of crop responses to water deficits and interaction with environmental variables, toward the development of model-based phenotyping. It includes and integrates the following steps:

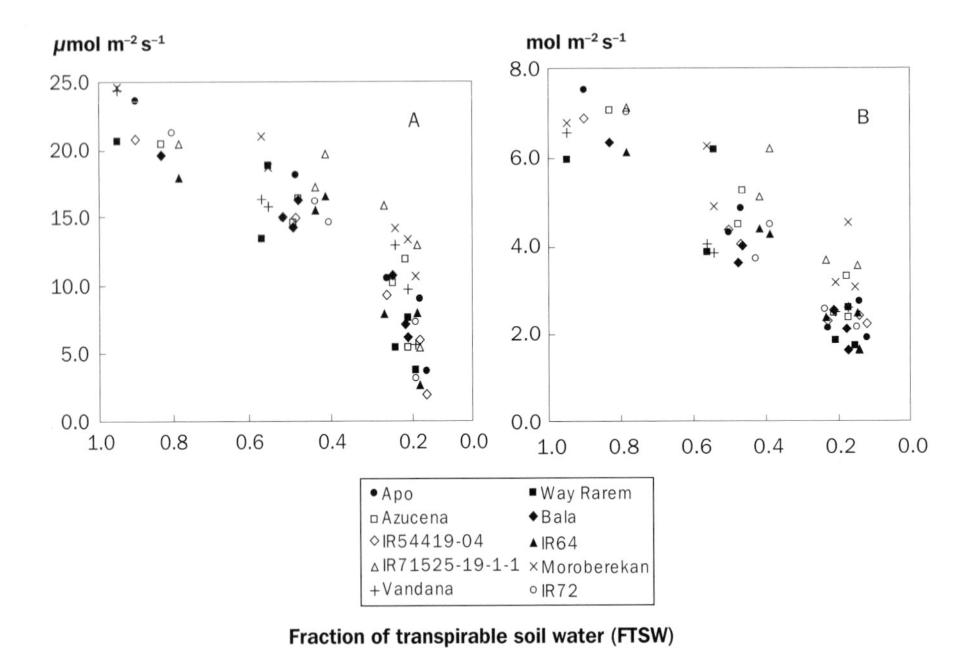

Fig. 2. Response of photosynthesis (A) and leaf transpiration (B) to FTSW of rice plants subjected to progressive soil drying in an upland field (IRRI, dry season 2007).

- Standard dry-season field-managed drought phenotyping, with control of stress timing, intensity, and duration
- Precise control of field irrigation and monitoring of soil moisture profiles and environmental parameters
- Continuous monitoring of crop phenology, canopy growth and development, and plant water status (destructive and nondestructive methods)
- Controlled FTSW dry-down methodology to confirm a phenotype and investigate its physiological basis (reverse physiology)
- Linking final crop yield, physiological traits, and putative molecular markers associated with resistance
- Field validation in the target environment (in wet season on-station and in drought-prone rainfed environments)

Control of field irrigation

Several irrigation regimes, including sprinkler, surface, furrow, and drip, have been used at IRRI to induce drought during specific periods and to investigate the response of rice varieties to drought stress (Lafitte and Courtois 2002). For many of the breeding experiments, water is withheld during the reproductive period—the most sensitive stage to drought stress—and flowering, grain filling, and yield are examined. Within rice, there is large variation in the time at which the reproductive stage is reached; therefore, experiments are separated into varieties with similar duration.

In upland fields, sprinkler irrigation is generally used to control soil water balance and plant water status, for managed-drought screening. Withholding irrigation allows targeted drought timing (reproductive or vegetative stage) and severity. For instance, a 2-week water deficit treatment during the dry season resulted in a dramatic decrease in FTSW (Fig. 2). If stress begins around 10 days before flowering, this can induce a yield reduction of 40% to 50% compared with the fully irrigated aerobic control treatment.

The line-source sprinkler irrigation system, first introduced at IRRI by O'Toole (Cruz and O'Toole 1984), is now used extensively to generate a differential gradient of soil moisture varying from well-watered to extremely dry, and thus allowing the side-by-side comparison of different amounts of drought stress on the same plots. A series of line-source experiments are currently being used with segregating backcross lines and near-isogenic lines (NILs) to dissect the physiological mechanisms and G × E interaction of major QTLs associated with yield under drought (Impa et al, unpublished).

A drip-irrigation system uses drip tapes to apply water to individual rice plants, allowing water management to be controlled in individual plots, and thus the application of variable drought-stress periods. This is preferably used to screen varieties with major differences in flowering time, by initiating the stress period at different times, to target a specific phenological stage for each variety.

In rainfed lowlands, drainage of flooded paddy fields at specific times (generally 2 to 4 weeks after transplanting) results in progressive soil drying with a stress intensity that varies with toposequence position and environmental conditions (rainfall and evaporative demand). This system has been successfully used during both dry and wet seasons, simulating intermittent dry spells, and to screen large collections of breeding lines and to identify several donor parents for drought-resistance breeding (Kumar et al 2007).

Monitoring of soil moisture profiles

The precise measurement of soil moisture is a prerequisite for the establishment of FTSW dry-down phenotyping in the field. A variety of devices are being tested for the optimization of soil moisture measurement precision, including classical home-made Hg tensiometers, gauge-type tensiometers, equitensiometers, thetaprobes, and Diviner-2000. A three-year international comparative study conducted by the International Atomic Energy Agency (IAEA) has carried out numerous soil moisture comparisons on a wide range of sensors, including the soil moisture neutron probe (SMNP), time domain reflectometry (TDR), capacitance probes (EnviroSCAN and Diviner-2000), Delta-T ThetaProbe, and profiling probe, tested under a wide range of soil types, vegetations, and experimental sites, under both irrigated and rainfed conditions in agricultural and field environments and in the laboratory (www.iaea.org/programmes/nafa/d1). The main conclusions indicated that all the devices, except the conventional TDR, required soil-specific calibration systems, and that success in using capacitance sensors such as the Sentek and Delta-T devices depends critically on proper calibration and careful installation of access tubes to avoid soil disturbance.

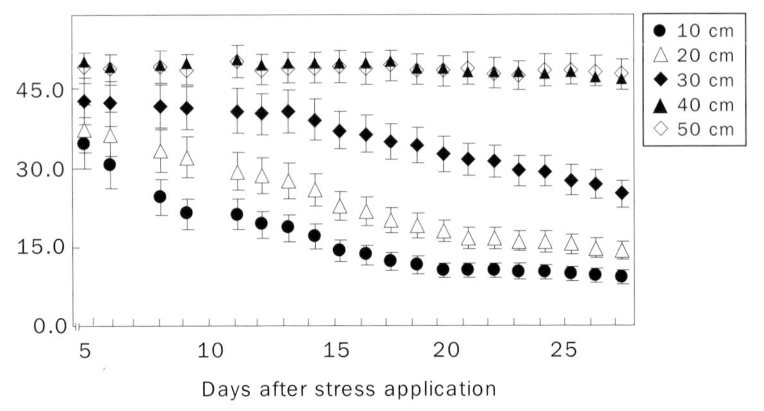

Fig. 3. Evolution of soil moisture at various soil depths after drought-stress application. Measurements were made using a Diviner-2000 capacitance probe.

Tensiometers measure soil water tension, which can be related to plant-stress levels, but with difficulty of conversion to soil moisture contents and a limited measurement range, between 0 and 80 kPa. The capacitance probe is a portable system that uses soil-water-content sensing technology to automatically record the SWC at each 10-cm depth increment up to 1.0-m depth, in about 2 seconds. It is designed to be moved from site to site in much the same way as a neutron-probe moisture meter is used, but with the advantage of not using a radioactive probe. The ThetaProbe avoids the limitations of an access tube by using steel spikes driven into the soil as the capacitance plates. The instrument can be either inserted into the soil surface to make one-time readings or buried for continuous *in situ* monitoring.

For drought phenotyping field experiments at IRRI, we are currently testing the combination of capacitance probes (Diviner-2000) for daily manual measurements of soil moisture profiles up to 1-m depth at 10-cm segments (Fig. 3) with the use of ThetaProbes (SM-200) connected to automatic dataloggers for continuous monitoring of soil moisture at a fixed (30 cm) soil depth. In addition, tensiometers are placed in all drought phenotyping and field screening experiments.

Nondestructive methods for plant growth and water status

The classical techniques used for measuring plant water use, water status, growth, and gas exchange are tedious and time-consuming, thus not suitable for precise high-throughput phenotyping. Recent efforts have focused on the development of automated and high-throughput phenotyping platforms, such as *Phenopsis*, which is based on automated pot weighing systems and digital imagery techniques (Granier et al 2006).

Imaging techniques including chlorophyll fluorescence, NDVI, infrared thermography, reflectance, autoluminescence, and multispectral fluorescence imagery have all been used for monitoring the effects of drought and other abiotic stresses on plants (see Chaerle et al 2007 for a review). Some of these techniques are specifically promising for application in high-throughput drought phenotyping. For instance, thermal imaging has proved useful in rapid screening for stomatal responses to stress; it can also be combined with fluorescence imaging to study photosynthesis (Chaerle et al 2007). Clear advantages of these imaging techniques are that they are nondestructive, as they allow large-scale, continuous, and possibly automated monitoring of dynamic spatial variations at whole-plant or canopy levels. They can also reveal the early signs of stress responses, thus making it possible to visualize the kinetics of stress responses in screening experiments and phenotyping dehydration-avoidance traits.

Digital imaging techniques are being tested at IRRI, based on conventional reflectance as a tool for monitoring plant growth and the dynamics of canopy development, which should permit predictive modeling of crop growth and performance (Kaminuma et al 2004). We are investigating the possible application of thermography in large-scale field screening of rice genotypes for stomatal behavior and water use under drought (Serraj and Cairns 2006). We are also using this technique as an indirect indicator for screening diverse germplasm accessions for drought-avoidance root traits under lowland conditions, before extracting roots for detailed morphological analyses. Thermography is showing promising results for capturing side-by-side differences in canopy temperature and plant water status, but large-scale extrapolation of the approach for field use is not straightforward, given the fluctuations in environmental conditions (Leinonen et al 2006).

The imaging approaches can also be combined with the use of carbon and oxygen isotope discrimination techniques (Condon et al 2004). An ongoing international research network coordinated by the IAEA has shown strong prospects for using these isotopic techniques in screening for water-use efficiency traits and yield stability in C_3 cereals (www-naweb.iaea.org/nafa/swmn/crp/2rcm-carbon-isotope.pdf). In addition to the potential use of leaf carbon isotope discrimination (CID), grain CID has been found to be positively associated with grain yield in wheat grown under both irrigated and drought conditions (Monneveux et al 2006). Recent experiments at IRRI have also confirmed this association in a set of rice breeding lines segregating for a major QTL for yield under drought (Bernier et al 2007). Leaf CID is based on the overall measurement of intercellular CO_2 concentration, without differentiating between the effects of stomatal closure and those of higher photosynthetic activity as potential mechanisms of enhanced water-use efficiency. A combination of CID with thermal and fluorescence imaging to measure stomatal conductance and photosynthetic activity, respectively, might allow a better resolution of stomatal and nonstomatal traits in plant responses to water deficits (Chaerle et al 2007).

The combination of visual, thermal, and chlorophyll fluorescence images with nondestructive isotopic techniques to extract canopy-specific parameters will lead the way to field-based phenotyping of crop performance as a function of FTSW. The key step in this development will be the improvement of automated algorithms for data capture and analysis, and model-based validation under a wide range of field situations.

Toward model-based phenotyping

The integration of FTSW dry-down methodology with nondestructive imaging and isotopic techniques for detailed measurements of plant growth and development parameters and yield components allows high-throughput, field-based, and precise drought phenotyping. This framework is now being used on a large scale on the IRRI experimental farm for detailed evaluation of potential drought-resistant donors and a large number of lines identified by drought breeding programs.

Simulation models have been used to understand and overcome the complexity of G × E × M interactions. Models can provide a tool to combine a mechanistic understanding of a drought trait with a range of weather scenarios. Models can also be instrumental in environmental characterization to support weighted selection for specific traits and adaptations. Modeling can play a role in enhancing the precision and integration of phenotyping either by linking model coefficients directly to QTLs (Tardieu 2003, Yin et al 1999, Chapman 2008) or more heuristically by guiding integrated phenotyping approaches (Hammer and Jordan, this volume).

Although there is some general agreement on the potential benefits of model-based phenotyping, pending questions remain regarding the general scope of the strategy, the link between crop simulation and genetic models, and the prototype of models suitable for phenotyping. Hammer and Jordan (this volume) concluded that, although an integrated systems approach to crop improvement is still in its infancy, rapid technology developments will speed its progress toward more relevance and a greater role in breeding for adaptation to drought.

IRRI has been involved in rice simulation modeling since the 1970s, with the development of several models, including a simple three-equation model (Sheehy et al 2004) and a more complex one, ORYZA (Bouman et al 2001). ORYZA2000 simulates the water balance and crop growth and development of lowland rice under both optimal and water-limited conditions. The parameters that characterize the rice plant's response to drought stress, although derived from pot experiments, could also simulate growth of IR72 under field drought conditions (Wopereis 1993). However, the model parameter values might not be valid across rice genotypes and growing ecosystems (Lilley and Fukai 1994a,b), as ORYZA 2000 did not simulate well crop performance under severe drought conditions. Current efforts at IRRI are for strengthening modeling capacity in rainfed environments and integrating systems analysis, drought physiology, and phenotyping with crop improvement strategies.

Phenotyping for gene expression and profiling

Phenotypic traits in gene expression experiments have been generally assumed to be fixed. However, crop growth and development traits are dynamic processes controlled by a complex network of genes. Analyzing phenotypic data measured at a single point in time in mapping and gene expression studies would not be suitable to capture the genetic control of developmental changes in response to stress. The concept of functional mapping has been recently proposed to characterize the QTL or nucleotides that underlie complex dynamic traits (Wu and Lin 2006).

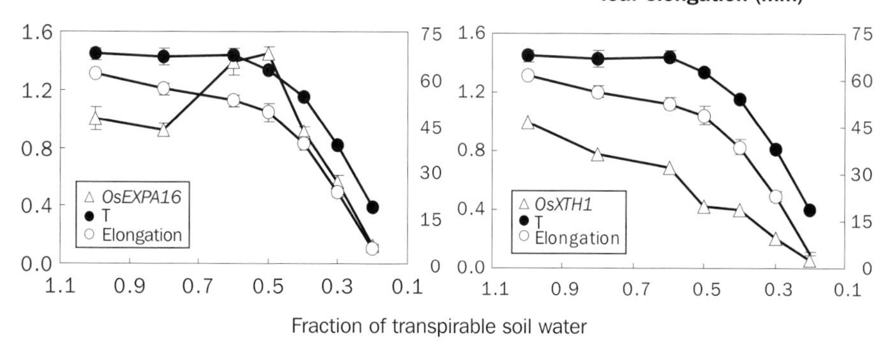

Fig. 4. Responses of leaf elongation, transpiration, and relative expression of one expansin and one XTH gene along with FTSW in rice variety IR64. Elongation was measured on newly emerged leaf #7, and transpiration was measured in 24 hours in the phytotron. The relative expression level of each gene (*OsEXPA16, OsXTH1*) was measured by real-time reverse transcription (RT)-PCR (1 = expression data from the well-watered control, FTSW = 1.0) in the elongating zone (3 cm) of leaf 7. Data are means (± S.E.) of at least 3 replicates.

We recently evaluated genotypic variation of dynamic responses of transpiration and leaf growth to soil water deficit under controlled and field conditions. The relationship between relative transpiration and soil drying, as measured by FTSW, was well described by linear-plateau functions that allowed the determination of the soil-water thresholds at which transpiration of drought-stressed plants began to decrease compared with the well-watered treatment. FTSW soil-water thresholds varied significantly among genotypes, suggesting a link between the kinetics of transpiration response to water deficits, leaf gas exchange, and plant growth parameters under drought. We used the FTSW dry-down phenotyping approach for analyzing gene expression and stress-response mechanisms of leaf elongation rate (LER). Transpiration and stomatal conductance both declined with increasing soil moisture deficit (Serraj et al 2008). However, there was a significant interaction of stress intensity with a greater reduction in stomatal conductance than in transpiration during severe stress (FTSW = 0.3 and below). LER was highly sensitive to water deficit, declining at a higher FTSW threshold value than transpiration and stomatal conductance (data not shown). Expression profiles of expansin and XTH genes were determined in the leaf-elongating zone at specific soil water deficits. Several genes, including *OsEXPA16*, *OsXTH1*, *OsXTH14*, and *OsXTH28*, showed similar changes in expression and consistent association with changes in leaf elongation (Fig. 4). Work is under way to profile large-scale expression of LER and QTL mapping of leaf and root growth under drought in controlled and field conditions.

A similar phenotyping framework has been used to analyze the physiological mechanisms and genetic basis of spikelet sterility and grain failure under reproductive-stage drought. Drought-induced spikelet sterility and inhibition of panicle exsertion in rice were recently related to the inhibition of peduncle elongation and the down-regulation of cell-wall invertase genes (Ji et al 2005). The physiological response of peduncle elongation rate to drought stress and the genetic variability of this trait have been recently analyzed in a set of diverse rice parental lines using the FTSW soil dry-down. Strong correlations were found between peduncle elongation rate, spikelet sterility, and yield under drought. QTL mapping for these traits is now under way in both upland and rainfed lowland field conditions.

Germplasm screening

Success in breeding for improved drought resistance depends essentially on the choice of parents, selection criteria, and robustness of the managed-screening protocols. The objectives of a screening system are to focus on TPEs and adaptation to major stress occurrence scenarios, and to minimize field variability for detecting heritable differences in drought resistance. Because of high rainfall and the unreliability of weather scenarios during the wet growing season at Los Baños, drought-screening at IRRI is mainly carried out during the dry season. Comparing several drought-screening protocols in upland or in drained lowland paddies, Lafitte and Courtois (2002) found that intermittent stress, imposed by withholding irrigation during the period bracketing the entire flowering and grain-filling stages, is generally reliable for ranking cultivar performance under drought, similarly to stress targeted precisely at the flowering period of individual cultivars.

Recent research findings at IRRI have demonstrated the feasibility of direct selection for yield under drought (Kumar et al 2008, Venuprasad et al 2007a). Since yield under stress is a function of yield potential, escape, and drought response, the use of the drought resistance index (DRI) can help to distinguish drought resistance from escape and yield potential (Bidinger et al 1987, Ouk et al 2006), and therefore further enhance the precision and reproducibility of drought screening.

Although breeding for upland and aerobic rice has recently made significant progress in developing new rice cultivars for water-short environments (Bernier et al 2008a), progress in rainfed lowlands has been relatively slow. Most improved cultivars grown in drought-prone rainfed lowlands were originally bred for irrigated conditions, and were never selected for drought tolerance (Kumar et al 2008). Drought escape has been exploited in the drought-prone areas of eastern India and Bangladesh through short-duration varieties, mainly of the aus germplasm group. But most of these varieties are not necessarily drought-resistant. The slow progress in the genetic improvement of grain yield in rainfed lowlands was explained by two major factors: the complexity of the target genotype × environment system and the insufficiency of genetic resources available to breeding programs (Cooper et al 1999b). Several studies have previously discussed the biophysical characteristics of the rainfed lowland ecosystem and their implications for breeding for yield and improved adaptation (Mackill et al 1996, Fukai et al 2001, Wade et al 1999).

Large genetic variation exists within rice and its wild *Oryza* relatives for performance under drought stress, but progress in developing improved cultivars has been relatively slow. Many parental lines and donors have been identified for drought resistance in upland (Bernier et al 2007), but only a few have been reported for the more extensive rainfed lowland system (Atlin et al 2006). The identification of parental materials and development of new populations were a major target for the IRRI rainfed lowland breeding program in the 1990s, focusing on the major target environments in eastern India and northeast Thailand (Sarkarung and Pantuwan 1999). Breeding populations were developed in the backgrounds of Mahsuri, Safri17, and Sabita for eastern India and KDML105 for northeast Thailand. An extensive G × E study in rainfed lowland by Wade et al (1999b) analyzed the interactions of 37 genotypes across 36 environments in India, Bangladesh, Thailand, Indonesia, and the Philippines from 1994 to 1997. Only a small group of genotypes were stable across environments. The cultivar NSG19 was found to be adapted to environments with rapid-onset late drought, whereas Sabita and KDML105 showed adaptation to environments with late maturity or recovery after drought.

Stress-sensitive mega-varieties are still widespread across South and Southeast Asian rainfed rice production systems, including Swarna, Sambha Mahsuri, IR36, IR64, BR11, and MTU 1010. These varieties generally preferred by farmers for their yield potential and quality traits are not tolerant of drought. As they were bred for the irrigated ecosystem, these varieties provide high yield in nondrought years, but they show a high yield reduction in mild to moderate drought years and collapse completely in severe drought-stress years (Kumar et al 2008).

In field experiments conducted at IRRI during the dry seasons of 2006 through 2008, large-scale field-managed drought screening has been focusing on the confirmation of drought-resistant breeding lines and identification of new potential donors of drought resistance within genebank germplasm collections, molecular breeding lines developed by backcrossing a series of donors to one of three elite recurrent parents (Lafitte et al 2006), *Oryza glaberrima* introgression lines, and hybrids and their parental lines, in addition to mutants and transgenic lines (Hervé and Serraj, this volume).

Genebank mini-core collections

A mini-core collection of 1,536 accessions from the six isozyme groups was selected to represent the diversity within *O. sativa* and was screened for drought resistance over three dry seasons (2005-07). As phenology varies widely in these sets, with days to 50% flowering varying between 56 and 145, yield components could not be used directly as selection criteria for screening diverse germplasm. Biomass accumulation under drought during the vegetative stage was used as an integrative trait, in addition to a series of morpho-physiological traits for the quantification of drought resistance. Biomass accumulation under drought varied between 10 and 580 g m^{-2}. Based on these data, association analysis studies are under way to identify correlations between allele haplotypes and drought-resistance phenotypic traits (Cairns et al, unpublished). In addition, selections have been made based on biomass performance and phenology under drought in upland and lowland conditions. Promising lines with potential drought-resistance traits include Kelee, Gul Murali, Gopal, Dumai, DA28, and Dular. New donors have also been identified within the mini-core for lowland drought resistance such as Basmati 334, Basmati 370, Habiganj boro, and Chikon sori (IRRI 2007).

Oryza glaberrima introgression lines

Two populations of introgressed lines of *O. glaberrima* crossed with IR64 and Apo were screened during the dry seasons of 2007 and 2008, under lowland and upland conditions, respectively. A total of 120 introgression lines from two sets of advanced backcross populations (BC_2F_3, BC_3F_4, and BC_4F_5) derived from *O. sativa* × *O. glaberrima* using *O. sativa* as a recurrent parent were used under two stress conditions: (1) severe vegetative and (2) reproductive-stage drought stress. Some of the high-yielding introgression lines yielded about 50% better than IR55423-01 (Apo), the recurrent parent, and 100% more than Vandana and PSBRc80, which were used as checks mainly associated with ability to maintain both high biomass production in dry soils and high harvest index (Bimpong et al, unpublished).

New crosses have been recently made using *O. glaberrima* lines originating from Mali (RAM) with IR64 and Apo. The introgression lines have been phenotyped under lowland and upland drought during the 2008 dry season, for measuring the genetic variability of drought responses and for mapping QTLs associated with drought resistance during the reproductive stage (Bimpong et al, unpublished).

Hybrid rice

The hybrid rice program at IRRI is currently dedicating substantial resources for screening breeding lines for performance under drought. In 2007 dry-season trials, several hybrids yielded high under both irrigated and drought-stressed environments; lines IR82372H and IR80228H yielded around 8.5 t ha^{-1} and 2.4 t ha^{-1} under irrigated and severe lowland drought, respectively. Similarly, line 82378H ranked among the top high yielders under both lowland and upland drought conditions. However, the correlation of yield performance of hybrids between the irrigated, upland, and lowland environments was generally low, which would suggest that selection of hybrids for performance has to be targeted at specific environments.

Overall, hybrid rice appears to offer an opportunity for combining improved drought resistance with tolerance of delayed planting and high yield potential in favorable environments (Atlin et al, this volume). This has resulted in promoting its adoption in the drought-prone shallow lowland areas across the eastern Indian states of Jarkhand, Bihar, Uttar Pradesh, and Chhattisgarh, where hybrid rice is slowly replacing drought-susceptible mega-varieties such as IR64 and IR36 (Atlin, personal communication).

Dehydration-avoidance traits

Given the low potential for improving rice performance under drought based on survival mechanisms of stage III, dehydration-avoidance mechanisms offer more promising avenues by improving plant characteristics to prolong stages I and II during progressive soil drying. Based on long-term multilocation drought studies in Southeast Asia, Fukai et al (this volume) concluded that rainfed lowland rice is mostly a drought avoider, with those genotypes that produce higher grain yield under drought being those able to maintain better plant water status, especially when stress occurs around flowering and grain setting. Dehydration avoidance is defined as the ability to maintain plant water status under soil water deficits, mainly through increased water uptake and/or reduced transpiration (Levitt 1980).

Plant water status

Leaf relative water content (RWC) and water potential (LWP) have long been associated with rice performance under water deficit (O'Toole and Moya 1978) and were found to be correlated with yield under drought stress when applied around flowering stage (Lafitte 2002, Jongdee et al 2006). Several QTLs have been mapped for RWC (Courtois et al 2000), LWP (Liu et al 2006), stomatal conductance (Khowaja and Price 2008), and several other traits putatively linked to drought resistance. Some inconclusive attempts have been made to exploit such QTLs in marker-assisted selection (MAS) schemes, but most QTLs identified have not been repeatable across environments and/or populations, or have not consistently affected either the target trait (Steele et al 2006) or grain yield under stress when introgressed into a susceptible cultivar (Steele et al 2007).

We have recently taken the reverse physiological approach, which consists of analyzing plant water status and related component traits in lines that have been confirmed with QTLs controlling grain yield under drought. We investigated the physiological basis of a major QTL for yield under drought (qtl12.1), which was recently mapped in the Vandana/Way Rarem population (Bernier et al 2007), by analyzing plant water status traits under both field and controlled environments (Bernier et al 2008b). The large effect of qtl12.1 on grain yield under drought stress in the Vandana/Way Rarem F_3-derived population was confirmed in this study. When exposed to severe drought stress at the reproductive stage, lines with the Way Rarem allele of qtl12.1 had improved grain yield, harvest index, and water status, and a higher spikelet number and weight. The lines with the Way Rarem allele of the QTL exhibited a reduced rate of leaf rolling and leaf drying, a higher LWP, and significantly increased stomatal conductance as well as a higher RWC. These results clearly indicated that the major effect of qtl12.1 on grain yield under drought was associated with an improvement in plant water status under stress, most likely through a dehydration-avoidance mechanism. A similar approach has also been used for dissecting the dehydration-avoidance mechanisms associated with major loci for performance under drought in near-isogenic rice lines derived from IR64/Aday Sel (Venuprasad et al 2007b).

Root traits

Significant progress has been made over the past decade in identifying QTLs associated with several root traits, including rooting depth, length, and thickness, which have long been emphasized as important adaptations to water deficits in rice (Nguyen et al 1997). Several QTLs have been mapped in rice for various root traits (see Price et al 2002 for a review), but it has not yet been possible to translate these results into a successful marker-assisted breeding program for drought avoidance. One of the main reasons for the lack of success is probably related to logistical and methodological problems of root phenotyping. The heritability of root-related traits is generally low and field screening for these traits is laborious and complicated by the high plasticity of root growth in response to even small changes in the environment and the complexity of G × E interactions. Therefore, most QTLs identified for root traits were detected in controlled environments, which are very different from the rooting environment plants experience in the field. The proof of concept in the ability to introgress root QTLs in rice was recently shown by Steele et al (2006). If validated with accurate phenotyping protocols and properly integrated into marker-assisted breeding programs, root QTL mapping and MAS strategies could then allow pyramiding of drought-avoidance mechanisms in rice and wheat cultivars and accelerate the development of locally adapted drought-resistant genotypes.

To investigate the physiological basis of a major QTL for yield under drought, which was previously mapped in the Vandana/Way Rarem population (Bernier et al 2007), we recently analyzed root traits under both field and controlled environments (Bernier et al 2008b). Traits related to deep root growth were positively correlated to water consumption in the drought-stress treatment. Based on regression analysis, the two most important traits were deep root length (below 30 cm) and maximum rooting depth. The lines with the Way Rarem allele of the QTL had an 18% higher deep root length than lines with the Vandana allele at the locus, which appears to be the most important difference in explaining the increased water uptake in the lines with the favorable allele of the QTL. Another candidate root trait to explain the improved water uptake would be better/faster root growth at depth. This hypothesis was also supported by the fact that the chromosome region where qtl12.1 is located has previously been identified as affecting rooting depth in a different population (Yue et al 2006). A further indication that qtl12.1 would be involved in improving deep root growth is that this trait is known to be important in conferring drought resistance in upland rice (where qtl12.1 is effective in improving drought resistance), but is not advantageous under lowland drought conditions, where root penetration ability would probably be more critical (Fukai and Cooper 1995). The overall conclusion of this study is that the Way Rarem allele of qtl12.1 improved root architecture (Bernier et al 2008b), which has resulted in increased water uptake, although this was relatively small (7%). However, our field root sampling methods were not precise enough to clearly capture these differences in root architecture among the recombinant inbred lines (RILs). Better evidence as to the nature of the effect of the QTL on roots can probably be obtained by conducting a precise root phenotyping experiment using NILs rather than the RIL used in this study. BC_3F_2 NILs with and without the Way Rarem allele of qtl12.1 in the Vandana genetic background are currently being developed at IRRI to clarify the effect of this locus. Transcript profiling experiments using these NILs will also be performed to identify differences in genes that are up-regulated and down-regulated due to the presence of qtl12.1. A similar approach is also being pursued for dissecting the dehydration-avoidance mechanisms and root traits associated with a major QTL for yield under drought in NILs derived from IR64/Aday Sel (Venuprasad et al 2007b) under lowland conditions.

Photosynthesis and stomata aperture traits

Stomatal regulation in response to soil drying is triggered by root-shoot chemical and/or hydraulic signaling (Tardieu and Davies 1993) as a key adaptation strategy to avoid tissue dehydration under drought. Various stomatal characteristics such as density, low conductance, high sensitivity to leaf water status, and ABA accumulation have been suggested as desirable traits in crop improvement for water-limited conditions, although, depending on the environment and occurrence of stress, they may also have trade-offs in terms of limiting carbon assimilation and crop yield. Significant genetic variation for the sensitivity of stomata to leaf water status was reported in rice (Price et al 1997), and the process of plant water conservation through stomatal regulation has been associated with reduced spikelet sterility and increased grain yield under reproductive-stage drought (Pantuwan et al 2002). The sensitivity of stomata to leaf water status is thought to be a highly heritable trait (Ludlow and Muchow 1990) and it has long been shown to have significant genetic variation in rice (Dingkuhn et al 1991, Price et al 1997).

Dry-down experiments carried out in pot studies under controlled conditions indicated that FTSW soil-water thresholds vary significantly among rice genotypes (Fig. 1). Based on this comparison, cultivars with lower threshold values (e.g., Apo) closed their stomata at lower FTSW (drier soil) and, hence, were able to maintain higher transpiration rates during the drying cycle than those with higher thresholds (e.g., IR72). These differences in transpiration response to FTSW are most likely associated with genetic differences in the control of transpiration and drought-avoidance mechanisms. However, transpiration response to soil drying is also influenced by VPD (Sinclair et al 2008), and the interactive effects of FTSW and VPD need to be resolved to better understand stomatal regulation and transpiration response to soil water deficits. The role of hormonal regulation and ABA is also still debated. Recent studies of stomatal responses to partial soil drying in rainfed lowland rice suggested a possible role of root signals (Siopongco et al 2008). A drought-induced increase in leaf ABA concentration under field conditions and strong association with soil moisture tension and gs suggested its involvement in mediating stomatal responses during early steps of drought responses in rice. However, the kinetics of the stress recovery after severing of droughted roots in the greenhouse could be attributed to increased hydraulic conductance. These responses imply a role for both chemical and hydraulic signals in rice (Siopongco et al 2008).

Instantaneous leaf gas exchange and chlorophyll fluorescence measurements were taken in dry-season upland screening experiments to compare the physiological stress responses of contrasting cultivars. Photosynthesis was inhibited by water deficits but to a variable extent in the genotypes tested. In general, the highest inhibition was observed in the genotypes that had inherently lower A, gs, and gm. Differences in plant water status among genotypes, probably related to variations in root systems, were associated with variations in photosynthesis and diffusional conductances (Centritto et al, unpublished). But, there were no lasting metabolic limitations to photosynthesis in response to the different levels and intensities of drought imposed in the genotypes tested. Furthermore, strong correlations between photosynthesis and the CO_2 diffusional conductances were found when taking into account not only stomatal but also mesophyll conductance. This is a clear indication that the chloroplast CO_2 concentration, as set by the combination of stomatal and mesophyll resistances to CO_2 diffusion, is the main limitation of photosynthesis for well-watered varieties as well as for varieties strongly affected by water deficit. A very strong correlation was found between low photosynthesis and low mesophyll conductance, both in genotypes with inherently low photosynthesis and in water-stressed plants of all varieties, indicating that the main limitation of photosynthesis in rice is the low chloroplastic CO_2 concentration. This suggests that the stress effect may be reversed if stomatal and/or mesophyll conductance to CO_2 diffusion is restored. It also indicates that rice genotypes with inherently high mesophyll conductance probably will be less affected by water deficit. This direct evidence of correlation between photosynthesis and diffusional conductances could provide a mechanistic basis for distinguishing photosynthetic causes of differences in WUE from causes related to stomatal regulation.

Carbon isotope discrimination and WUE traits

As an integrative parameter of stomatal aperture traits, carbon isotope discrimination (CID) has been successfully used as a selection criterion for water-use efficiency and yield under drought (Condon et al 2004). The negative correlation between CID and WUE, consistent with theoretical predictions, was confirmed in several crop plants, including rice (Dingkuhn et al 1991, Peng et al 1998, Impa et al 2005). However, field studies in wheat have also revealed that the correlation between CID and grain yield was often positive rather than negative (Condon et al 2004, Monneveux et al 2006), leading to the conclusion that the anticipated association was hidden because correlations between CID and yield can be influenced by several physiological differences among genotypes (Rebetzke et al 2002). A successful breeding program began in Australia, aiming at the introduction of enhanced WUE into germplasm within similar genetic backgrounds. Two hundred F_2:$_3$ families were first analyzed for CID under field conditions, and six families with extreme CID values were used in the backcross-breeding program. CID analysis was used again to select genotypes in the BC_2F_2:$_4$ families for testing in multilocation trials. Those lines selected for increased WUE using CID produced significantly greater yields in eight of the nine field environments, with relative yield increase reaching 11% in the driest environment (Rebetzke et al 2002). Two wheat varieties have been released so far from this program, which makes this one of the few success stories of trait-based selection for improved crop performance under drought (Sinclair et al 2004).

An international coordinated research project and consortium led by the IAEA has been validating this methodology, showing strong prospects for using these traits in screening for high WUE and yield stability in C_3 cereals under drought and salinity conditions (www-naweb.iaea.org/nafa/swmn/crp/2rcm-carbon-isotope.pdf). In rice, grain CID was also found to be strongly associated with grain yield under both irrigated and drought conditions. Two years of dry-season field testing at IRRI have recently confirmed the association between grain yield and CID among rice lines segregating for a major QTL of yield under drought (Bernier et al, unpublished).

Canopy temperature

Canopy temperature is a sensitive indicator of plant stress level, and has long been associated with stomatal conductance at the leaf level. This has led to the development of infrared sensing of canopy temperature as a technique for the estimation of gs and plant water status (Jones 1992). During a period of water shortage, plants conserve water by closing their stomata, thus raising their internal temperature. Leaf temperature is thus correlated with the stress level of the plant. However, leaf temperature is influenced by stomatal and boundary layer resistances as well as by meteorological conditions. The rate of leaf transpiration is only one of many components of the canopy energy balance that affect canopy temperature: factors such as radiation, wind speed, air temperature, humidity, and VPD all have major effects (Jones 1992).

Thermal sensing has been used successfully in field screening for monitoring differences in stomatal responses to drought in crops such as rice and wheat (Garrity and O'Toole 1995, Amani et al 1996). Using an infrared thermometer gun on single rice leaves, Garrity and O'Toole (1995) were able to screen rice varieties for reproductive-stage drought-avoidance traits. Their major findings were that mean canopy temperature increased from 28 to 37 °C during the drought stress period, and both grain yield and spikelet fertility were significantly correlated to midday canopy temperature on the day of flowering. They also found highly significant differences in canopy temperature among rice cultivars, reporting that lines with high drought-avoidance scores consistently remained coolest under stress (Garrity and O'Toole 1995). However, using an infrared gun on single plants is often subject to high field variation and it has limited scope as a high-throughput and reproducible screening method.

The recent development of infrared thermal imagery has revived the interest in using canopy temperature as a screening tool, especially if up-scaled to plot or field levels (Leinonen and Jones 2004). Infrared cameras can now be used to detect whole-plant, or even plot, temperature within seconds. IR thermal imaging has been used successfully to screen for stomatal and other mutants in the laboratory (Merlot et al 2002), and for precision phenotyping of plant water status under water deficits, especially when combined with automated image analysis to facilitate analysis (Leinonen and Jones 2004). Based on the energy balance theory (Monteith 1973), Horie et al (2006) developed a methodology based on field measurements of sunlit and suddenly shaded canopy surface temperatures. Simultaneous recording of microclimate data allowed the estimation of the evapotranspiration rate, aerodynamic resistance, and canopy diffusive resistance under field conditions. This study further demonstrated the possibility of relating the quantitative estimation of rice canopy physiological characteristics under field conditions with crop growth rate during the period preceding heading and final grain yield (Horie et al 2006). This remote-sensing technique could be useful as a rapid screening tool for effective selection of high-yielding and drought-adapted rice genotypes under field conditions.

Fig. 5. View of a drought phenotyping trial at IRRI upland farm (dry season 2006), with digital (above) and infrared thermal (below) images of the plots. Well-watered plots are mostly in blue (low canopy temperature) and drought-stressed plots in yellow/red (high canopy temperature).

Our recent collaborative work with H.G. Jones has demonstrated that there is potential for larger-scale field drought phenotyping with thermal imaging, even in the humid climate of the Philippines (Serraj and Cairns 2006). Studies were made during the dry season of 2006 to test the precision and sensitivity of thermal imaging for phenotyping rice water status under drought. Thermal images were obtained from a 6-m-high tower, using a Thermacam P25 long-wave thermal imager with a sensitivity of 0.1 °C and accuracy of ±2 °C. Parallel visible images were obtained either with the onboard digital imager or a normal digital camera. Measurements were made from about 5 m above the ground using a mobile tower that could be moved around the plot. The experiments included three replicates of 50 lines with well-irrigated and droughted plots for each accession in alternate rows. Figure 5 shows a view of the experiment, together with a comparable thermal image. The average temperature differences due to drought were between 1 °C and 1.8 °C, whereas the temperatures of individual lines varied by as much as 1.9 °C for the well-irrigated plots and 3.7 °C for the dry plots. These results confirmed the expectation that thermal sensors are more sensitive to differences when the crops are drought-stressed. A second experiment analyzed in more detail the response of 12 rice genotypes to soil moisture in a line-source irrigation gradient experiment (data not shown).

The sensitivity of the thermal imaging approach is determined by a combination of incident radiation, wind speed, and humidity. A useful measure of the sensitivity was given by the difference in temperature between a wet and a dry reference surface (e.g., discs of wet and dry filter paper are convenient). Given that the camera has a sensitivity of 0.1 °C, we found that, even on the most humid day during the experimental period, the actual resolution allowed good discrimination of differing stomatal conductances among genotypes. This result was in contrast to the conclusion of Araus et al (2002), suggesting that canopy temperature measurement was useful only under invariable conditions in hot and dry environments (i.e., high VPD). However, no attempt was made in our study to relate canopy temperature to crop yield, or to calibrate the thermal images in terms of absolute stomatal conductance (Leinonen et al 2006) as we aimed to concentrate on the more readily usable relative differences in canopy temperature. Therefore, direct comparisons of the two sets of results were not easy to make in this particular study. Nevertheless, several previous studies have shown that there is normally a close relationship between porometer and thermal measurements, with the errors in each method being comparable (e.g., Jones et al 2002). In this study, there was also a good correlation between the general responses for the two methods. Further work has begun for the calibration of infrared thermal imaging in routine screening of rice genotypes for dehydration-avoidance traits and plant water status under drought.

Conclusions: an integrated strategy for improving drought resistance

Although the need for precise characterization of the drought-prone TPE has long been emphasized, this effort is yet to be carried out systematically across the rainfed rice environments in South and Southeast Asia and sub-Saharan Africa. Simulation models can play a role in both the characterization and enhancement of the precision and integration of phenotyping either by linking model coefficients directly to or more heuristically guiding integrated phenotyping approaches. Increased crop yield and water productivity require the optimization of the physiological processes involved in the initial critical stages of plant response to soil drying, water-use efficiency, and dehydration-avoidance mechanisms. Overall, it is now well accepted that the complexity of the drought syndrome can be tackled only with a holistic approach integrating plant breeding with physiological dissection of the resistance traits and molecular genetic tools together with agronomic practices that lead to better conservation and use of soil moisture and matching crop genotypes with the environment. Some of the steps involved in this multidisciplinary approach (Fig. 6) are described below:

1. Define the target drought-prone environment(s) and identify the predominant type(s) of drought stress and the rice varieties preferred by farmers. Define the phenological and morphological traits that contribute substantially toward adaptation to drought stress(es) in the target environment(s).
2. Use simulation modeling and systems analysis to evaluate crop response to the major drought patterns, and assess the value of candidate physiological traits in the target environment.

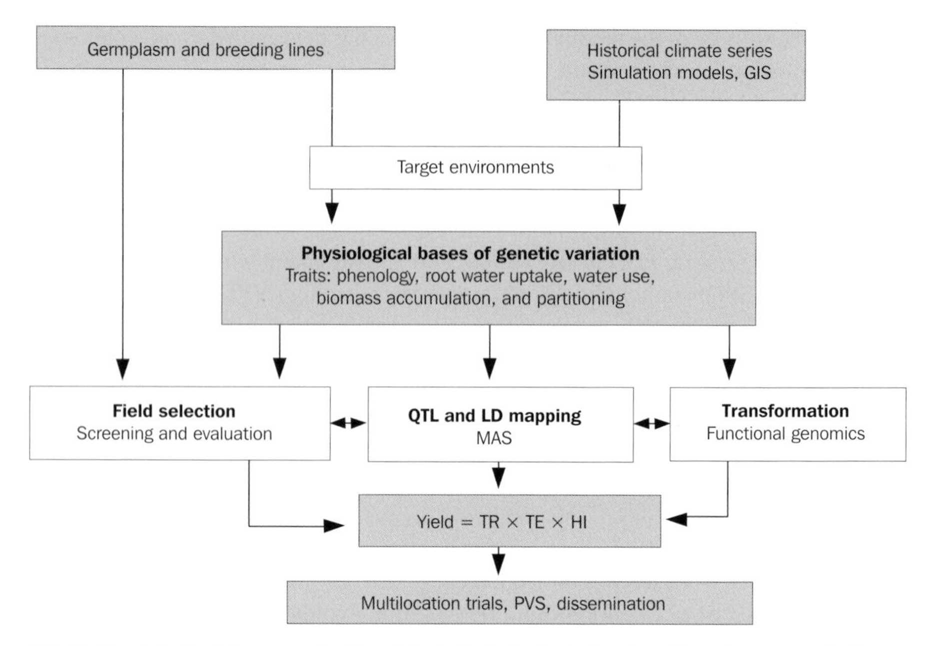

Fig. 6. Physiological framework of an integrated strategy for genetic enhancement of crop grain yield and its components under drought. TR = total plant water uptake; TE = transpiration efficiency; HI = harvest index.

3. Develop and refine appropriate screening methodologies for characterizing genetic stocks that could serve as donor parents for the traits of interest.
4. Identify the genetic stocks for various putative constitutive and inducible traits in the germplasm and establish genetic correlations between the traits of interest and the degree of adaptation to the targeted drought stress.
5. Harness functional genomics, transgenics, and reverse genetics tools to understand the genetic control of the relevant traits.
6. Use mapping populations and/or linkage disequilibrium mapping to identify genetic markers and QTLs for traits that are critical for drought resistance.
7. Incorporate some of the components of relevant physiological traits into various agronomic genetic backgrounds to provide a range of materials with specific traits of interest (i.e., developing NILs, RILs, and BC populations) for improving drought adaptation of locally adapted varieties.
8. Test the MAS products under well-managed field screening and in farmer-participatory multilocation trials.

This framework is also discussed in more detail in relation to the strategy for the Drought Frontier Project (Serraj and Atlin, this volume).

References

Amani I, Fischer RA, Reynolds MP. 1996. Canopy temperature depression association with yield of irrigated spring wheat cultivars in a hot climate. J. Agron. Crop Sci. 176:119-129.

Araus JL, Slafer GA, Reynolds MP, Royo C. 2002. Plant breeding and drought in C_3 cereals: what should we breed for? Ann. Bot. 89:925-940.

Atlin GN, Lafitte HR, Tao D, Laza M, Amante M, Courtois B. 2006. Developing rice cultivars for high-fertility upland systems in the Asian tropics. Field Crops Res. 97:43-52.

Bänziger M, Setimela PS, Hodson D, Vivek B. 2006. Breeding for improved abiotic stress tolerance in maize adapted to southern Africa. Agric. Water Manage. 80:212-224.

Bernier J, Atlin GN, Serraj R, Kumar A, Spaner D. 2008a. Breeding upland rice for drought resistance (a review). J. Sci. Food Agric. 88:927-939.

Bernier J, Kumar A, Ramaiah V, Spaner D, Atlin G. 2007. A large-effect QTL for grain yield under reproductive-stage drought stress in upland rice. Crop Sci. 47:507-516.

Bernier J, Serraj R, Impa S, Gowda V, Owane R, Bennett J, Kumar A, Spaner D, Atlin G. 2008b. Increased water uptake explains the effect of qtl12.1, a large-effect drought-resistance QTL in upland rice. Field Crops Res. (In press.)

Bhatnagar-Mathur P, Jyostna-Devi M, Reddy DS, Lavanya M, Vadez V, Serraj R, Yamaguchi-Shinozaki K, Sharma KK. 2007. Stress-inducible expression of *Arabidopsis thaliana* DREB1A in transgenic peanut (*Arachis hypogaea* L.) improves transpiration efficiency under water limiting conditions. Plant Cell Rep. 26:2071-2082.

Bidinger FR, Mahalakshmi V, Rao GDP. 1987. Assessment of drought resistance in pearl millet [*Pennisetum americanum* (L.) Leeke]. II. Estimation of genotype response to stress. Aust. J. Agric. Res. 38:49-59.

Boote KJ, Sinclair TR. 2006. Crop physiology: significant discoveries and our changing perspective on research. Crop Sci. 46:2270-2277.

Bouman BAM, Kropff MJ, Tuong TP, Wopereis MCS, ten Berge HFM, Van Laar HH. 2001. ORYZA2000: modeling lowland rice. International Rice Research Institute, Los Baños, Philippines, and Wageningen University and Research Centre, Netherlands.

Chaerle L, Leinonen I, Jones HG, Van der Straeten D. 2007. Monitoring and screening plant populations with combined thermal and chlorophyll imaging. J. Exp. Bot. 58:773-784.

Chapman SC. 2008. Use of crop models to understand genotype by environment interactions for drought in real-world and simulated plant breeding trials. Euphytica 161:195-208.

Condon AG, Richards RA, Rebetzke GJ, Farquhar GD. 2004. Breeding for high water-use efficiency. J. Exp. Bot. 55:2447-2460.

Cooper M, Fukai S, Wade LJ. 1999b. How can breeding contribute to more productive and sustainable rainfed lowland rice systems? Field Crops Res. 64:199-209.

Cooper M, Rajatasereekul S, Immark S, Fukai S, Basnayake J. 1999a. Rainfed lowland rice breeding strategies for Northeast Thailand. I. Genotypic variation and genotype × environment interactions for grain yield. Field Crops Res. 64:131-151.

Courtois B, Lafitte HR.1999. Improving rice for drought-prone upland environments. Proceedings of the Workshop on Genetic Improvement of Rice for Water-Limited Environments, 1-3 December 1998, Los Baños, Philippines.

Crouch JH, Serraj R. 2002. DNA marker technology as a tool for genetic enhancement of drought tolerance at ICRISAT. In: Saxena NP, editor. International Workshop on Field Screening for Drought Tolerance in Rice. Patancheru (India): International Crops Research Institute for the Semi-Arid Tropics.

Cruz RT, O'Toole JC. 1984. Dryland rice response to an irrigation gradient at flowering stage. Agron. J. 76:178-183.

Ding S, Pandey S, Chen C, Bhandari H. 2005. Drought and rice farmers' coping mechanisms in the rice production systems of southern China. A final research report submitted to the Rockefeller Foundation. Department of Agricultural Economics, Huazhong Agricultural University, Hubei, China.

Dingkuhn M, Farquhar GD, De Datta SK, O'Toole JC. 1991. Discrimination of ^{13}C among upland rices having different water use efficiencies. Aust. J. Agric. Res. 42:1123-1131.

Edmeades GO, Bolanos J, Chapman SC, Lafitte HR, Banziger M. 1999. Selection improves drought tolerance in tropical maize populations. I. Gains in biomass, grain yield, and harvest index. Crop Sci. 39:1306-1315.

Fukai S, Basnayake J, Cooper M. 2001. Modeling water availability, crop growth, and yield of rainfed lowland rice genotypes in northeast Thailand. In: Tuong TP, Kam SP, Wade L, Pandey S, Bouman BAM, Hardy B, editors. Proceedings of the International Workshop on Characterizing and Understanding Rainfed Environments, Bali, Indonesia, 5-9 December 1999. Los Baños (Philippines). International Rice Research Institute p 111-130.

Fukai S, Cooper M. 1995. Development of drought-resistant cultivars using physio-morphological traits in rice. Field Crops Res. 40:67-86.

Garrity DP, Oldeman LR, Morris RA. 1986. Rainfed lowland rice ecosystems: characterization and distribution. In: Progress in rainfed lowland rice. Los Baños (Philippines): International Rice Research Institute. p 3-23.

Garrity DP, O'Toole JC. 1995. Selection for reproductive stage drought avoidance in rice, using infrared thermometry. Agron. J. 87:773-779.

Granier C, Aguirrezabal L, Chenu K, Cookson SJ, Dauzat M, Hamard P, Thioux JJ, Rolland G, Bouchier-Combaud S, Lebaudy A, Muller B, Simonneau T, Tardieu F. 2006. Phenopsis, an automated platform for reproducible phenotyping of plant responses to soil water deficit in *Arabidopsis thaliana* permitted the identification of an accession with low sensitivity to soil water deficit. New Phytol. 169:623-635.

Heinemann AB, Dingkuhn M, Luquet D, Combres JC, Chapman S. 2008. Characterization of drought stress environments for upland rice and maize in central Brazil. Euphytica 162(3):395-410.

Horie T, Matsuura S, Takai T, Kuwasaki K, Ohsumi A, Shiraiwa T. 2006. Genotypic difference in canopy diffusive conductance measured by a new remote-sensing method and its association with the difference in rice yield potential. Plant Cell Environ. 29:653-660.

Impa SM, Nadaradjan S, Boominathan P, Shashidhar G, Bindumadhava H, Sheshshayee MS. 2005. Carbon isotope discrimination accurately reflects variability in WUE measured at a whole plant level in rice. Crop Sci. 45:2517-2522.

Ji XM, Raveendran M, Oane R, Ismail A, Lafitte R, Bruskiewich R, Cheng SH, Bennett J. 2005. Tissue-specific expression and drought responsiveness of cell-wall invertase genes of rice at flowering. Plant Mol. Biol. 59(6):945-964.

Jones HG. 1992. Plants and microclimate. 2nd Edition. Cambridge (UK): Cambridge University Press.

Jones HG, Stoll M, Santos T, de Sousa C, Chaves MM, Grant OM. 2002. Use of infrared thermography for monitoring stomatal closure in the field: application to grapevine. J. Exp. Bot. 53:2249-2260.

Jongdee B, Pantuwan G, Fukai S, Fischer K. 2006. Improving drought tolerance in rainfed lowland rice: an example from Thailand. Agric. Water Manage. 80(1-3):225-240.

Kaminuma E, Heida N, Tsumoto Y, Yamamoto N, Goto N, Okamoto N, Konagaya A, Matsui M, Toyoda T. 2004. Automatic quantification of morphological traits via three-dimensional measurement of *Arabidopsis*. Plant J. 38(2):358-365.

Khowaja FS, Price AH. 2008. QTL mapping rolling, stomatal conductance and dimension traits of excised leaves in the Bala × Azucena recombinant inbred population of rice. Field Crops Res. 106(3):248-257.

Kumar A, Bernier J, Verulkar S, Lafitte HR, Atlin GN. 2008. Breeding for drought tolerance: direct selection for yield, response to selection and use of drought-tolerant donors in upland and lowland-adapted populations. Field Crops Res. 107:221-231.

Kumar R, Venuprasad R, Atlin GN. 2007. Genetic analysis of rainfed lowland rice drought tolerance under naturally-occurring stress in eastern India: heritability and QTL effects. Field Crops Res. 103:42-52.

Lafitte HR. 2002. Relationship between leaf relative water content during reproductive stage water deficit and grain formation in rice. Field Crops Res. 76:165-174.

Lafitte HR, Courtois B. 2002. Interpreting cultivar × environment interactions for yield in upland rice: assigning value to drought-adaptive traits. Crop Sci. 42:1409-1420.

Lafitte HR, Li ZK, Vijayakumar CHM, Gao YM, Shi Y, Xu JL, Fu BY, Ali AJ, Domingo J, Maghirang R, Torres R, Mackill D. 2006. Improvement of rice drought tolerance through backcross breeding: evaluation of donors and selection in drought nurseries. Field Crops Res. 97:77-86.

Leinonen I, Grant OM, Tagliavia CPP, Chaves MM, Jones HG. 2006. Estimating stomatal conductance with thermal imagery. Plant Cell Environ. 29:1508-1518.

Leinonen I, Jones HG. 2004. Combining thermal and visible imagery for estimating canopy temperature and identifying plant stress. J. Exp. Bot. 55:1423-1431.

Leung H. 2008. Stressed genomics—bringing relief to rice fields. Curr. Opin. Plant Biol. 11:201-208.

Levitt J. 1980. Water stress, dehydration and drought injury. In: Kozlowski TT, editor. Responses of plants to environmental stresses. Vol II. London: Academic Press.

Lilley JM, Fukai S. 1994a. Effect of timing and severity of water-deficit on 4 diverse rice cultivars. 1. Rooting pattern and soil-water extraction. Field Crops Res. 37:205-213.

Lilley JM, Fukai S. 1994b. Effect of timing and severity of water-deficit on 4 diverse rice cultivars. 2. Physiological responses to soil-water deficit. Field Crops Res. 37:215-223.

Liu H, Mei H, Yu X, Zou G, Liu G, Luo L. 2006. Towards improving the drought tolerance of rice in China. Plant Genet. Res. 4(1):47-53.

Ludlow MM, Muchow RC. 1990. A critical-evaluation of traits for improving crop yields in water-limited environments. Adv. Agron. 43:107-153.

Mackill, DJ, Coffman WR, Garrity DP. 1996. Rainfed lowland rice improvement. Los Baños (Philippines): International Rice Research Institute. 242 p.

Merlot S, Mustilli AC, Genty B, North H, Lefebvre V, Sotta B, Vavasseur A, Giraudat J. 2002. Use of infrared thermal imaging to isolate *Arabidopsis* mutants defective in stomatal regulation. Plant J. 30:601-609.

Miflin B. 2000. Crop improvement in the 21st century. J. Exp. Bot. 51:1-8.

Monneveux P, Rekika D, Acevedo E, Merah O. 2006. Effect of drought on leaf gas exchange, carbon isotope discrimination, transpiration efficiency and productivity in field grown durum wheat genotypes. Plant Sci. 170:867-872.

Monteith JL. 1973. Principles of environmental physics. London (UK): Edward Arnold.

Nguyen HT, Babu RC, Blum A. 1997. Breeding for drought resistance in rice: physiological and molecular genetics considerations. Crop Sci. 37:1426-1434.

Nigam SN, Chandra S, Sridevi KR, Bhukta M, Reddy AGS, Rachaputi NR, Wright GC, Reddy PV, Deshmukh MP, Mathur RK, Basu MS, Vasundhara S, Varman PV, Nagda AK. 2005. Efficiency of physiological trait-based and empirical selection approaches for drought tolerance in groundnut. Ann. Appl. Biol. 146:433-439.

O'Toole JC. 2004. Rice and water: the final frontier. First International Conference on Rice for the Future. 31 August-2 Sept. 2004, Bangkok, Thailand.

O'Toole JC, Moya TB. 1978. Genotypic variation in maintenance of leaf water potential in rice. Crop Sci. 18:873-876.

Ouk M, Basnayake J, Tsubo M, Fukai S, Fischer KS, Cooper M, Nesbitt H. 2006. Use of drought response index for identification of drought tolerant genotypes in rainfed lowland rice. Field Crops Res. 99:48-58.

Pandey S, Bhandari H, Hardy B, editors. 2007. Economic costs of drought and rice farmers' coping mechanisms: a cross-country comparative analysis from Asia. Los Baños (Philippines): International Rice Research Institute. 203 p.

Pantuwan G, Fukai S, Cooper M, Rajatasereekul S, O'Toole JC. 2002. Yield response of rice (*Oryza sativa* L.) genotypes to different types of drought under rainfed lowlands. Part 3. Plant factors contributing to drought resistance. Field Crops Res. 73:181-200.

Passioura JB. 1977. Grain yield, harvest index and water use of wheat. J. Aust. Inst. Agric. Sci. 43:21.

Peng S, Laza RC, Khush GS, Sanico AL, Visperas RM, Garcia FV. 1998. Transpiration efficiencies of indica and improved tropical japonica rice grown under irrigated conditions. Euphytica 103:103-108.

Price AH, Cairns JE, Horton P, Jones HG, Griffiths H. 2002. Linking drought-resistance mechanisms in upland rice using a QTL approach: progress and new opportunities to integrate stomatal and mesophyll responses. J. Exp. Bot. 53:989-1004.

Price AH, Young EM, Tomos AD. 1997. Quantitative trait loci associated with stomatal conductance, leaf rolling and heading date mapped in upland rice (*Oryza sativa*). New Phytol. 137:83-91.

Ray JD, Sinclair TR. 1997. Stomatal closure of maize hybrids in response to drying soil. Crop Sci. 37:803-807.

Rebetzke GJ, Condon AG, Richards RA, Farquhar GD. 2002. Selection for reduced carbon isotope discrimination increases aerial biomass and grain yield of rainfed bread wheat. Crop Sci. 42:739-745.

Richards RA. 2006. Physiological traits used in the breeding of new cultivars for water-scarce environments. Agric. Water Manage. 80(1-3):197-211.

Sadras VO, Milroy SP. 1996. Soil-water thresholds for the responses of leaf expansion and gas exchange. Field Crops Res. 47:253-266.

Sarkarung S, Pantuwan G. 1999. Improving rice for drought-prone rainfed lowland environments. In: Ito O, O'Toole JC, Hardy B, editors. Genetic improvement of rice for water-limited environments. Los Baños (Philippines): International Rice Research Institute. p 57-70.

Serraj R, Allen HL, Sinclair TR. 1999. Soybean leaf growth and gas exchange response to drought under carbon dioxide enrichment. Global Change Biol. 5:283-292.

Serraj R, Cairns JE. 2006 Diagnosing drought. Rice Today 5(3):32-33.

Serraj R, Sinclair TR. 1997. Variation among soybean cultivars in dinitrogen fixation response to drought. Agron. J. 89:963-969.
Serraj R, Sinclair TR. 2002. Osmolyte accumulation: can it really help increase crop yield under drought conditions? Plant Cell Environ. 25:335-341.
Sheehy JE, Mitchell PL, Ferrer AB. 2004. Bi-phasic growth patterns in rice. Ann. Bot. 94:811-817.
Sinclair TR. 2005. Theoretical analysis of soil and plant traits influencing daily plant water flux on drying soils. Agron. J. 97:1148-1152.
Sinclair TR, Ludlow MM. 1986. Influence of soil water supply on the plant water balance of four tropical grain legumes. Aust. J. Plant Physiol. 13:329-341.
Sinclair TR, Muchow RC. 2001. System analysis of plant traits to increase grain yield on limited water supplies. Agron. J. 93:263-270.
Sinclair TR, Purcell LC, Sneller CH. 2004. Crop transformation and the challenge to increase yield potential. Trends Plant Sci. 9:70-75.
Sinclair TR, Zwieniecki MA, Holbrook NM. 2008. Low leaf hydraulic conductance associated with drought tolerance in soybean. Physiol. Plant. 132:446-451.
Siopongco JDLC, Sekiya K, Yamauchi A, Egdane J, Ismail AM, Wade LJ. 2008. Stomatal responses in rainfed lowland rice to partial soil drying: evidence for root signals. Plant Prod. Sci. 11:28-41.
Steele KA, Price AH, Shashidhar HE, Witcombe JR. 2006. Marker-assisted selection to introgress rice QTLs controlling root traits and aroma into an Indian upland rice variety. Theor. Appl. Genet. 112:208-221.
Steele KA, Virk DS, Kumar R, Prasad SC, Witcombe JR. 2007. Field evaluation of upland rice lines selected for QTLs controlling root traits. Field Crops Res. 101:180-186.
Tardieu F. 2003. Virtual plants: modeling as a tool for the genomics of tolerance to water deficit. Trends Plant Sci. 8:9-14.
Tardieu F, Davies WJ. 1993. Integration of hydraulic and chemical signaling in the control of stomatal conductance and water status of droughted plants. Plant Cell Environ. 16:341-349.
Towfiqul Islam M. 2008. Modeling of drought for aman rice in the northwest region of Bangladesh. PhD thesis. Bangladesh Agricultural University, Mymensingh. 194 p.
Venuprasad R, Lafitte HR, Atlin GN. 2007a. Response to direct selection for grain yield under drought stress. Crop Sci. 47:285-293.
Venuprasad R, Zenna N, Choi IR, Amante M, Virk PS, Kumar A, Atlin GN. 2007b. Identification of marker loci associated with tungro and drought tolerance in near-isogenic rice lines derived from IR64/Aday Sel. Int. Rice Res. Notes 32(1):27-29.
Wade LJ, Fukai S, Samson BK, Ali A, Mazid MA. 1999a. Rainfed lowland rice: physical environment and cultivar requirements. Field Crops Res. 64:3-12.
Wade LJ, McLaren CG, Quintana L, Harnpichitvitaya D, Rajatasereekul S, Sarawgi AK, Kumar A, Ahmed HU, Singh SAK, Rodriguez R, Siopongco J, Sarkarung S. 1999b. Genotype by environment interactions across diverse rainfed lowland rice environments. Field Crops Res. 64:35-50.
Wopereis MCS. 1993. Quantifying the impact of soil and climate variability on rainfed rice production. PhD thesis. Wageningen (Netherlands): Wageningen Agricultural University. 188 p.
Wu RL, Lin M. 2006. Functional mapping: how to map and study the genetic architecture of dynamic complex traits. Nat. Rev. Genet. 7(3):229-237.

Yin X, Kropff M, Stam P. 1999. The role of ecophysiological models in QTL analysis: the example of specific leaf area in barley. Heredity 82:415-421.
Yue B, Xue W, Xiong L, Yu X, Luo L, Cui K, Jin D, Xing Y, Zhang Q. 2006. Genetic basis of drought resistance at reproductive stage in rice: separation of drought tolerance from drought avoidance. Genetics 172:1213-1228.

Notes

Authors' address: International Rice Research Institute, Los Baños, Philippines.

The rice root system: from QTLs to genes to alleles

Brigitte Courtois, Nourollah Ahmadi, Christophe Perin, Delphine Luquet, and Emmanuel Guiderdoni

In rice, a deep and thick root system is generally considered to be a useful trait for maintaining yield under water stress in a broad range of conditions, notably in rainfed ecosystems. A large natural variation in root architecture is observed in rice due to the specific adaptation of groups of rice cultivars to contrasting hydrological conditions (irrigated, rainfed, or upland). Understanding the genetic and molecular mechanisms controlling the development of the root system and its adaptive plasticity in response to water availability and soil environment would help in improving yield stability of rice crops confronting drought situations. This paper reviews current knowledge on QTLs detected for constitutive and adaptive root development, the present status of QTL integration on a consensus QTL map linked to the rice physical map through sequenced markers, and QTL validation in near-isogenic or substitution lines. A key step is now to link QTLs to genes, which can be done through several approaches. We are presently exploring some of them: looking for QTLs for physiological parameters derived from models, fine-mapping QTLs for root depth using meta-analyses and recombinant genetics, searching for known root architecture and stress-response genes by direct genetics in an insertion mutant collection, and searching for orthologs in rice of genes involved in root development in other species. We will combine these approaches with association studies between polymorphism within validated candidate genes and phenotypic variation that will confirm the interest of the genes in the target plant material, and, in addition, will give access to a range of alleles at the genes. We think that these combined approaches could hasten the discovery of important genes and alleles involved in root traits, provided that a high-throughput reliable phenotyping technique linked to field performance, which is presently missing, is developed. This will provide breeding programs with markers allowing the combination of favorable alleles at key loci for root traits. Although genotype building involving a few loci is now well under control, we are still lacking experience in how to accumulate alleles at many loci in the most efficient way. A marker-aided recurrent selection strategy could help achieve this goal.

It is generally agreed among the scientific community that no universal trait can confer drought resistance in all water-limited situations. A clear diagnosis of water-deficit situations in target environments is key to orienting the choice of traits and focusing our research. An avoidance strategy, however, seems a reasonable option in most agronomic situations in which the objective is not just survival but growth maintenance and final crop productivity.

The root system through its architectural characteristics (distribution in the soil layers, depth, and thickness) and dynamics of growth is considered as the most consensual of all traits contributing to drought avoidance. The root system plays an essential role in plant growth and development through its function of water and mineral extraction, anchorage, and competition with weeds, and it is an important site of hormonal synthesis.

If the role of root systems is no longer questioned in some water-deficit situations, notably those faced by upland rice in Asia (Courtois and Lafitte 1999), its consideration in breeding programs is still modest. Several elements are required to develop an efficient breeding program for a given trait: genetic variability for the target traits(s); efficient screening techniques, preferably high-throughput; some understanding of the physiology of the traits(s); some understanding of the genetic control of the trait(s) implying knowledge of the many constitutive and inducible quantitative trait loci (QTLs) or genes involved, alleles at these QTLs or genes, and markers strongly linked to the QTLs or in the genes themselves; and, lastly, a good strategy to pyramid the favorable alleles at the QTLs or genes. The complexity of the breeding work increases with the number of genes and alleles involved, the small size of their effects, and the importance of epistatic and/or pleiotropic effects of these genes. From the initial studies (see the paragraph on QTLs controlling root morphology), the genetic control of the rice root system is already known to be complex, involving a large number of genes with small effects and low heritability that influence the whole-plant architecture. It makes sense, therefore, to start by reducing this complexity through genetic dissection of the traits into elementary components whose role, individual or in interaction, needs to be understood. However, once this is achieved, the choice of the right combinations of traits, genes, and alleles to construct an elite cultivar with the phenotype required for the target environment will still represent a serious challenge. New tools such as simulation and approaches such as recurrent selection, adapted to this complexity, will be needed.

This paper reviews the present status of research on rice root systems in relation to these points, with the exception of the physiological aspects that are tackled elsewhere in this volume, and presents the work on-going at Cirad.

Genetic diversity of root systems in rice

Rice is characterized by a shallow and limited root system compared with other cereal crops such as maize or sorghum. Nevertheless, significant genetic variation is observed among rice varieties for morphological traits such as root number, diameter, depth, branching, vertical density distribution, and root-to-shoot ratio, or traits linked to root activity such as water extraction pattern, activity at depth, and root pulling force (O'Toole and Bland 1987). Differences are also observed for root plasticity in response to water stress, with some varieties being able to adjust root growth to maintain water supply (O'Toole and Bland 1987). However, the existence of genetic variation for a given root trait, while necessary for genetic improvement, does not imply that the trait has an adaptive value for the target situations. It remains to be demonstrated that the trait has a positive effect on the phenotype in the conditions likely to be experienced by the crop. A quantification of its effect may be useful to assess whether a breeding effort is worthwhile, given the range of variability existing in the species.

The organization of the variability of root morphology in *Oryza sativa* reflects primarily the ecosystem adaptation of the different varietal groups and, secondarily, their genetic structure (Courtois et al 1996, Lafitte et al 2001). Irrigated varieties have thin, highly branched superficial roots with narrow vessels and a low root-to-shoot ratio. Irrigated varieties encompass indica types (enzymatic group 1), the temperate component of japonicas and the bulus from Indonesia (both belonging to group 6), and boro accessions from South Asia (group 2) with unusually thin roots. Upland aerobic varieties generally have thicker less-branched longer roots and a larger root-to-shoot ratio. Upland varieties regroup the tropical japonica component from Southeast Asia (group 6) and the short-duration aus group from South Asia (group 2), which is intermediate with a root distribution profile similar to that of japonica, but with thinner roots. The other groups, including deepwater and floating rice (groups 3 and 4) and sadri and basmati irrigated rice originating from the Himalayan border (group 5), have root thickness and root distribution profiles closer to those of indicas. Genetic variability is mainly distributed between groups, while the variability within groups is more limited.

For most groups, improvement of the root system is possible, but, for most traditional upland japonica genotypes, which are already deep-rooted, it may be difficult to find donors in the *O. sativa* species to improve them further.

The root development of *Oryza glaberrima* and wild species mostly belonging to the primary gene pool has also been studied (Liu et al 2004). Although some accessions have large constitutive root mass at depth, neither *O. glaberrima* nor wild species appear to be a better source of alleles for improved constitutive or adaptive root distribution than the japonica varieties from *O. sativa*.

Screening techniques

The root system has been appropriately coined the "hidden half." Its analysis is tedious due to methodological problems: difficulties to observe it *in situ*, large sensitivity to environmental variations, and low heritability under field conditions.

The most commonly used evaluation techniques in recent years were PVC pipe experiments under greenhouse conditions and hydroponic culture in growth chambers. Very few experiments have been conducted under field conditions. The rare attempts showed that, using auger sampling, environmental variation could be higher than genetic variation for root parameters linked to root depth (Yadav et al, unpublished results).

The choice of methods such as pipe experiments or hydroponic culture derives from increased ease and speed to phenotype the root system and possibilities to study its dynamics. This also translates into an increased level of artificiality. An issue to address is how performance varies among different experimental conditions and relates to yield under stress. Some preliminary evidence exists showing that field and pipe performance can be well correlated (Breseghello et al, this volume), but this issue needs more thorough investigation.

An experiment has been conducted at Cirad to assess the links between results found in PVC pipes under well-watered aerobic conditions (Shen et al 2001) and those found in a hydroponic experiment. The lines evaluated were 23 near-isogenic lines (NILs) of IR64 introgressed with one or two QTLs from Azucena. The range of values in root attributes under hydroponic conditions was much narrower than in PVC pipes, which could be a problem when the objective is to evaluate varietal differences. The correlations between the two conditions were 0.668 for maximum root length, 0.410 for total root weight, and 0.515 for number of roots. The correlations were positive and reasonably high considering that genetic variation within this set of NILs was generated by only one or two QTLs. Hence, this generates some confidence in the use of hydroponics.

Nevertheless, none of these methods can be considered as high-throughput. New methods are under development at Cirad, derived from methods developed for *Arabidopsis thaliana*. The first method, nondestructive, involves following the root and shoot growth of pregerminated sterilized seeds transferred on 0.8% agarose gel containing half-strength MS medium with half-strength MS vitamins in 0.2 × 0.2-m square Petri plates, placed at an angle of 15% from the vertical in a growth chamber under a 12-h photoperiod. Another option is to grow excised root meristem on solid medium containing active charcoal in the same Petri plates and follow root development. These techniques are adapted to assess early root growth for a few days and are amenable to software-assisted follow-up of root development. The first experiments with small sets of accessions showed that the varietal ranking conformed to what was known on the performance of the varieties, provided attention was paid to a range of small experimental details (Perin et al, unpublished results). As with other methods, however, the link between root growth under such conditions and field performance under stress remains to be established.

QTLs controlling root morphology

With the development of molecular marker use in rice, the detection of QTLs controlling root traits was undertaken in recombinant inbred or doubled-haploid line mapping populations, mostly derived from indica × japonica crosses. Both constitutive and adaptive traits were studied. An inventory of the literature dealing with QTL mapping for drought resistance in rice records 36 articles between 1992 and 2006. Together, they identify about 1,100 main-effect QTLs. These studies, however, involved different experimental conditions, populations, molecular marker types, analysis methods, or thresholds, as well as differences in the information provided (e.g., indication or not of the peak position or the confidence interval). We felt that tools facilitating comparisons between studies and allowing an easy passage to the genes underlying the QTLs were needed. The Gramene database (www.gramene.org/) fills this role to some extent but, in this database, the information on QTLs is restricted to the QTL position, which is not satisfactory for traits strongly influenced by experimental conditions. We extracted the QTL information (experimental conditions, population size, LOD, r^2, etc.) from the published papers. We defined the position of each QTL on an integrative genetic map derived from the reference map of the IR64 × Azucena population as well as on the physical map (TIGR v4). Maps and QTL data will be available soon in TropgeneDB (http://tropgenedb.cirad.fr/) and can be visualized through OrygenesDB (http://orygenesdb.cirad.fr/). From these databases, it will be possible to extract QTLs by trait, by experimental conditions (e.g., aerobic versus anaerobic, well-watered versus stressed), or by any other recorded information.

Constitutive traits

Root morphological traits such as maximum root length, root thickness, and vertical distribution of root mass were studied in five different populations, under well-drained conditions representative of upland conditions (Champoux et al 1995, Yadav et al 1997, Hemamalini et al 2000, Courtois et al 2003), under anaerobic conditions representative of the rainfed lowland situation (Kamoshita et al 2002a,b), or under hydroponic conditions (Price and Tomos 1997, Xu et al 2004, Yue et al 2006). Figure 1 shows an example of the position of the QTL for maximum root length for the three conditions, obtained using the information from our database.

Root penetration through a 5-mm-thick wax-petrolatum layer simulating the impedance of compacted rice fields was studied in the same populations, focusing on root number, penetrated root number, penetrated root length, penetrated root thickness, and root penetration index (Ray et al 1996, Ali et al 2000, Price et al 2000, Zheng et al 2000, Zhang et al 2001). Figure 2 shows the synthesis of these studies for root penetration index.

A set of constitutive QTLs seems to be regularly identified on the long arm of chromosomes 1, 2, 4, 7, and 9, and, to a lesser extent, on the short arm of chromosome 3 and the long arm of chromosomes 3, 6, and 11 (Périn et al 2006). Those are generally the QTLs with the highest LOD scores. Most QTLs, however, are specific for a study. This moderate reproducibility can be due to G × E interactions, although experimental conditions were relatively homogeneous for a group of traits, or to threshold effects.

These studies also highlighted the links between root and shoot development and the influence on root trait expression of the semidwarfism gene *sd1* located on the long arm of chromosome 1.

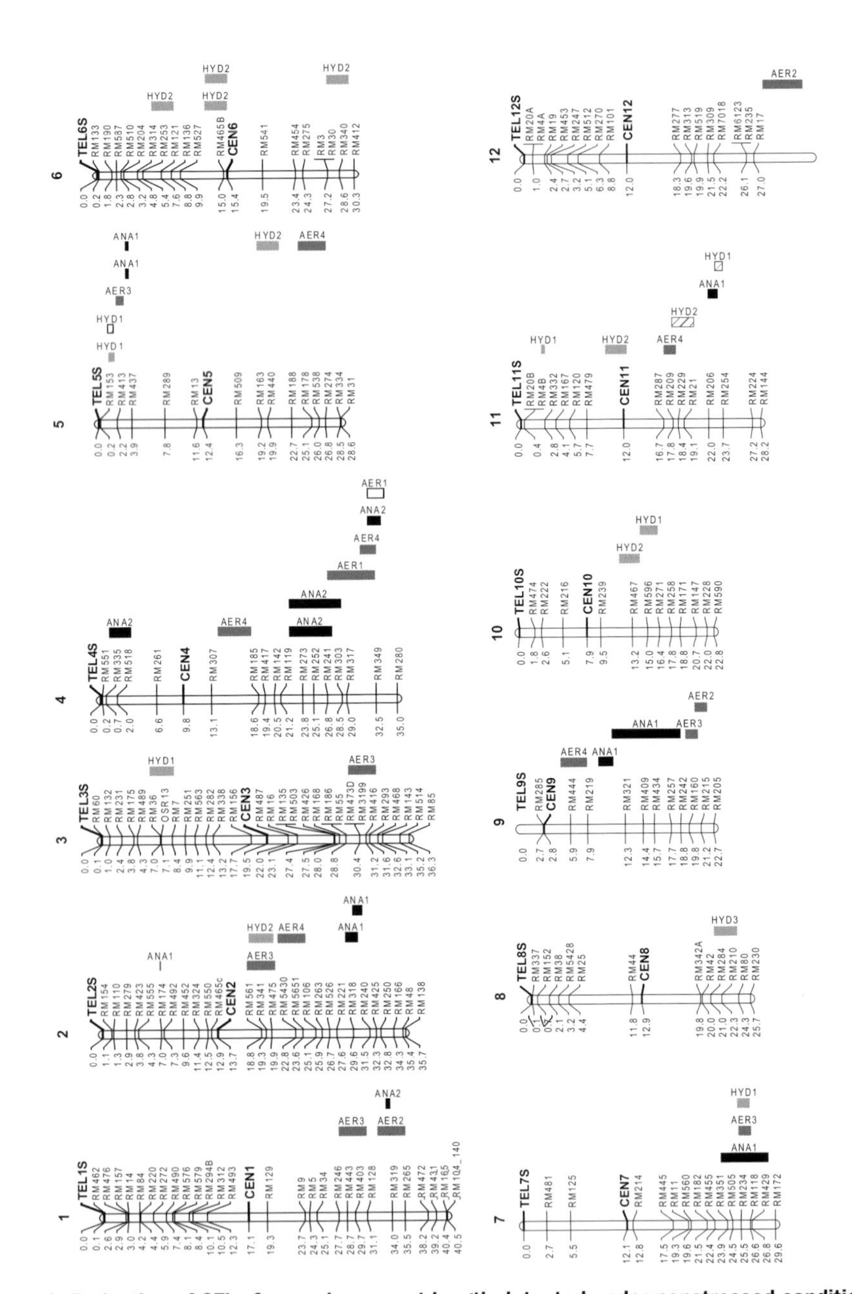

Fig. 1. Projection of QTLs for maximum root length detected under nonstressed conditions on the IR64 × Azucena SSD physical framework map (accumulated distances between markers are in Mb). Black segments correspond to QTLs obtained under anaerobic conditions (Kamoshita et al 2002a,b). Strongly shaded segments correspond to QTLs detected under aerobic conditions (Yadav et al 1997, Hemamalini et al 2000, Courtois et al 2003). Lightly shaded segments correspond to QTLs detected under hydroponic conditions (Price and Tomos 1997, Xu et al 2004, Yue et al 2006).

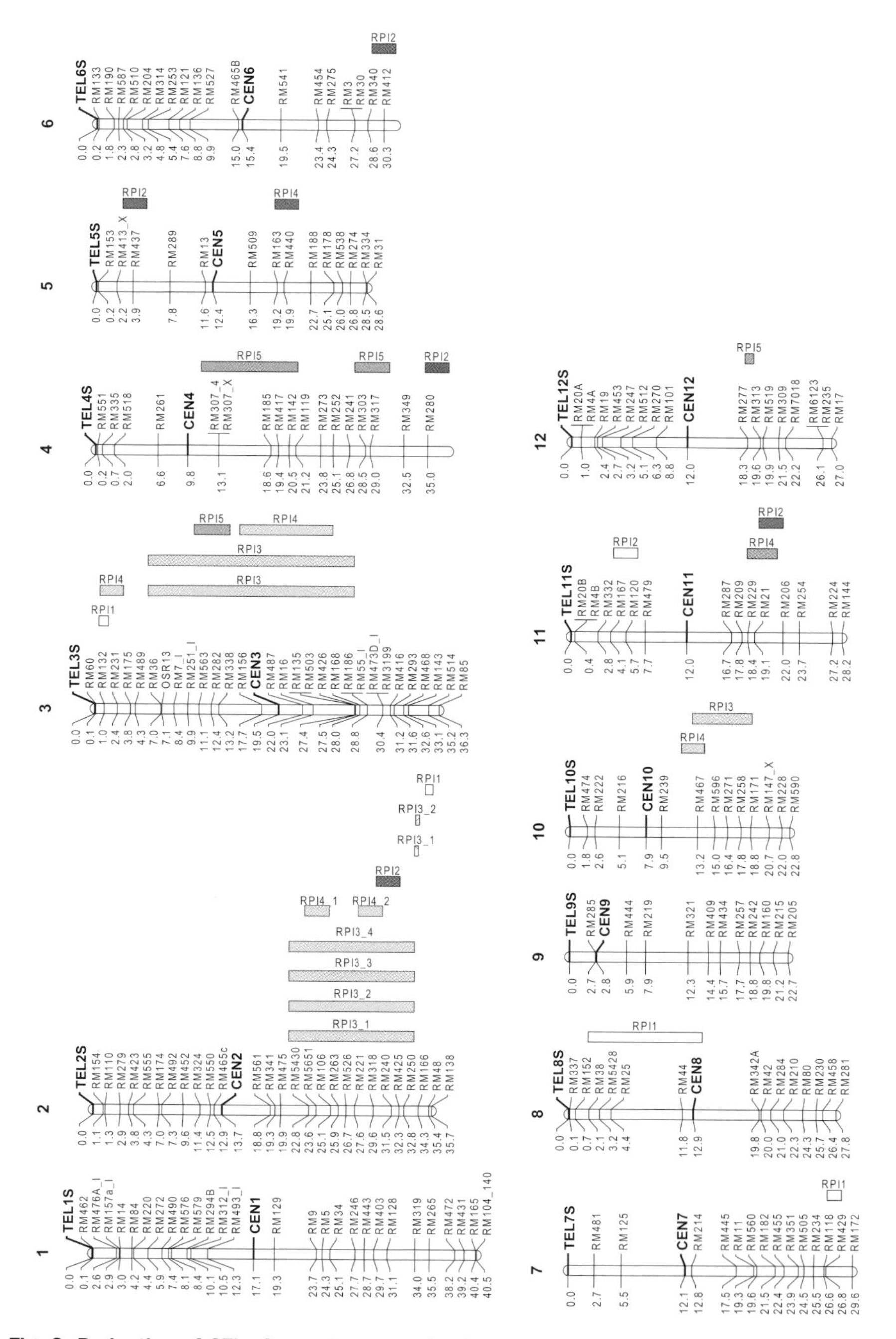

Fig. 2. Projection of QTLs for root penetration index on the IR64 × Azucena SSD physical framework map (accumulated distances between markers are in Mb). QTLs obtained by Ray et al (1996), Ali et al (2000), Price et al (2000), Zheng et al (2000), Zhang et al (2001).

Adaptive traits

The root system adapts under stress. This adjustment includes morphological and architectural changes and modifications in root/shoot partitioning. Environmental stimuli, below and above ground, induce a response of the rice root system through passive plasticity in response to carbon assimilate availability and active plasticity through signals for the regulatory systems involved in root morphogenesis. Such signals can be induced by nitrogen or phosphorus application, or water stress.

Searches for QTLs involved in root system plasticity (Hemamalini et al 2000, Price et al 2002, Zhang et al 2001, Zheng et al 2003) have been undertaken mainly under contrasting water supply conditions. Most authors reported an absence or a relatively small number of root trait QTLs common across different growing conditions, which indicated strong QTL × environment interactions, suggesting that both constitutive and plasticity QTLs may contribute to root development under water deficit.

More recently, using 178 lines of the IR64/Azucena SSD population, our team analyzed rice phenotypic plasticity during early vegetative growth by comparing root/shoot biomass partitioning, plant architecture, and development under three growth conditions: phosphorus deficiency, osmotic stress, and a control hydroponic culture. The two stimuli applied after 2 weeks' growth led to a similar expression of phenotypic plasticity: higher allocation of biomass to roots versus shoots and enhanced root growth and development when the ontogenetic differences between environments were taken into account. Five QTLs were detected for root/shoot ratio. Among them, two had a constitutive expression while two others had inducible expression and the last one was detected only for relative response variables. Several other inducible QTLs were detected controlling maximum root length, root biomass, and the number of nodal roots. Inducible QTLs related to either phosphorus deficiency or osmotic stress did not co-segregate. These results suggest that rice response to these stimuli involves different regulation and signaling networks. Looking for common points between these networks, we are currently investigating the relationships between biomass and sugar allocation to roots and the activities of some enzymes (e.g., invertases, sucrose synthase) involved in sugar metabolism that could provide a generic explanation for root phenotypic plasticity.

QTL validation through NIL development

QTL validation through NIL production in the studied background

Building on the genetic dissection of constitutive root traits and identification of QTLs with general effect, an attempt was made to transfer these QTLs into elite material. Using the results of the QTL analysis from Yadav et al (1997), Shen et al (2001) introgressed alleles for root depth from Azucena, a deep-rooted variety, into IR64, a shallow-rooted variety used as the recurrent parent. DH lines of the IR64 × Azucena population carrying the favorable alleles at markers surrounding the QTLs were chosen as donor parents. Four QTLs on chromosomes 1, 2, 7, and 9 were manipulated to produce near-isogenic lines (NILs) introgressed with one or two QTLs. Three cycles of marker-assisted backcrosses were performed with selection based on molecular information alone. The phenotypic evaluation of BC_3F_3 families showed that, except in the case of the QTL on chromosome 2, the introgressed QTLs were expressed in the recipient background with the expected effects. Comparisons of the NIL allelic profiles, however, did not allow more precise location of the QTLs. The quality of the initial QTL analysis appeared to be crucial for final progress. The work highlighted the importance of localizing QTLs in the smallest possible confidence intervals, correctly detecting QTLs linked in the repulsion phase, and taking into account epistasis.

These lines were tested in the field, under both well-watered aerobic and drought-stress conditions. A couple of lines were performing better than IR64 and lines with increased root depth showed better yield than lines without (Lafitte et al, unpublished results). Large segments had been introgressed to take into account the uncertainty on QTL position and an important linkage drag was observed for tillering.

QTL validation through NIL production in an unknown background

In normal breeding programs, the situation would be different, however, from that just reported. Preliminary QTL detection studies would probably have been conducted only on a small fraction of all possible parents.

Steele et al (2005) explored this avenue by undertaking the introgression of QTLs from Azucena into an elite Indian line (Kalinga III) for which no QTL detection had been previously attempted. They manipulated four QTLs on chromosomes 2, 7, 9, and 11 and pyramided them to obtain a line carrying the four QTLs in a 6-year marker-aided backcross program. They concluded that the QTL from chromosome 9, the same used in the previous study, significantly increased root depth under both irrigated and drought-stress treatments while the other QTLs did not show significant effects in their trials.

These studies, which involved a long effort, showed that QTL introgression is feasible but not without a loss of efficiency due to the limited precision in QTL location. Identifying the genes underlying the QTLs and developing better markers based on the genes themselves to manipulate them efficiently would considerably help in this context.

Linking QTLs to genes

Our team is presently exploring several ways to link QTLs to genes more tightly: modeling via QTL detection on model-based phenotypes, statistics via meta QTL analyses for candidate gene identification, recombinant genetics via fine-mapping of QTLs, and functional genomics using mutant collections and orthology relationships.

Detection of QTLs using model-based phenotypes

It is hypothesized that, by using eco-physiological models, it could be possible to extract hidden process-based traits (e.g., meristem growth rate) reflecting better gene expression than integrative traits (e.g., organ size, leaf area index) because part of the environmental variation would have been overcome by the model. Examples in other species have demonstrated the validity of this hypothesis (Tardieu et al 2005).

To pursue this approach in rice, we used Ecomeristem, a plant growth model based on heuristic approaches to extract plant genetic parameters from phenotypic data and address the adaptive responses of morphogenesis to environmental changes or stresses (Luquet et al 2006, Dingkuhn et al 2006). The model dynamically simulates rice phenotypes under genotypic and environmental control. The internal engine of the model is based on a state variable named internal competition index (Ic), determined for each time step as the ratio between carbon supply, provided mainly by the environment (radiation, temperature), and carbon demand, linked to organ initiation and expansion. The genotype-dependent parameters of the model, in particular, meristem growth rate, Ic threshold for tillering, or plastochron, control the response of one genotype to environment and to Ic. These parameters can be optimized for each genotype and then used as traits to characterize the lines from a mapping population, for example.

This model was applied to the IR64 × Azucena SSD population of 178 lines and plant growth was monitored for 5 weeks. QTLs for integrative traits and for model parameters mostly co-localized but some model-specific QTLs were also identified that could be interesting to analyze (Ahmadi et al, unpublished results).

So far, the model describes only the shoot morphogenetic behavior at the vegetative stage, with the root being only a sink for biomass. Extension of the model to the entire life cycle is under way. It is also planned to introduce a root module. In rice, root and shoot emissions are synchronized. Our intention is to exploit this mirror effect and assess whether it is possible to derive root parameters from the shoot ones (much easier to measure) once the effects of stress on synchrony are better understood.

Meta-analyses for candidate gene identification

Numerous QTL detection studies have been conducted on root system characteristics. An increase in the precision of QTL location and a decrease in confidence interval size can be obtained by running meta-analysis on independent experiments (Goffinet and Gerber 2000). The meta-analysis developed in Biomercator is based on the principles described by Goffinet and Gerber and makes use of the confidence interval size and QTL position. It computes an Akaike criterion to choose the most likely possibility between models with one to n QTLs. Examples of the power of this approach are given in Chardon et al (2004) for flowering time in maize and in Guo et al (2006) for resistance to cyst nematode in soybean.

When the QTL confidence interval is not known, it is possible to estimate it using the size of the population and the r^2 of the QTL as described by Darvasi and Soller (1997) and Weller and Soller (2004) for various types of mapping populations.

Using this approach on QTLs for root depth positioned on the physical map of the long arm of chromosome 9 (15 independent QTLs from 6 studies), covering 5 Mb, we were able to find two meta-QTLs, each of 1 Mb confidence interval. Using a search tool available in our genome browser OrygenesDB, we extracted all the genes underlying these two physical intervals and were able to recover several candidate genes in these areas.

Fine-mapping QTLs: toward QTL cloning

The recombinant genetic approach is powerful and has a proven record of efficiency from fine-mapping to QTL cloning, but it requires the development of new plant material complementing the mapping populations. To fine-map constitutive and adaptive root QTLs, we are using two different sets of materials. The first one is the set of chromosome segment substitution lines (CSSLs) derived from the Nipponbare × Kasalath cross, developed by Dr. Yano, RGP, NIAS, Tsukuba, Japan. This set of around 40 lines represents the whole Kasalath genome decomposed in small segments individually introgressed into a Nipponbare background and constitutes an extremely powerful tool to locate QTLs in 40-cM-size chromosomal segments without perturbations due to differences in genetic background. This material has been used to clone the first QTLs in rice (Yano et al 2000). We have verified that there were large differences between the two parents in terms of root growth. Interspecific populations of CSSL are also available but they are often affected by sterility problems that drive the segments carrying the fertility alleles to be fixed. In comparison, the Nipponbare × Kasalath set has a particularly clean background. The CSSLs will be evaluated with various methods and fine-mapping of selected target QTLs will be undertaken.

We are also using the NILs introgressed with constitutive QTLs for root depth developed in the QTL validation experiment presented above (Shen et al 2001) to fine-map the detected QTLs. The limits of the introgression segments in the BC_3F_5 near-isogenic lines of IR64 were more precisely defined (Courtois et al, unpublished results). The lines were evaluated under hydroponic conditions to make a final choice of a couple of target QTLs.

Our results led us to focus on the QTLs of chromosomes 2 and 9. We have undertaken a backcross program starting from the advanced BC_3F_5 lines with the intention of producing large recombinant populations for each QTL target. We are presently harvesting the F_2 generation of these crosses that will be screened for recombinants in the target areas before evaluating the recombinants under hydroponic conditions.

Genomic approaches to root system improvement

Among the large range of genomic resources and tools now available for rice root improvement, we choose to use mutant collections and orthology relationships.

Direct and reverse genetics based on mutant collections. Knockout mutant collections can be used to identify or validate genes involved in root system development. World rice resources amount to around 250,000 mutants, all types pooled (T-DNA, *Tos17*, *Ac-Ds*), among which 70,000 have a flanking sequence tag (FST) attached.

We screened a collection of 2,700 mutants carrying a Gal4:UAS:GFP cassette (Johnson et al 2005). The construct involving a reporter gene coding for a green fluorescent protein (GFP) permits visualization under UV of the location(s) of gene expression. This nondestructive technique allowed identification of 75 mutants with specific expression in roots (vessels, epidermis, lateral root emergence sites, apex, quiescent center), among which 65 were confirmed in the next generation and 13 showed a clear modification of their root phenotype (Rebouillat et al, unpublished data). One of these mutants, *hzl*, characterized by early death of the seminal root meristem, was chosen for a more detailed physiological, histological, and molecular characterization to better understand root morphogenesis (Rebouillat 2006).

Mutants can also be used for reverse genetics. Knowing the genes involved in root development pathways, it is possible to check in FST databases such as OrygenesDB whether mutants with insertions in these genes are available. The next step should be to identify mutants with a phenotype different from that of the wild type, and assess whether the phenotype co-segregates with the insert. Mutants have been identified for promising genes co-segregating with the root QTL on chromosome 9 and are presently undergoing this sequence of validation steps (Guiderdoni et al, unpublished data).

Orthology relationships. A large body of information on constitutive root development genes derived from mutant analysis is available in *Arabidopsis thaliana* (Van den Berg et al 1997, Casimiro et al 2003, Schiefelbein 2003). Since both *Arabidopsis* and rice genomes are sequenced and the full proteomes are available, the establishment of clear orthology relationships between the genes of the two species is possible. We are currently using Greenphyl, a proteome-based pipeline of identification of putative rice orthologs of *Arabidopsis* genes based on phylogenomics approaches (Conte et al 2008) to identify genes for root development orthologous to those detected in *Arabidopsis*.

The limitation of this approach lies in the fact that the root system of monocots has specificities, notably in root morphology and anatomy, and root development in coordination with the aerial organs, which makes it different from the root system of *Arabidopsis*. *Arabidopsis* has a simple pivotal root system with no exodermis, no sclerenchyma, a single cortical layer, and no aerenchyma, whereas rice has a complex fibrous root system with several root types from embryonic or postembryonic origin, each originating from different tissues. These differences, which have been extensively reviewed by Périn et al (2006), will likely relate to rice- or monocot-specific master genes, some of which are listed in Hochholdinger et al (2004).

From genes to alleles through association studies

Once candidate genes are identified by any of these methods and validated, the next step is to identify alleles of primary interest in the studied species, to provide tools for practical application of genomics in breeding programs. This objective can be addressed by investigating the organization of nucleotide diversity and the associations between gene polymorphism and trait variation among large collections of genetically diverse materials. This approach, referred to as "association studies," was initially developed in human genetics and is becoming increasingly important in animal and plant genetics (Thornsberry et al 2001, Remington et al 2001).

In comparison with classical QTL studies using mapping populations, association studies have several advantages. There is no need to produce specific material since they use the natural diversity. They compare several alleles at once. They permit making use of existing information on phenotypes of genetic resources. They normally improve the precision of the localization of the genes affecting the phenotype because local linkage disequilibrium (LD) is low in such collections, lower than in mapping populations in most cases and, therefore, the resolution of these approaches is higher. The comparative advantage of association studies, however, and the way to run them (whole-genome or candidate gene targeted approach) will depend on local LD magnitude.

This method solves some of the problems encountered in QTL analysis but has drawbacks of its own. Since local LD linked to physical distance is difficult to distinguish from LD due to population history, it is important to avoid structured collections. Sampling is therefore particularly important and the approach is quite restrictive in the type of populations that can be used. Since it implies a high marker density, it supposes that the number of candidate genes has been reduced to a manageable level; otherwise, the genotyping burden and cost can be too high.

Once validated candidate genes are available in a reasonable number, the principle is simple. Primers are designed to amplify candidate genes in a small panel representative of the diversity of the species. For fragments in which single nucleotide polymorphisms (SNPs) are detected, the SNPs are genotyped in the target larger collection. The same collection is phenotyped for the trait of interest. The genetic structure of the collection also has to be determined using a small number of neutral markers. Bayesian methods are generally used to determine the number of populations composing the sample, and a percentage of genome allocation (admixture) to each population can be computed for each accession. The association between SNP and variation of the trait is determined by ANOVA, with the percentage of admixture used as a covariate to correct from the structure if one is observed. This method allows determination of whether a gene is involved in the control of a trait, and, if the resolution is fine enough, can identify a range of SNPs or the specific SNP responsible for the phenotype.

This approach has been chosen by the Generation Challenge Program. In this framework, we have started to analyze nucleotide diversity in genes known to play a role in drought resistance in cereals such as invertases, auxin efflux carriers, ASR, or ERECTA (ADOC and Haploryza projects).

Strategies to accumulate favorable alleles

Once favorable alleles at relevant genes have been identified, the next issue is how to combine them to produce an elite genotype. The approach will vary according to the number of genes involved, which can be very high for traits as complex as those related to roots and adaptation to water limitation.

Genotype building

Genotype building is a solution well adapted for a limited number of genes. The first option explored was the introgression of favorable alleles coming from a line with limited agronomic value into a good genetic background. Several papers based on simulation explained how to optimize the methodology (Hospital and Charcosset 1997, Hospital 2001). More recently, Servin et al (2004) proposed efficient ways to do genotype building by pyramiding several alleles from different parents into an elite genotype. Positive results have already been obtained with the first approach in rice (Shen et al 2001).

Marker-aided recurrent selection

When many QTLs/genes are involved, either because the target trait is highly quantitative or complex or because several traits have to be combined, the genotype building approach is no longer feasible and the path forward is not as obvious or simple. One possibility could be to shift to marker-assisted recurrent selection (Hospital et al 1997). Recurrent selection works through successive cycles of selection and recombination and achieves long-term genetic progress though accumulation of numerous favorable alleles of loci with small individual effect. This has been widely used in rice breeding in Latin America (Guimaraes and Chatel 2004), relying on a recessive male sterility gene to facilitate recombination. No use of molecular markers in this framework has been attempted yet.

Theoretical papers have indicated approaches for optimization of the approach (Hospital et al 2000, Charmet et al 2001, Bernardo and Charcosset 2006).

A molecular score is computed for each individual, combining alleles at markers and phenotype or based on markers alone, depending on ease and cost of phenotyping. To reach maximum efficiency, the selection involves two steps: choice of the best individuals to recombine and then choice of the best pairs to cross on the basis of complementarity.

Experimental results using this strategy were obtained in maize (Blanc 2006) following QTL detection in a multiparental design (Blanc et al 2006). The results were ambivalent. Lines accumulating the favorable alleles at all QTLs were obtained but the progress observed was less than that expected, probably because of an overestimation of QTL effects as is classically observed (Melchinger et al 1998) and the fact that epistasis was not taken into account. The study gave interesting guidelines on how to weigh the alleles to compute the molecular score and avoid the very fast fixation of alleles with a major effect. The factors influencing success were the initial quality of QTL detection and the distribution of the favorable alleles among the parental lines, with the influence of lines carrying only a few interesting alleles in an otherwise unfavorable background being detrimental.

We have started developing a new breeding population, using already available information on loci, alleles, and phenotypes of interest for drought resistance. This population will be employed in our proposed marker-aided recurrent selection.

To optimize the process when a large choice of genes and alleles is available, it may be useful to seek support from crop modeling and integrate simulation tools in breeding programs to predict plant behavior in the target environment on the basis of model parameters and/or plant allelic composition. The recent review of Hammer et al (2006) highlights the possibilities of this modeling approach for complex adaptive traits.

Concluding remarks

The results obtained to date permit some optimism in the use of this range of molecular and model-based methods to improve the rice root system in a way that could translate into an advantage for yield under stress. However, two caveats need to be clearly spelled out to prevent overestimating expectations.

What if the variability existing in the species for root morphology and plasticity is not sufficient to cause differences in the field? If this occurs, this means that we will have to go for genetic engineering to optimize native expression or to introduce new mechanisms in the species. Such a strategy presupposes a good understanding of the genes involved and their interaction with other genes so all the knowledge accumulated to this point is still going to be extremely useful.

What if the genetic potential reached is good but difficulties are encountered in realizing it in the target environment because of soil compaction, acid subsoil, nematodes, or termites? This emphasizes the need for strong interaction with agronomists and the importance of managing the environment properly.

References

Ali ML, Pathan MS, Zhang J, Bai G, Sarkarung S, Nguyen HT. 2000. Mapping QTLs for root traits in a recombinant inbred population from two indica ecotypes in rice. Theor. Appl. Genet. 101:756-766.

Bernardo R, Charcosset A. 2006. Usefulness of gene information in marker-assisted recurrent selection: a simulation appraisal. Crop Sci. 46:614-621.

Blanc G. 2006. Sélection assistée par marqueurs dans un dispositif multiparental connecté: application au maïs et approche par simulations. PhD thesis. INA-PG, 150 p.

Blanc G, Charcosset A, Mangin B, Gallais A, Moreau L. 2006. Connected populations for detecting QTL and testing for epistasis: an application in maize. Theor. Appl. Genet. 113:206-224.

Casimiro I, Beeckman T, Graham N, Bhalerao R, Zhang H, Casero P, Sandberg G, Bennett MJ. 2003. Dissecting *Arabidopsis* lateral root development. Trends Plant Sci. 8:165-171.

Champoux MC, Wang G, Sarkarung S, Mackill DJ, O'Toole JC, Huang N, McCouch SR. 1995. Locating genes associated with root morphology and drought avoidance in rice via linkage to molecular markers. Theor. Appl. Genet. 90:969-981.

Chardon F, Virlon B, Moreau L, Falque M, Joets J, Decousset L, Murigneux A, Charcosset A. 2004. Genetic architecture of flowering time in maize as inferred from quantitative trait loci meta-analysis and synteny conservation with the rice genome. Genetics 168:2169-2185.

Charmet G, Robert N, Perretant MR, Gay G, Sourdille P, Groos C, Bernard S, Bernard M. 2001. Marker-assisted recurrent selection for accumulating QTLs for bread-making related traits. Euphytica 119:89-93.

Conte MG, Gaillard S, Lanau N, Rouard M, Périn C. 2008. GreenPhylDB: a database for plant comparative genomics. Nucl. Acids Res. 36:D991-D998.

Courtois B, Chaitep W, Moolsri S, Sinha PK, Trébuil G, Yadav R. 1996. Drought resistance and germplasm improvement: ongoing research in the Upland Rice Research Consortium. In: Piggin C, Courtois B, Schmit V, editors. Upland rice research in partnership. IRRI Discussion Paper Series No. 16. Los Baños (Philippines): International Rice Research Institute. p 154-175.

Courtois B, Lafitte RH. 1999. Rice improvement for the drought-prone upland environments of Asia. In: O'Toole J, Ito O, Hardy B, editors. Genetic improvement of rice for water-limited environments. Los Baños (Philippines): International Rice Research Institute. p 35-56.

Courtois B, Shen L, Petalcorin W, Carandang S, Mauleon R, Li Z. 2003. Locating QTLs controlling constitutive root traits in the rice population IAC 165 × Co39. Euphytica 134:335-345.

Darvasi A, Soller M.1997. A simple method to calculate resolving power and confidence interval of QTL map location. Behav. Genet. 27:125-132.

Dingkuhn M, Luquet D, Kim HK, Tambour L, Clément-Vidal A. 2006. EcoMeristem, a model of morphogenesis and competition among sinks in rice. 2. Simulating genotype responses to phosphorus deficiency. Funct. Plant Biol. 33:325-337.

Goffinet B, Gerber S. 2000. Quantitative trait loci: a meta-analysis. Genetics 155:436-473.

Guimaraes E, Chatel M. 2004. Exploiting rice genetic resources through population improvement. In: Guimaraes EP, editor. Population improvement: a way of exploiting the rice genetic resources of Latin America. CIAT, Cali, Colombia. p 53-74.

Guo B, Sleper DA, Lu P, Shannon JG, Nguyen HT, Arelli PR. 2006. QTLs associated with resistance to soybean cyst nematode in soybean: meta-analysis of QTL locations. Crop Sci. 46:595-602.

Hammer G, Cooper M, Tardieu F, Welch S, Walsh B, van Eeuwijk F, Chapman S, Podlich D. 2006. Models for navigating biological complexity in breeding improved crop plants. Trends Plant Sci. 11:587-593.

Hemamalini GS, Shashidhar HE, Hittalmani S. 2000. Molecular marker assisted tagging of morphological and physiological traits under two contrasting moisture regimes at peak vegetative stage in rice. Euphytica 112:69-78.

Hochholdinger F, Park WJ, Sauer M, Woll K. 2004. From weeds to crops: genetic analysis of root development in cereals. Trends Plant Sci. 9(1):42-48.

Hospital F. 2001. Size of donor chromosome segments around introgressed loci and reduction of linkage drag in marker-assisted back-cross programs. Genetics 158:1363-1379.

Hospital F, Charcosset A. 1997. Marker assisted introgression of quantitative trait loci. Genetics 147:1469-1485.

Hospital F, Goldringer I, Openshaw S. 2000. Efficient marker-based recurrent selection for multiple quantitative trait loci. Genet. Res. 75:357-368.

Hospital F, Moreau L, Lacoudre A, Charcosset A, Gallais A. 1997. More on the efficiency of marker-assisted selection. Theor. Appl. Genet. 95:1181-1189.

Johnson AA, Hibberd JM, Gay C, Essah PA, Haseloff J, Tester M, Guiderdoni E. 2005. Spatial control of transgene expression in rice using the Gal4 enhancer trapping system. Plant J. 41:779-789.

Kamoshita A, Wade LJ, Ali ML, Pathan MS, Zhang J, Sarkarung S, Nguyen HT. 2002a. Mapping QTLs for root morphology of a rice population adapted to rainfed lowland conditions. Theor. Appl. Genet. 104:880-893.

Kamoshita A, Zhang J, Sipongco J, Sarkarung S, Nguyen HT, Wade LJ. 2002b. Effect of phenotyping environment on identification of QTLs for rice root morphology under anaerobic conditions. Crop Sci. 42:255-265.

Lafitte RH, Champoux MC, McLaren G, O'Toole JC. 2001. Rice root morphological traits are related to isozyme groups and adaptation. Field Crops Res. 71:57-70.
Liu L, Lafitte R, Guan D. 2004. Wild *Oryza* species as potential sources of drought-adaptive traits. Euphytica 138:149-161.
Luquet D, Dingkuhn M, Kim H, Tambour L, Clément-Vidal A. 2006. EcoMeristem, a model of morphogenesis and competition among sinks in rice. 1. Concept, validation and sensitivity analysis. Funct. Plant Biol. 33:309-323.
Melchinger AE, Utz HF, Schon CC. 1998. Quantitative trait locus mapping using different testers and independent population samples in maize reveals low power of QTL detection and large bias in estimating QTL effects. Genetics 149:383-403.
O'Toole JC, Bland WL. 1987. Genotypic variation in crop plant root systems. Adv. Agron. 41:91-145.
Périn C, Breitler JC, Diévart A, Gantet P, Courtois B, Ahmadi N, de Raissac M, This D, Le QH, Brasileiro AMC, Meynard D, Verdeil JL, Guiderdoni E. 2006. Novel insights in rice root adaptive development. In: Brar DS, Mackill DJ, Hardy B, editors. Rice genetics V. Proceedings of the Fifth International Rice Genetics Symposium, 19-23 November 2005, Manila, Philippines. Singapore: World Scientific Publishing and Los Baños (Philippines): International Rice Research Institute. p 117-141.
Price AH, Steele KA, Moore BJ, Barraclough PB, Clark LJ. 2000. A combined RFLP and AFLP linkage map of upland rice used to identify QTLs for root penetration ability. Theor. Appl. Genet. 100:49-56.
Price AH, Tomos AD. 1997. Genetic dissection of root growth in rice. II. Mapping quantitative trait loci using molecular markers. Theor. Appl. Genet. 95:143-152.
Price AH, Townend J, Jones MP, Audebert A, Courtois B. 2002. Mapping QTLs associated with drought avoidance in upland rice grown in the Philippines and West Africa. Plant Mol. Biol. 48:683-695.
Ray JD, Yu LX, McCouch SR, Champoux MC, Wang G, Nguyen HT. 1996. Mapping quantitative trait loci associated with root penetration ability in rice. Theor. Appl. Genet. 92:627-636.
Rebouillat J. 2006. Etude cellulaire et moléculaire du développement racinaire chez le riz (cv Nipponbare): criblage et caractérisation d'une collection de lignées d'insertion enhancer trap GAL4:UAS:GFP. PhD thesis. University of Montpellier II, France. 98 p.
Remington DL, Thornsberry JM, Matsuoka Y, Wilson LM, Whitt SR, Doebley J, Kresovich S, Goodman MM, Buckler ES. 2001. Structure of linkage disequilibrium and phenotypic associations in the maize genome. Proc. Natl. Acad. Sci. USA 20:11479-11484.
Schielfelbein JW. 2003. Cell-fate specification of the epidermis: a common patterning mechanism in the root and shoot. Curr. Opin. Plant Biol. 6:74-78.
Servin B, Martin OC, Mezard M, Hospital F. 2004. Toward a theory of marker-assisted gene pyramiding. Genetics 168:513-523.
Shen L, Courtois B, McNally KL, Robin S, Li Z. 2001. Evaluation of near-isogenic lines introgressed with QTLs for root depth through marker-aided selection. Theor. Appl. Genet. 103:75-83.
Steele KA, Price AH, Sashidhar HE, Witcombe JR. 2005. Marker-assisted selection to introgress rice QTLs controlling root traits into an Indian upland rice variety. Theor. Appl. Genet. 112(2):208-221.
Tardieu F, Reymond M, Muller B, Simonneau T, Sadok W, Welcker C. 2005. Linking physiological and genetic analyses of the control of leaf growth under changing environmental conditions. Aust. J. Agric. Res. 56:937-946.

Thornsberry JM, Goodman MM, Doebley J, Kresovich S, Nielsen D, Buckler ES. 2001. *Dwarf8* polymorphisms associate with variation in flowering time. Nat. Genet. 28:286-289.

Van den Berg C, Willemsen V, Hendricks G, Weisbeek P, Scheres B. 1997. Short-range control of cell differentiation in the *Arabidopsis* root meristem. Nature 390:287-289.

Weller J, Soller M. 2004. An analytical formula to estimate confidence interval of QTL location with a saturated genetic map as a function of experimental design. Theor. Appl. Genet. 109:1224-1229.

Xu CG, Li XQ, Xue Y, Huang YW, Gao J, Xing YZ. 2004. Comparison of quantitative trait loci controlling seedling characteristics at two seedling stages using rice recombinant inbred lines. Theor. Appl. Genet. 109:640-647.

Yadav R, Courtois B, Huang N, McLaren G. 1997. Mapping genes controlling root morphology and root distribution in a double-haploid population of rice. Theor. Appl. Genet. 94:619-632.

Yano M, Katayose Y, Ashikari M, Yamanouchi U, Monna L, Fuse T, Baba T, Yamamoto K, Umehara Y, Nagamura Y, Sasaki T. 2000. *Hd1*, a major photoperiod sensitivity quantitative trait locus in rice, is closely related to the *Arabidopsis* flowering time gene *CONSTANS*. Plant Cell 12:2473-2483.

Yue B, Xue W, Xiong L, Yu X, Luo L, Cui K, Jin D, Xing Y, Zhang Q. 2006. Genetic basis of drought resistance at reproductive stage in rice: separation of drought tolerance from drought avoidance. Genetics 172:1213-1228.

Zhang J, Zheng HG, Aarti A, Pantuwan G, Nguyen TT, Tripathy JN, Sarial AK, Robin S, Babu RC, Nguyen BD, Sarkarung S, Blum A, Nguyen HT. 2001. Locating genomic regions associated with components of drought resistance in rice: comparative mapping within and across species. Theor. Appl. Genet. 103:19-29.

Zheng HG, Babu CR, Pathan MS, Ali L, Huang N, Courtois B, Nguyen HT. 2000. Quantitative trait loci for root penetration ability and root thickness: comparison of genetic backgrounds. Genome 43:53-61.

Zheng BS, Yang L, Zhang WP, Mao CZ, Wu YR, Yi KK, Liu FY, Wu P. 2003. Mapping QTLs and candidate genes for rice root traits under different water-supply conditions and comparative analysis across three populations. Theor. Appl. Genet. 107:1505-1515.

Notes

Authors' address: Cirad, UMR DAP, Avenue Agropolis, Montpellier 34398, France.

An integrated systems approach to crop improvement[1]

Graeme L. Hammer and David Jordan

Progress in crop improvement is limited by the ability to identify favorable combinations of genotypes (G) and management practices (M) given the resources available to search among possible combinations in the target population of environments (E). Crop improvement can be viewed as a search strategy on a complex G × M × E adaptation or fitness landscape. Here we consider the design of an integrated systems approach to crop improvement that incorporates advanced technologies in molecular markers, statistics, bioinformatics, and crop physiology and modeling. We suggest that such an approach can enhance the efficiency of crop improvement relative to conventional phenotypic selection by changing the focus from the paradigm of *identifying superior varieties* to a focus on *identifying superior combinations of genetic regions and management systems*. A comprehensive information system to support decisions on identifying target combinations is the critical core of the approach. We discuss the role of ecophysiology and modeling in this integrated systems approach by reviewing (1) applications in environmental characterization to underpin weighted selection, (2) complex trait physiology and genetics to enhance the stability of QTL models by linking the vector of coefficients defining the dynamic model to the genetic regions generating variability, and (3) phenotypic prediction in the target population of environments to assess the value of putative combinations of traits and management systems and enhance the utility of QTL models in selection. We examine *in silico* evidence of the value of ecophysiology and modeling to crop improvement for complex traits and note that, although

[1]This paper was published originally as Hammer GL, Jordan DR. 2007. An integrated systems approach to crop improvement. In: Spiertz JHJ, Struik PC, van Laar HH, editors. *Scale and Complexity in Plant Systems Research: Gene-Plant-Crop Relations.* Wageningen UR - Frontis Series No. 21, Springer, Dordecht, The Netherlands. ISBN: 978-1-4020-5904-9 (hard cover) and ISBN: 978-1-4020-5905-6 (soft cover). p 45-61. It is reproduced here in an updated form with the kind permission of Springer Science and Business Media.

there is no definitive position, it seems clear that there is sufficient promise to warrant continued effort. We discuss criteria determining the nature of models required and argue that a greater degree of biological robustness is required for modeling the physiology and genetics of complex traits. We conclude that, although an integrated systems approach to crop improvement is in its infancy, we expect that the potential benefits and further technology developments will likely enhance its rate of development and that this approach will be particularly relevant in breeding for adaptation to water limitation.

Progress in crop improvement depends on identifying favorable combinations of genotypes and management practices from among innumerable possible combinations. Available resources and variability in target environments limit this search process. Crop improvement can be viewed as a search strategy on a complex adaptation or fitness landscape, which consists of the phenotypic consequences of genotype (G) and management (M) combinations in target environments (E) (Cooper and Hammer 1996). The phenotypic consequences of only a very small fraction of all possible G × M × E combinations can be evaluated experimentally. Hence, most of the fitness landscape remains hidden to its explorer, even if the experiments remain simple and measure only yield of as many combinations as resources allow, as in standard multienvironment trials. Despite this, conventional breeding strategies based on phenotypic selection and principles of statistical quantitative genetics (Lynch and Walsh 1998) have been able to achieve sustained levels of yield improvement (Duvick et al 2004). But, to maintain this rate of advance requires increasing resources. Can an integrated approach incorporating advanced technologies in molecular markers, statistics, bioinformatics, and crop physiology and modeling enhance the efficiency of crop improvement?

The complexity of the phenotypic fitness landscape arises from G × E, M × E, G × G, and G × M × E interactions. Traits associated with genetic variation (e.g., maturity, tillering) may rank differently for yield depending on environment (Hammer and Vanderlip 1989, van Oosterom et al 2003), management interventions (e.g., row configuration, density) may rank differently depending on environment (Whish et al 2005), and combinations of traits and management (e.g., maturity × density) may rank differently in different environments (Wade et al 1993). In addition, the genetic architecture of the gene network underpinning complex multigenic adaptive traits is likely to involve varying degrees of epistatic interactions. In such situations, trait expression is governed by context-dependent gene effects, that is, interaction with other genes (Podlich et al 2004). Such G × G interactions add substantially to genetic architecture complexity, with major implications for G × M × E interactions and rate of progress in crop improvement (Cooper et al 2005).

It has been more than a decade since the 1994 international symposium at which Cooper and Hammer (1996) advanced the concept of crop improvement as a search strategy on a G × M × E adaptation landscape and outlined a general framework for an integrated systems approach to crop improvement. Their framework incorporated simultaneous manipulation of plant genetics and crop management and considered how crop physiological understanding and modeling might add value to existing plant breeding methodologies. Plant breeding requires the prediction of phenotype based on genotype to underpin yield advance and this provided the logical entry for advances in quantitative functional physiology.

Since the 1994 symposium, there has been considerable development in these concepts and methodologies. Advances in understanding the complexities of gene-to-phenotype and phenotype-to-genotype associations for traits, and the potential to use this knowledge in plant breeding, were the subject of a symposium at the most recent International Crop Science Congress (Brisbane, Australia, 2004: www.cropscience.org.au). Revised versions of invited papers to that symposium, which set out the current state of knowledge, were published subsequently in a special issue of the *Australian Journal of Agricultural Research* introduced by Cooper and Hammer (2005). Several other key review papers (Cooper et al 2002, Hammer et al 2002, 2006, Chapman et al 2002, Tardieu 2003, Yin et al 2004) cover developments in linking physiological and genetic modeling for crop improvement and in pursuing the G × M × E concept to enhance molecular breeding. In addition, there have been continuing advances in capacity for molecular genotyping and genomics approaches (Somerville and Dangl 2000, Jaccoud et al 2001) and in statistics and bioinformatics (van Eeuwijk et al 2005, Verbyla et al 2003). These advances have enhanced the possibility for an integrated systems approach to crop improvement to link to the genomic region level for complex traits. This is despite the limited progress of molecular breeding for complex traits to date due to gene and environment context dependencies (Podlich et al 2004).

Here, we consider the design and implementation of such an integrated systems paradigm for crop improvement. We assess progress from the initial concept construction in 1994 (Cooper and Hammer 1996) and focus on the linking role of crop ecophysiology and modeling to enhance the potential of molecular breeding and the efficiency of crop improvement in general (Hammer et al 2006). We use sorghum as a case-study species, not only because it is the central focus of our crop improvement research and is relevant to the issue of adaptation to water limitation but also because there is advanced physiological understanding, well-developed modeling capability, and a mature set of molecular technologies and genome resources, all linked to an operational breeding and crop improvement program (Henzell and Jordan 2006, Jordan et al 2006).

Design of an integrated crop improvement program

The central tenet of the integrated systems approach to crop improvement proposed is to change the focus from the paradigm of *identifying superior varieties* to a focus on *identifying superior combinations of genetic regions and packaging these regions into varieties*. Beyond this, it can change the focus from the breeding paradigm of *only developing superior varieties* to a crop improvement paradigm of *developing superior combinations of genetic regions and management systems to optimize resource capture and sustainability in particular cropping environments.* Key decisions in the integrated program relate to the selection of genotypes, management practices, and

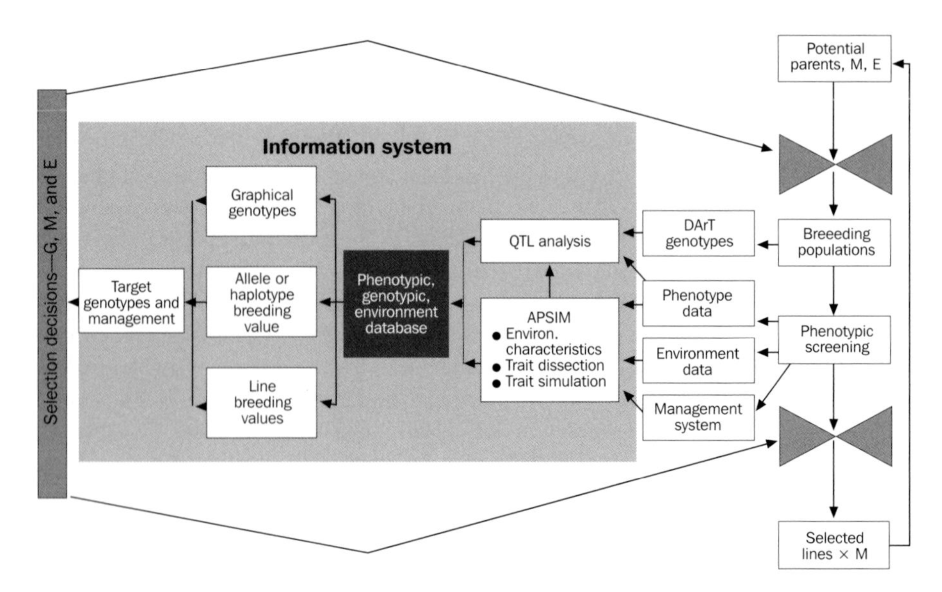

Fig. 1. Overview of an integrated crop improvement program.

test environments (Fig. 1). A comprehensive information system supporting these decisions is the critical core of the approach.

The design of the program involves a novel approach to integrating four relatively new technologies to enhance effectiveness in crop improvement:

1. Enhanced marker technology—low-cost, high-throughput genotyping allowing all of the genotypes tested in a breeding program to be genotyped with relatively high marker density (e.g., using DArT technology (Jaccoud et al 2001)).
2. Enhanced QTL detection methods—novel statistical approaches, pedigree-based methods, and associative genetics to allow marker detection directly in breeding populations (Verbyla et al 2003, Jordan et al 2004, van Eeuwijk et al 2005).
3. Enhanced gene-to-phenotype linkages—dynamic physiology and modeling frameworks to dissect complex traits to functional components to enhance association of phenotype with marker profiles (e.g., Leon et al 2001, Reymond et al 2003, Tardieu 2003, Yin et al 2004, Messina et al 2006, Manschadi et al 2006).
4. *In silico* evaluation—advanced modeling frameworks to characterize environments and to evaluate the utility of trait and management combinations in target environments (Chapman et al 2000a,b, Hammer et al 2005).

The proposed integration (Fig. 1) provides the means to work across levels of biological organization from genetic regions to plant growth, development, and yield while retaining the scale of a functional breeding and crop improvement program. The physiology and modeling provide a "knowledge bank" of process understanding.

Modeling can generate benchmarks within the breeding program trialing system against which the degree of advance associated with new genetic recombinations and management systems can be assessed, despite genotype-by-management-by-environment interactions. Valuable novel combinations of regions can be identified and linked to dense marker profiles, which will be available across the breeding program via the enhanced marker technology. The advanced statistical procedures will identify patterns of desirable genomic regions. The information accumulated in the breeding program over time will enable the identification of key genomic regions and their value in breeding, as the genetic associations among lines will be known. Existing phenotypic information and populations generated in the breeding program will be used in contrast to the conventional approach of developing populations specifically for mapping or validating markers. Key regions of unknown function can then be targeted for physiological analysis and modeling to build the information base. Modeling can be deployed to add value to conventional field testing by examining potential combinations of traits and management systems in a range of production environments (sites, soils, season types) via simulation analysis. Such projections of genotype and management combinations onto target environments contribute to the measures of breeding value.

Role of ecophysiology and modeling in integrated crop improvement

There are three general areas in which crop ecophysiology and modeling can play a role in the integrated approach to crop improvement (Fig. 1): (1) characterizing environments to define the nature and frequency of challenges in the target population of environments (TPE), (2) understanding and dissecting the physiology and genetics of complex traits, and (3) predicting phenotypes of G × M combinations in the TPE.

Environmental characterization

Using modeling to characterize environments in the TPE can assist in unraveling G × E interactions in a manner that aids selection decisions and improves the rate of yield gain in crop improvement programs. Muchow et al (1996) demonstrated that a sorghum simulation model (Hammer and Muchow 1994) could be used to characterize water-limited environments more effectively than indices based only on climatic data. The time course of a relative transpiration (RT) index was derived from the dynamic interactions implicit in the model. It was used to define the nature of the water limitation experienced by the crop throughout the growing season. Chapman et al (2000a) classified environments in the TPE for sorghum in Australia based on the time course of RT and identified three distinct environment types (Fig. 2). They found that the frequency of environment types at specific locations correlated with patterns of discrimination among hybrids detected in multienvironment trials (MET) at those locations. When the same simulation and classification procedure was applied to the TPE using historical climate data (Chapman et al 2000b), they noted that changes in frequency of environment types over time periods relevant to a breeding program affected yield likelihood and generated differing patterns of G × E (Fig. 3).

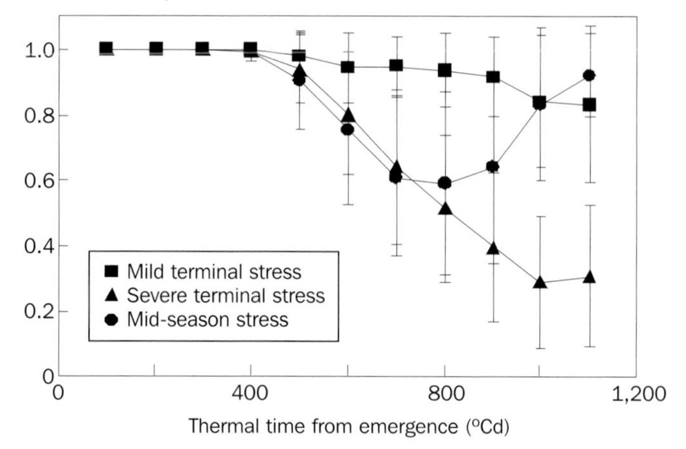

Fig. 2. Environmental characterization of sorghum production environments in Australia based on the time course of simulated relative transpiration throughout the crop life cycle (adapted from Chapman et al 2000a).

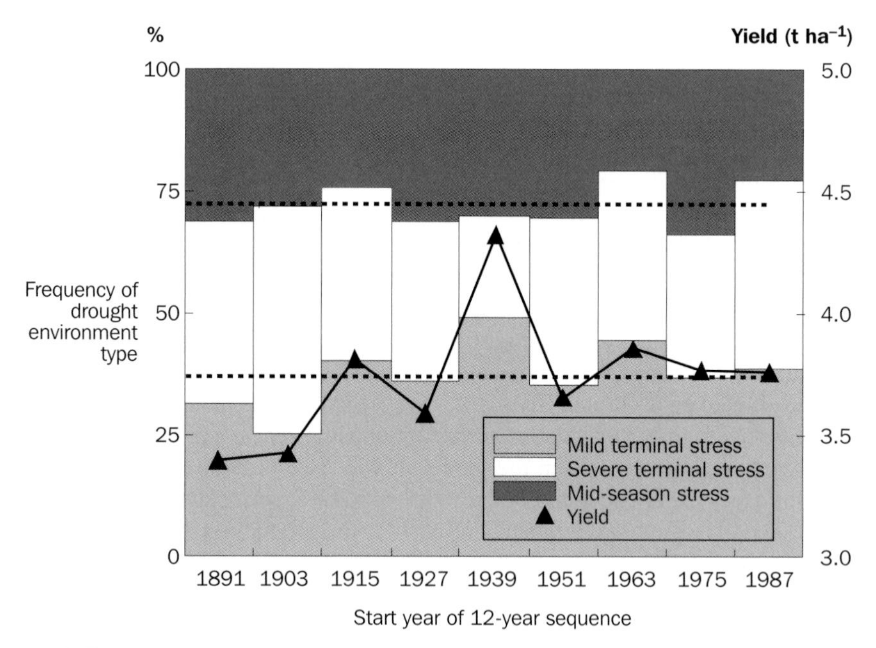

Fig. 3. Frequencies of environment types in consecutive 12-year periods during the 20th century for sorghum in Australia (adapted from Chapman et al 2000b).

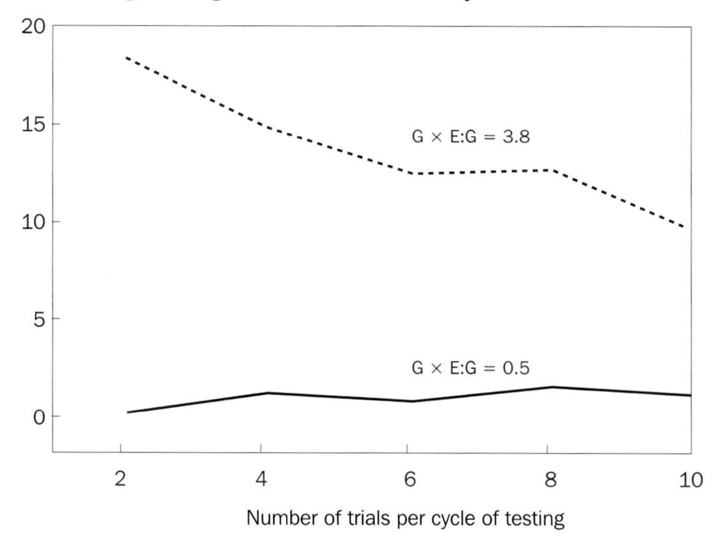

Fig. 4. Percent advantage of weighted selection versus number of trials per cycle of testing in a breeding program with high or low G × E (adapted from Podlich et al 1999).

They suggested that weighting genotype performance by the representativeness of the selection environment in each MET with respect to the TPE would be advantageous in breeding programs in these variable environments. Podlich et al (1999) used breeding system simulation to demonstrate the advantage of such a weighted selection strategy in variable environments, especially when G × E was high (Fig. 4). Löffler et al (2005) used a simulation-based environment classification of the TPE for the corn belt in the U.S. to improve cultivar performance predictability for a maize breeding program.

Complex trait physiology and genetics

A dynamic crop modeling framework can aid understanding of the physiology and genetics of complex traits in a way that has potential to enhance efficiency in crop improvement programs. The model provides an analytical framework to specify the functional basis underpinning phenotypic variation in a complex trait. The vector of coefficients that specifies the functional and process control equations in the model is the basis for the link to genetic modes of action (i.e., QTLs or genes). The notion of using a virtual or *in silico* plant for this purpose has been discussed by Hammer et al (2002, 2004, 2006), Tardieu (2003), Yin et al (2004), and Dingkuhn et al (2005).

Dissection of phenotypic variability in complex traits requires detailed experimental studies in controlled genetic backgrounds to unravel the functional biology underpinning the variability. In sorghum, studies on genotypes differing in their ability to retain green leaves during grain-filling under terminal drought, known as the "stay green" trait (Borrell et al 2000, 2001), have suggested that the trait arises as an emergent consequence of differences in underlying factors such as leaf size, specific leaf nitrogen, dry matter partitioning, nitrogen uptake, and transpiration or transpiration efficiency. This understanding is being used in fine-mapping studies with near-isogenic lines (NILs) to isolate target genes in the genomic regions associated with the stay green trait (Tao et al 2000). In other studies on genotypes from a population differing in tillering (Kim et al 2006), size of leaves on the main culm and the consequent dynamics of internal plant competition for assimilate have been identified as a likely causal factor. This is consistent with the concepts presented by Luquet et al (2006) in modeling morphogenesis and competition among sinks in rice. In wheat, Manschadi et al (2006) used large root observation chambers to quantify differences in root system architecture between lines differing in adaptation to water limitation. They used their experimental findings to modify a wheat model and could demonstrate by simulation that consequences of the root system architecture differences on the dynamics of soil water extraction during the crop cycle could generate the nature of the adaptation differences observed between the lines in the field.

To date, however, a modeling approach to connecting trait physiology to underpinning genetics has been demonstrated comprehensively only at the organ or component level for traits such as expansive growth of leaves (Reymond et al 2003, Tardieu 2003, Chenu et al 2008) and crop development (Leon et al 2001, Yin et al 2005, Messina et al 2006). In these cases, coefficients defining differences among lines in process responses to environmental influences have been linked with QTL analyses. Reymond et al (2003) combined QTL analysis with an ecophysiological model of the response of maize leaf elongation rate to temperature and water deficit by phenotyping a population and conducting the QTL analysis on the fitted model parameters. Using the derived relationships between model coefficients and QTLs, they were able to predict responses of lines with novel combinations of QTLs in a range of environments. Chenu et al (2008) have now linked this organ-level capability into a whole-plant modeling framework to provide the basis for gene-to-phenotype simulation at the crop level. Messina et al (2006) achieved similar results in predicting soybean development by linking temperature and photoperiod responsiveness coefficients of a photo-thermal phenology model to allelic variants at known regulating loci. They used a study on NILs varying at these loci to derive the relationships and then applied them successfully in predicting the development of other genotypes in a range of environments.

It may be possible to use a modeling approach to link more directly with gene networks controlling growth and development processes (Welch et al 2005). Knowledge is emerging rapidly from studies on model plants to support modeling frameworks based on experimental evidence for understanding the action of gene networks at the biochemical level (e.g., Blazquez 2000). For example, Koornneef et al (1998) presented a working model for the genetic control of flowering time in *Arabidopsis* based on extensive molecular-genetic studies to dissect this process. These studies employed a large number of mutant genotypes of *Arabidopsis* varying in time to flowering. The genetic, molecular, and physiological analyses have led to the elucidation of components and pathways involved. Welch et al (2003) adapted the qualitative understanding reported for *Arabidopsis* to a quantitative predictive model of transition to flowering using a genetic neural network approach. Morgan and Finlayson (2000) have presented a similar qualitative model for flowering in sorghum, based on their extensive studies with mutant genotypes. Beyond this, Dong (2003) developed a dynamic flowering-time model of the gene network in *Arabidopsis* that simulated the temperature- and photoperiod-dependent dynamics of mRNA expression for key genes in the network. He used controlled environment and gene expression studies for a range of mutants to develop the model and was able to successfully predict transition to flowering for a far wider range of G × E combinations than used in model development.

The scientific insights gained from this approach at the organ or component level could be connected to more conventional crop models to explore interactions among development and growth and yield processes, thus providing an effective bridge between genetic architecture and phenotypic expression. Messina et al (2006) connected their prediction of development in soybean based on the presence of specific genetic loci to cultivar performance in breeding trials. van Oosterom et al (2006) connected a simplified gene network model for photoperiod control of transition to flowering in sorghum to the APSIM generic crop modeling platform (Wang et al 2002) to demonstrate that an input of allelic variability could generate G × E for yield as an emergent consequence of the model dynamics. The use of modeling technologies in support of understanding the consequences of alterations of specific genes can occur via validated QTL models linked to model coefficients or via direct linkages to gene networks when underpinning knowledge is sufficient. These approaches provide the major opportunities to effectively use modeling in an integrated approach to crop improvement.

Phenotypic prediction in the TPE

Using modeling to project consequences of G × M combinations in the TPE can generate information that aids in selection decisions and improves the rate of yield gain in crop improvement programs. Numerous studies have approached this by exploring the putative value of potential trait variation in a range of species using a diversity of crop models (e.g., Spitters and Schapendonk 1990, Muchow et al 1991, Aggarwal et al 1997, Boote et al 2001, Sinclair and Muchow 2001, Asseng and van Herwaarden 2003, Sinclair et al 2005) or by exploring the optimization of trait and management combinations (e.g., Hammer et al 1996). Using this approach requires confidence in the adequacy of the crop model to simulate effects of trait variation. This aspect is discussed below in considering the nature of models required to support the integrated systems approach to crop improvement. It also requires rigorous specification of soil (e.g., water-holding capacity) and climate (e.g., daily temperature and radiation) conditions for relevant production zones of the TPE as inputs to the simulation analysis.

An example of a model-generated G × M × E interaction relates to manipulation of tillering (G) and row spacing (M) in dryland grain sorghum production systems in Australia (Fig. 5). Canopy development and consequent demand for water are affected by the extent of tillering (Kim et al 2006) and row configuration (Whish et al 2005). Figure 5 shows the results of a 50-year simulation using historical climate data for

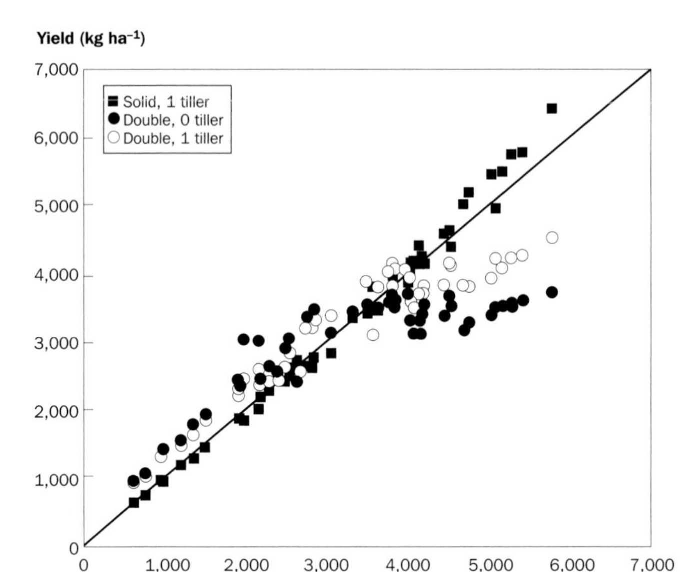

Fig. 5. Simulated yield of sorghum varying in tiller number (0 or 1) and row configuration (solid or double skip) for a 50-year simulation at Emerald, Australia (for details, see text). Yield for each combination each year is plotted against yield of the standard treatment (plants with no tillers grown in a solid 1-m row configuration) in that year.

Emerald in central Queensland, Australia, with the sorghum model implementation in APSIM (www.apsim.info/apsim/; Wang et al 2002). The simulation was conducted for a medium-maturing hybrid planted in early January each year assuming 80 mm of available water in a 120-cm-deep vertosol soil that held a maximum of 130 mm of plant-available water. In wetter higher-yielding years, the greater cover associated with solid row configuration and tillering is advantageous. But, in drier low-yielding years, the lower cover associated with uniculm plants grown in a double skip row configuration is advantageous. There is a crossover at a yield of about 3.5 t ha^{-1} in the standard treatment (solid row configuration and no tillers). When a random error component is added by assuming a coefficient of variation of 12% (as per Hammer et al 1996) and three replicates stochastically generated, the resultant 50-year MET has a highly significant G (tillering) × M (row configuration) × E (year) interaction (data not shown). Hence, the interaction was an emergent consequence of the model dynamics generated by a change in one plant attribute and one management factor. The study of Manschadi et al (2006), noted earlier, on root system architecture and adaptation to water limitation in wheat provides a good example of model-generated G × E.

Value of ecophysiology and modeling in integrated crop improvement

As noted in the introduction, the key question to resolve is whether incorporating ecophysiological understanding and modeling in an integrated approach can enhance the efficiency of crop improvement. Is it possible to achieve a rate of yield improvement better than can be obtained by continued conventional empirical breeding based on phenotypic selection?

Beyond the demonstrated value of using models for environment characterization noted earlier, there is now some *in silico* evidence supporting a tentatively positive response to these questions in relation to crop improvement for complex traits (Cooper et al 2002, 2005, Chapman et al 2002, 2003, Hammer et al 2005, 2006). In those studies, sorghum phenotypes were simulated for a broad range of production environments in Australia based on assumed levels of variation in 15 genes controlling four adaptive traits. "Virtual genotypes" were created by deriving combinations of expression states that depended on the number of positive alleles present for each trait. Expression states were then linked with crop model coefficients that quantified their physiological effects. By simulating a range of such virtual genotypes over a range of production environments, a database of simulated phenotypes was generated. The database of simulated phenotypes was linked to the QU-GENE breeding system simulation platform (Podlich and Cooper 1998) to explore effects of cycles of selection on yield gain for a range of breeding strategies. When marker-assisted selection (MAS) breeding strategies were simulated, the inclusion of marker-trait associations based on physiological knowledge and marker weights based on simulated trait value in the TPE significantly increased the average rate of yield gain over MAS strategies without such knowledge and modeling capability (Fig. 6). This result was dependent on (1) the assumed existence of stable QTL models that linked regions to model coefficients and (2) the lessening of gene and environment context dependencies of the QTLs via inherent interactions in the model dynamics that allowed robust projection of consequences of combinations onto the performance landscape in the TPE.

In a more comprehensive simulation analysis of response to breeding strategies, Cooper et al (2005) examined a range of genetic models incorporating varying degrees of additive, epistatic, and G × E effects that generated a spectrum of complexity in the resultant performance landscape. They quantified the qualitative expectation that response to phenotypic selection (PS) decreased as complexity of the genetic architecture of the trait increased. They also quantified the relative advantage of MAS over PS by simulating differences in response after 5 cycles of selection for the same range of genetic models. They placed the sorghum example above in the context of this diverse set of situations. Although the performance landscape generated in that case demonstrated a relatively high complexity, their analysis indicated that G × E was the major component of genetic architecture influencing complexity, and that there was only a modest advantage of the MAS strategy proposed over PS. The previous analyses (Chapman et al 2003, Hammer et al 2005) had emphasized that value generated from the inclusion of physiological knowledge and modeling resulted from the enhanced ability to deal with environment context dependencies

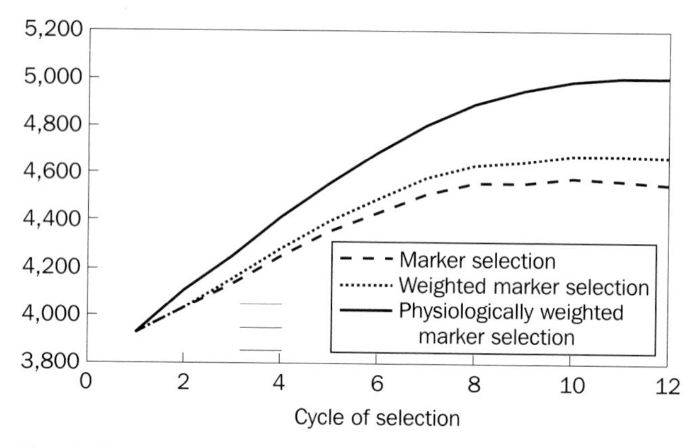

Fig. 6. Average yield in the target population of environments (TPE) over 12 cycles of a sorghum crop improvement program. The trajectories for marker selection and weighted marker selection represent the average result over individual breeding system simulations based on a QTL analysis from single environments. Marker selection incorporates only detection of markers, whereas weighted marker selection includes the weighting associated with each marker in the single detection environment. The trajectory for physiologically weighted marker selection represents the average result over simulations in which markers have been assigned to physiological traits and marker weights have been derived from the simulated value of that trait in the entire TPE (after Hammer et al 2005).

(i.e., QTL × E interaction) in the use of markers. However, an optimistic view of overall value may have been presented.

Hence, while there is as yet no definitive answer to our question, it seems clear that there is sufficient promise to warrant continued effort in pursuing approaches to using physiological knowledge and modeling to enhance crop improvement for complex traits.

What is the nature of the models required?

Many conventional agronomic models are adequate and suitable for characterizing the abiotic stress patterns experienced by crops. The major requirements for such environmental characterization are reliable predictions of ontogeny, canopy dynamics, and water use. This aligns with pressures on the development of agronomic models in which predictive capacity for growth and yield outputs has often been more to the fore than biological robustness or mechanistic rigor in components. Predicting ontogeny establishes the developmental time base relevant to perception and effects of stresses (e.g., water limitation around flowering, high temperature during grain-filling, etc.). Predicting canopy dynamics well is important in capturing the patterns of potential water use throughout the season. Canopy leaf area and the associated cover, in conjunction with environmental factors (e.g., radiation, vapor pressure deficit), determine demand for water. Ability of the crop to meet this demand can be related to root depth, soil water content, and extraction capacity in each soil layer occupied by roots. The sunflower model of Chapman et al (1993) details a generic water supply/demand framework of this nature but numerous other models with varying approaches (see review of Hammer et al 2002) would be adequate for environmental characterization.

A greater degree of biological robustness is required for modeling the physiology and genetics of complex traits (Hammer et al 2002, 2006). Dissection of the underlying components associated with function and control of complex traits and projection of their effects onto the TPE requires biological realism. Models need to be sufficiently detailed so that important physiological linkages and interactions are simulated implicitly. They should incorporate a hierarchy of physiological processes and input variables based on experimental analyses (Tardieu 2003). The phenotype becomes an emergent consequence of variation in system architecture and control and its interaction with the environment. This requires that (1) the physiological modes of action of the traits are understood and quantified and (2) the model is sufficiently detailed and robust to realistically simulate the interactions with crop growth and development generated by expression of the trait in any particular environment (Hammer et al 1996). Tests of the integrative ability of the crop model to project consequences onto the TPE can range from qualitative sensibility testing of responses based on biological knowledge through to formal validation, when appropriate data are available (e.g., Messina et al 2006). Robust models could add significant value to discussions on the likely value of putative traits as breeding targets for indirect selection (e.g., Richards et al 2002, Morgan et al 2002) and to considerations of simpler targets/measures for the high-throughput phenotyping required for QTL modeling and forward selection in a breeding program (e.g., Reynolds et al 1998).

It is necessary to gain an understanding of the physiology and genetics of complex traits from studies in controlled genetic backgrounds. Tardieu (2003) used transgenic plants to link genetic responses to coefficients of a model of water flux through the plant. Messina et al (2006) used near-isogenic lines varying at specific loci to derive coefficients for a phenology model that could then be estimated via linear functions of the alleles present. This contrasts with initial attempts to use agronomic crop models by optimizing a range of model coefficients to best fit observed phenotypic variation among sets of diverse genotypes, which had limited success (White and Hoogenboom 1996). The modest predictive capabilities found highlighted the need to better understand the physiological basis of the genetic variation involved via studies with controlled genetic backgrounds before seeking such predictive capability across diverse material.

The ability to generate stable associations between model coefficients and QTLs provides another criterion for model realism and adequacy to deal with the physiology and genetics of complex traits (Welch et al 2005). Reymond et al (2003) were able to achieve stable QTLs for their ecophysiological model of leaf elongation rate in maize. Similarly, Messina et al (2006) and Yin et al (2005) found stable QTLs for photo-thermal phenology models for soybean and barley, respectively. However, Yin et al (1999) were unable to find stable associations with QTLs for a study on specific leaf area (SLA, cm^2 g^{-1}) in barley. This suggested a lack of validity with which the crop model architecture and associated coefficients captured and integrated the physiological basis of the genetic variation. The barley model used in their study simulated leaf expansion as the product of carbohydrate partitioning to the leaf and SLA. Tardieu et al (1999) presented a modeling framework to explore whether leaf expansion was a consequence of specific leaf area or vice-versa. They were able to conclude the latter and argued that leaf expansion should be modeled independently of the plant carbon budget and that it was largely driven by temperature. Despite now having this enhanced understanding of control of leaf expansion in cereals, many crop simulation models continue to erroneously use the SLA-driven approach. However, this would likely have few consequences when using such models for agronomic or environmental characterization purposes.

Kitano (2004) discusses robustness as a fundamental feature of complex evolvable systems, such as biological organisms. He notes that system controls and modularity are basic features providing system robustness and that system control is the prime mechanism for coping with environmental perturbations. Attention to these aspects is likely to be important in the progression to the type of models most suited to study of the physiology and genetics of complex traits. This notion accords with the separation of physical and control equations in plant models (Tardieu 2003) and with the motivation behind the ongoing development of the APSIM modular generic crop routines (Wang et al 2002). The latter is designed to capture advances in knowledge as they occur, while retaining parsimony in approach to the G × M × E modeling objective. Our current research is designed to test the ability of this type of model to generate more stable associations between model coefficients and QTLs.

Concluding remarks

We suggest that an integrated systems approach to crop improvement that incorporates advanced technologies in molecular markers, statistics, bioinformatics, and crop physiology and modeling is likely to significantly enhance the efficiency of crop improvement. We discuss the design of such a system and consider the linking role of crop ecophysiology and modeling. A role of modeling in environmental characterization to support weighted selection is clear. It also seems clear that physiology and modeling will contribute significantly in the area of complex traits. The exact nature of this contribution is still emerging and is the focus of ongoing research. Attention to biological robustness in modeling will likely assist in this regard. While an integrated systems approach is in its infancy, we expect that the potential benefits and further technology developments will likely enhance its rate of development. To this end, we are simultaneously pursuing the development and implementation of an integrated systems approach to crop improvement in the Australian sorghum program.

References

Aggarwal PK, Kropff MJ, Cassman KG, Ten Berge HFM. 1997. Simulating genotypic strategies for increasing rice yield potential in irrigated, tropical environments. Field Crops Res. 51:5-17.

Asseng S, van Herwaarden AF. 2003. Analysis of the benefits to wheat yield from assimilates stored prior to grain filling in a range of environments. Plant Soil 256:217-229.

Blazquez M. 2000. Flower development pathways. J. Cell Sci. 113:3547-3548.

Boote KJ, Kropff MJ, Bindraban PS. 2001. Physiology and modelling of traits in crop plants: implications for genetic improvement. Agric. Syst. 70:395-420.

Borrell AK, Hammer GL, Henzell RG. 2000. Does maintaining green leaf area in sorghum improve yield under drought? 2. Dry matter production and yield. Crop Sci. 40:1037-1048.

Borrell AK, Hammer GL, van Oosterom EJ. 2001. Stay-green: a consequence of the balance between supply and demand for nitrogen during grain filling? Ann. Appl. Biol. 138:91-95.

Chapman SC, Cooper MC, Hammer GL, Butler D. 2000a. Genotype by environment interactions affecting grain sorghum. II. Frequencies of different seasonal patterns of drought stress are related to location effects on hybrid yields. Aust. J. Agric. Res. 50:209-222.

Chapman SC, Cooper M, Podlich D, Hammer GL. 2003. Evaluating plant breeding strategies by simulating gene action and dryland environment effects. Agron. J. 95:99-113.

Chapman SC, Hammer GL, Butler D, Cooper M. 2000b. Genotype by environment interactions affecting grain sorghum. III. Temporal sequences and spatial patterns in the target population of environments. Aust. J. Agric. Res. 50:223-234.

Chapman SC, Hammer GL, Meinke H. 1993. A sunflower simulation model. I. Model development. Agron. J. 85:725-735.

Chapman SC, Hammer GL, Podlich DW, Cooper M. 2002. Linking bio-physical and genetic models to integrate physiology, molecular biology and plant breeding. In: Kang M, editor. Quantitative genetics, genomics, and plant breeding. Wallingford (UK): CAB International. p 167-187.

Chenu K, Chapman SC, Hammer GL, McLean G, Tardieu F. 2008. Short term responses of leaf growth rate to water deficit scale up to whole plant and crop levels: an integrated modelling approach in maize. Plant Cell Environ. 31:378-391.

Condon AG, Richards RA, Rebetzke GJ, Farquhar GD. 2002. Improving intrinsic water-use efficiency and crop yield. Crop Sci. 42:122-131.

Cooper M, Chapman SC, Podlich DW, Hammer GL. 2002. The GP problem: quantifying gene-to-phenotype relationships. In Silico Biol. 2:151-164.

Cooper M, Hammer GL. 1996. Synthesis of strategies for crop improvement. In: Cooper M, Hammer GL, editors. Plant adaptation and crop improvement. CAB International, ICRISAT, and IRRI. p 591-623.

Cooper M, Hammer GL. 2005. Complex traits and plant breeding: can we understand the complexities of gene-to-phenotype relationships and use such knowledge to enhance plant breeding outcomes? Aust. J. Agric. Res. 56:869-872.

Cooper M, Podlich DW, Smith OS. 2005. Gene-to-phenotype models and complex trait genetics. Aust. J. Agric. Res. 56:895-918.

Dingkuhn M, Luquet D, Quilot B, Reffye PD. 2005. Environmental and genetic control of morphogenesis in crops: towards models simulating phenotypic plasticity. Aust. J. Agric. Res. 56:1289-1302.

Dong Z. 2003. Incorporation of genomic information into the simulation of flowering time in *Arabidopsis thaliana*. Ph.D. dissertation. Kansas State University, Manhattan, Kan., USA.

Duvick DN, Smith JSC, Cooper M. 2004. Long-term selection in a commercial hybrid maize breeding program. Plant Breed. Rev. 24:109-151.

Hammer G, Chapman S, van Oosterom E, Podlich D. 2005. Trait physiology and crop modelling as a framework to link phenotypic complexity to underlying genetic systems. Aust. J. Agric. Res. 56:947-960.

Hammer G, Cooper M, Tardieu F, Welch S, Walsh B, van Eeuwijk F, Chapman S, Podlich D. 2006. Models for navigating biological complexity in breeding improved crop plants. Trends Plant Sci. 11:587-593.

Hammer GL, Butler D, Muchow RC, Meinke H. 1996. Integrating physiological understanding and plant breeding via crop modelling and optimization. In: Cooper M, Hammer GL, editors. Plant adaptation and crop improvement. CAB International, ICRISAT, and IRRI. p 419-441.

Hammer GL, Kropff MJ, Sinclair TR, Porter JR. 2002. Future contributions of crop modelling – from heuristics and supporting decision-making to understanding genetic regulation and aiding crop improvement. Eur. J. Agron. 18:15-31.

Hammer GL, Muchow RC. 1994. Assessing climatic risk to sorghum production in water-limited subtropical environments. I. Development and testing of a simulation model. Field Crops Res. 36:221-234.

Hammer GL, Sinclair TR, Chapman S, van Oosterom E. 2004. On systems thinking, systems biology and the in silico plant. Plant Physiol. 134:909-911.

Hammer GL, Vanderlip RL. 1989. Genotype by environment interaction in grain sorghum. III. Modeling the impact in field environments. Crop Sci. 29:385-391.

Henzell RG, Jordan DR. 2006. History of grain sorghum breeding in Australia, including the development of resistances to midge, drought and ergot. Invited paper, 5th Australian Sorghum Conference, 30 Jan.-2 Feb. 2006, Gold Coast, Australia.

Jaccoud D, Peng K, Feinstein D, Kilian A. 2001. Diversity arrays: a solid state technology for sequence information independent genotyping. Nucl. Acids Res. 29(4): e25.

Jordan DR, Hammer GL, Henzell RG. 2006. Breeding for yield in the DPI&F breeding program. Invited paper, 5th Australian Sorghum Conference, 30 Jan.-2 Feb. 2006, Gold Coast, Australia.

Jordan DR, Tao YZ, Godwin ID, Henzell RG, Cooper M, McIntyre CL. 2004. Comparison of identity by descent and identity by state for detecting genetic regions under selection in a sorghum pedigree breeding program. Mol. Breed. 14:441-454.

Kim HK, van Oosterom E, Luquet D, Dingkuhn M, Hammer GL. 2006. Physiology and genetics of tillering. Contributed paper, 5th Australian Sorghum Conference, 30 Jan.-2 Feb. 2006, Gold Coast, Australia.

Kitano H. 2004. Biological robustness. Nature Rev. Gen. 5:826-837.

Koornneef M, Alonso-Blanco C, Peeters AJM, Soppe W. 1998. Genetic control of flowering time in *Arabidopsis*. Ann. Rev. Plant Physiol. Plant Mol. Biol. 49:345-370.

Leon AJ, Lee M, Andrade FH. 2001. Quantitative trait loci for growing degree days to flowering and photoperiod response in sunflower (*Helianthus annuus* L.). Theor. Appl. Genet. 102:497-503.

Löffler C, Wei J, Fast T, Gogerty J, Langton S, Bergman M, Merrill B, Cooper M. 2005. Classification of maize environments using crop simulation and geographic information systems. Crop Sci. 45:1708-1716.

Luquet D, Dingkuhn D, Kim HK, Tambour L, Clement-Vidal A. 2006. *EcoMeristem*, a model of morphogenesis and competition among sinks in rice. 1. Concept, validation and sensitivity analysis. Funct. Plant Biol. 33:309-323.

Lynch M, Walsh B. 1998. Genetics and analysis of quantitative traits. Sunderland, Mass. (USA): Sinauer Associates, Inc.

Manschadi AM, Christopher J, deVoil P, Hammer GL. 2006. The role of root architectural traits in adaptation of wheat to water-limited environments. Funct. Plant Biol. 33:823-837.

Messina CD, Jones JW, Boote KJ, Vallejos CE. 2006. A gene-based model to simulate soybean development and yield response to environment. Crop Sci. 46:456-466.

Morgan PW, Finlayson SA. 2000. Physiology and genetics of maturity and height. In: Smith CW, Frederiksen RA, editors. Sorghum: origin, history, technology and production. John Wiley & Sons, New York. p 227-259.

Morgan PW, Finlayson SA, Childs KL, Mullet JE, Rooney WL. 2002. Opportunities to improve adaptability and yield in grasses: lessons from sorghum. Crop Sci. 42:1791-1799.

Muchow RC, Cooper M, Hammer GL. 1996. Characterising environmental challenges using models. In: Cooper M, Hammer GL, editors. Plant adaptation and crop improvement. CAB International, ICRISAT, and IRRI. p 349-364.

Muchow RC, Hammer GL, Carberry PS. 1991. Optimising crop and cultivar selection in response to climatic risk. In: Muchow RC, Bellamy JA, editors. Climatic risk in crop production: models and management for the semiarid tropics and subtropics. Wallingford (UK): CAB International. p 235-262.

Podlich D, Cooper M. 1998. QU-GENE: a simulation platform for quantitative analysis of genetic models. Bioinformatics 14:632-653.

Podlich DW, Cooper M, Basford KE 1999. Computer simulation of a selection strategy to accommodate genotype-environment interactions in a wheat recurrent selection programme. Plant Breed. 118:17-28.

Podlich DW, Winkler CR, Cooper M. 2004. Mapping as you go: an effective approach for marker-assisted selection of complex traits. Crop Sci. 44:1560-1571.

Reymond M, Muller B, Leonardi A, Charcosset A, Tardieu F. 2003. Combining quantitative trait loci analysis and an ecophysiological model to analyze the genetic variability of the responses of maize leaf growth to temperature and water deficit. Plant Physiol. 131:664-675.

Reynolds MP, Singh RP, Ibrahim A, Ageeb OAA, Larqué-Saavedra A, Quick JS. 1998. Evaluating physiological traits to complement empirical selection for wheat in warm environments. Euphytica 100:85-94.

Richards RA, Rebetzke GJ, Condon AG, van Herwaarden AF. 2002. Breeding opportunities for increasing the efficiency of water use and crop yield in temperate cereals. Crop Sci. 42:111-121.

Sinclair TR, Hammer GL, van Oosterom EJ. 2005. Potential yield and water-use efficiency benefits in sorghum from limited maximum transpiration rate. Funct. Plant Biol. 32: 945-952.

Sinclair TR, Muchow RC. 2001. System analysis of plant traits to increase grain yield on limited water supplies. Agron. J. 93:263-270.

Somerville C, Dangl J. 2000. Plant biology in 2010. Science 290:2077-2078.

Spitters CJT, Schapendonk AHCM. 1990. Evaluation of breeding strategies for drought tolerance in potato by means of crop growth simulation. Plant Soil 123:193-203.

Tao YZ, Henzell RG, Jordan DR, Butler DG, Kelly AM, McIntyre CL. 2000. Identification of genomic regions associated with staygreen in sorghum by testing RILs in multiple environments. Theor. Appl. Genet. 100:1225-1232.

Tardieu F. 2003. Virtual plants: modelling as a tool for the genomics of tolerance to water deficit. Trends Plant Sci. 8:9-14.

Tardieu F, Granier C, Muller B. 1999. Modelling leaf expansion in a fluctuating environment: are changes in specific leaf area a consequence of changes in expansion rate? New Phytol. 143:33-43.

van Eeuwijk FA, Malosetti M, Yin X, Struik PC, Stam P. 2005. Statistical models for genotype by environment data: from conventional ANOVA models to eco-physiological QTL models. Aust. J. Agric. Res. 56:883-894.

van Oosterom EJ, Bidinger FR, Weltzien ER. 2003. A yield architectural framework to explain adaptation of pearl millet to environmental stress. Field Crops Res. 80:33-56.

van Oosterom EJ, Weltzien ER, Yadav OP, Bidinger FR. 2006. Grain yield components of pearl millet under optimum conditions can be used to identify germplasm with adaptation to arid zones. Field Crops Res. 96:407-421.

van Oosterom EJ, Hammer GL, Chapman SC, Doherty A. 2006. A simple gene network model for photoperiodic response of floral transition in sorghum can generate genotype-by-environment interactions in grain yield at the crop level. In: Mercer CF, editor. Breeding for success: diversity in action. Proceedings of the 13th Australasian Plant Breeding Conference, Christchurch, New Zealand, 18-21 April 2006. p 687-691. CD.

Verbyla AP, Eckermann PJ, Thompson R, Cullis B. 2003. The analysis of quantitative trait loci in multi-environment trials using a multiplicative mixed model. Aust. J. Agric. Res. 54:1395-1408.

Wade LJ, Douglas ACL, Bell KL. 1993. Variation among sorghum hybrids in the plant density required to maximise grain yield over environments. Aust. J. Exp. Agric. 33:185-191.

Wang E, Robertson MJ, Hammer GL, Carberry PS, Holzworth D, Meinke H, Chapman SC, Hargreaves JNG, Huth NI, McLean G. 2002. Development of a generic crop model template in the cropping system model APSIM. Eur. J. Agron. 18:121-140.

Welch SM, Roe JL, Dong Z. 2003. A genetic neural network model for flowering time control in *Arabidopsis thaliana*. Agron. J. 95:71-81.

Welch SM, Dong Z, Roe JL, Das S. 2005. Flowering time control: gene network modelling and the link to quantitative genetics. Aust. J. Agric. Res. 56:919-936.

Whish J, Butler G, Castor M, Cawthray S, Broad I, Carberry P, Hammer G, McLean G, Routley R, Yeates S. 2005. Modelling the effects of row configuration on sorghum yield in north-eastern Australia. Aust. J. Agric. Res. 56:11-23.

White JW, Hoogenboom G. 1996. Simulating effects of genes for physiological traits in a process-oriented crop model. Agron. J. 88:416-422.

Yin X, Kropff M, Stam P. 1999. The role of ecophysiological models in QTL analysis: the example of specific leaf area in barley. Heredity 82:415-421.

Yin X, Struik PC, Kropff MJ. 2004. Role of crop physiology in predicting gene-to-phenotype relationships. Trends Plant Sci. 9:426-432.

Yin X, Struik PC, van Eeuwijk FA, Stam P, Tang J. 2005. QTL analysis and QTL-based prediction of flowering phenology in recombinant inbred lines of barley. J. Exp. Bot. 56:967-976.

Notes

Authors' addresses: Agricultural Production Systems Research Unit (APSRU), School of Land and Food Sciences, The University of Queensland, Brisbane, Queensland 4072, Australia; Hermitage Research Station, Queensland Department of Primary Industries, Warwick, Queensland 4370, Australia, e-mail: g.hammer@uq.edu.au.

Acknowledgments: The ideas summarized here have evolved over a number of years and have been aided by input during discussions with many others. In particular, ongoing discussions with Andrew Borrell, Scott Chapman, Mark Cooper, Bob Henzell, Dean Podlich, Francois Tardieu, Erik van Oosterom, Fred van Eeuwijk, Bruce Walsh, and Steve Welch following the 4th International Crop Science Congress symposium in Brisbane in 2004 have been particularly influential. We also thank Greg McLean and Brendan Power for assistance with sorghum G × M × E simulations and analyses.

Management of rainfed rice systems

Drought-prone rainfed lowland rice in Asia: limitations and management options

S.M. Haefele and B.A.M. Bouman

The rice-based rainfed lowland system in Asia covers about 45 million hectares, almost 30% of the total rice area worldwide. Rice (*Oryza sativa* L.) is the main crop in this system and it grows in bunded fields that are flooded for at least part of the season. Overall, drought stress is considered the most important limitation to production in rainfed lowlands and is estimated to frequently affect about 19 to 23 million hectares. Severe and regular droughts affect mainly rainfed lowlands in South and Southeast Asia, but regional weather patterns, topography, and soil characteristics cause considerable drought-risk variations within and beyond these regions. In addition, drought-prone environments are often simultaneously affected by other abiotic stresses such as submergence, adverse soil conditions, pests, and weeds. Two main management strategies for drought-stress alleviation in rice can be distinguished. The first strategy is based on management options that allow escaping drought by either avoiding dry periods or by providing access to additional water resources. The second strategy is to moderate drought by reducing unproductive water losses and thereby "saving" water for productive transpiration. Both strategies contain several management options that offer considerable scope for improving drought-prone rainfed lowlands; however, direct seeding and improved nutrient management are probably the most widely applicable options. "Aerobic rice" as a new system of rice cultivation is still under development but promises considerable opportunities in specific target environments within the rainfed lowlands of Asia.

The rainfed lowland system in Asia covers about 45 million hectares, almost 30% of the total rice area worldwide (Haefele and Hijmans 2007) (Fig. 1). In South Asia, rainfed lowlands are concentrated in India (16.1 million ha) and Bangladesh (5.1 million ha). In Southeast Asia, important areas of rainfed rice are located in Thailand, Indonesia, Vietnam, Myanmar, Cambodia, and the Philippines (8.2, 4.0, 2.9, 2.4, 1.6, and 1.3 million ha, respectively). In East and northeast Asia, only China has a larger area of rainfed lowland rice (1.8 million ha). Conventionally, two main categories of rainfed lowlands are distinguished: shallow rainfed lowlands where water depths usually fluctuate between 0 and 0.3 m and intermediate rainfed lowlands with water

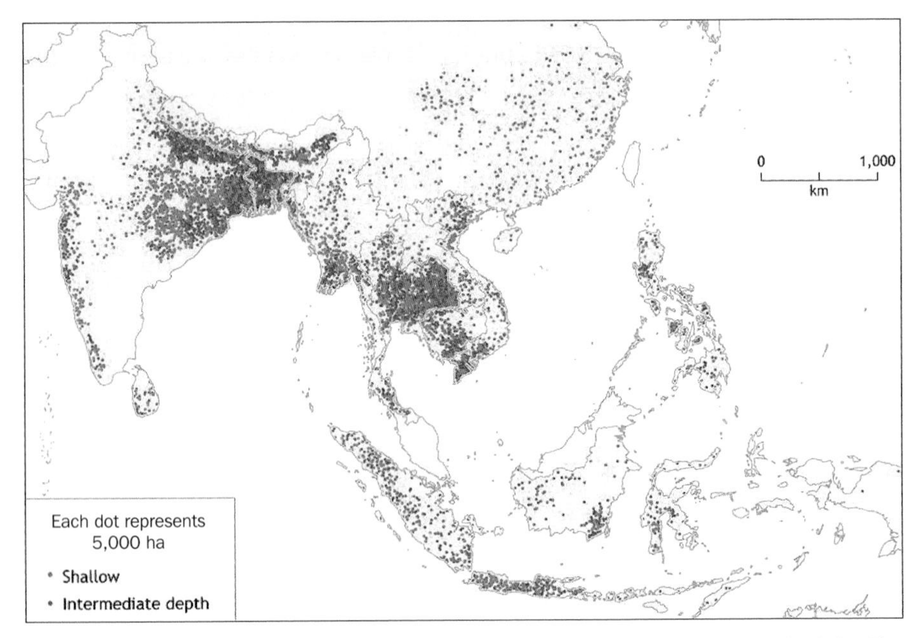

Fig. 1. Distribution of shallow and intermediate rainfed lowland ecosystems in Asia. The map is based on an updated and revised version of the database developed by Huke and Huke (1997).

depths between 0.3 and 1.0 m. Shallow rainfed lowlands are almost equally distributed between South and Southeast Asia (15.3 and 16.1 million ha, respectively), whereas intermediate rainfed lowlands are more frequent in South Asia than in Southeast Asia (6.9 and 3.7 million ha, respectively). Although rainfed lowlands include favorable environments with conditions similar to those of irrigated systems, most of the area in this agroecosystem faces various biophysical constraints to rice production. In addition, the main climatic factor in most of the region is the monsoon, which deposits >80% of the rainfall within just a few months, not allowing a rainfed crop in the dry season. In the rainy season, the options for crops other than rice are often limited because at least temporary flooding due to heavy rains does occur in most years. As a consequence of the resulting low and unstable system productivity, poverty is widespread and often severe in communities largely dependent on rainfed rice.

To improve the productivity and production of drought-prone systems, a combination of improved varieties and crop management is necessary. Ludlow (1989) defined three principal mechanisms of genotypic adaptations contributing to increased yields in water-limited environments: (1) drought escape, in which the crop completes its life in the time when no water limitation occurs; (2) drought avoidance, in which the crop maximizes its water uptake and minimizes its water loss; and (3) drought tolerance, in which the crop continues to grow and function at reduced water contents. Two other options difficult to group in this system are a higher yield potential and improved tolerance of other simultaneously occurring stresses (Jearakongman et al 1995, Mackill 1986). No widely accepted analogous terminology has been proposed for crop management options that can reduce crop losses due to water limitation. Nevertheless, a wide range of management options has been developed, although most of the work has been done for nonrice crops. Debaeke and Aboudrare (2004) described the main strategies in similar terms as drought escape or avoidance, crop rationing, and irrigation. Tuong (1999) summarized options for productive water use for irrigated rice and described the main strategies as increasing yield per unit of water evapotranspired, reducing nonbeneficial water losses, and using rainfall more effectively. And, although a considerable number of studies were conducted to evaluate management options for their potential to reduce drought stress in rainfed lowland rice, no equally comprehensive studies for this agroecosystem are available.

Improved germplasm is without doubt a necessary first step to significantly reduce risk and increase productivity in drought-prone rice-based lowlands. But optimal use of such new germplasm will require a combination with adequate and, if possible, improved management techniques. Therefore, our objective here is to present an overview of important and often limiting characteristics of rainfed lowlands in Asia and to describe important management options to reduce the effect of drought on rice. In addition, we will present and discuss some recent findings from "aerobic rice" research because they provide some new insights into rice production under water-limited conditions.

Important characteristics of the environment

According to the agroecological zone classification system (FAO), most rice-based rainfed lowlands in Asia are situated in the warm subhumid tropics (eastern India, Myanmar, Thailand) and the warm humid tropics (Laos, Cambodia, Vietnam, Bangladesh, the Philippines, Indonesia, Sri Lanka, Malaysia). Accordingly, regular and sometimes severe droughts affect mainly rainfed lowlands in eastern India, northeast Thailand, and parts of Myanmar and Laos (Fig. 2).

However, regional weather patterns, topography, and soil characteristics cause considerable drought-risk variations within and beyond these regions. In general, intermediate rainfed lowlands occur in the lower part of the landscape, usually in the vicinity of larger rivers. In contrast, shallow rainfed lowlands are mostly situated outside of the larger floodplains and the typical topomorphology is an undulating landscape with small to medium height differences within the toposequence. Depending on the slope and soil characteristics, this can have considerable effects on plant available water and nutrient resources (Fig. 3). On upper terraces, a coarser texture can contribute to lower water and nutrient retention capacity of the soil and lower levels of indigenous soil fertility. Water movement down the slope further reduces available water resources and removes nutrients. The groundwater level is often below the main rooting zone and contributes little to plant available water resources. On medium terraces, water and nutrient losses to lower positions can be balanced by inputs from upslope. On lower terraces and valley bottoms, water and nutrient losses are usually

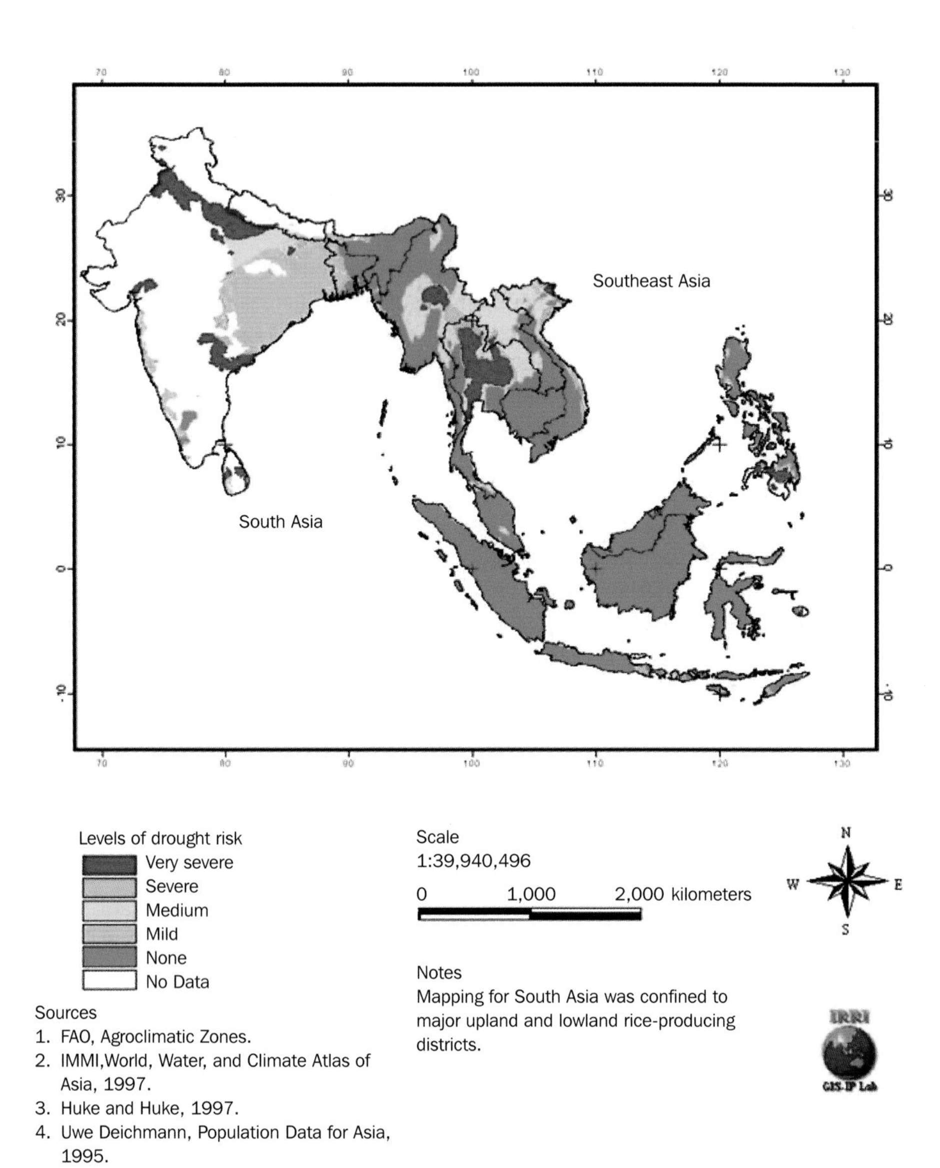

Fig. 2. Distribution and severity of drought risk for regions with significant areas of rainfed rice in Asia. Ranking of drought severity was developed based on number of humid months and critical thresholds of number of rainy days in the preceding and post months of the humid period. Adapted from Kam et al (2000).

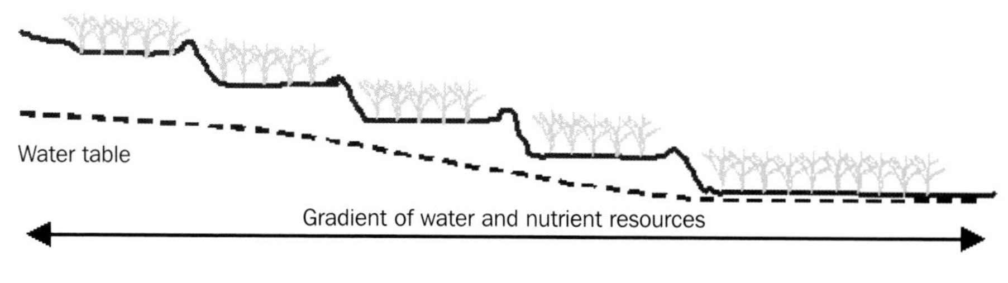

Upper terraces	Medium terraces	Lower terraces/valley bottom
Water		
High hydraulic head, higher percolation rate, lower soil water storage capacity, low water table	High hydraulic head, intermediate percolation rate/soil water storage capacity/water table	Low hydraulic head, lower percolation rate, higher soil water storage capacity, high water table
High lateral water losses, no/low water input from above, limited access to groundwater	Medium lateral water losses, medium water input from above, medium access to groundwater	Low lateral water losses, considerable water input from above, good access to groundwater
Soil		
Coarser texture, lower CEC, lower SOM concentrations	Intermediate	Finer texture, higher SOM concentrations, higher CEC
Lower water and nutrient retention, negative natural nutrient balance, lower soil fertility	Intermediate	Higher water and nutrient retention, positive natural nutrient balance, higher soil fertility
Effect on risk		
Higher drought risk, shorter favorable season, higher weed competition	Intermediate	Lower drought risk, longer season, better natural weed control, high submergence risk
Effect on management		
Direct seeding can be preferable, nutrients can be equally limiting as water; small fertilizer rates to make optimal use of available water resources, OM applications can be important to improve the soil	Intermediate	Transplanting can be preferable, fertilizer management can target higher yields, lower risk increases the incentive to use inorganic fertilizers
Preferable varietal traits		
Very short duration, drought tolerance, greater rooting depth	Drought tolerance	Drought and submergence tolerance, at least intermediate height, short to long duration

Fig. 3. Topographic effects in rainfed lowland environments on water and nutrient resources and related crop management issues.

smaller than inputs from above. The water table is often close to the surface and the main rooting horizon. Soils generally have a higher indigenous soil fertility because of a finer texture, nutrient inputs from above, and frequently higher soil organic matter contents.

The consequences of these resource gradients are obviously a higher drought risk and more severe nutrient limitations for upper terraces, and, due to runoff from the slopes and upstream, a higher submergence risk for lower terraces. Water accumulation in the lower parts of the landscape frequently enables/imposes an earlier crop establishment and may cause harvest delays. Weeds will often have a competitive advantage on upper terraces because of shorter durations of flooded conditions and drought stress, whereas they are better suppressed by the floodwater layer in the lower fields. These processes and consequences of toposequences are obviously not limited to unfavorable rainfed lowlands and can affect yields equally in favorable lowlands (Pane et al 2005, Vityakorn 1989, Homma et al 2003).

Overall, abiotic stresses constitute the most important constraints to productivity and intensification of rainfed lowland rice. Drought stress is the most important limitation to production and is estimated to frequently affect about 19 to 23 million ha of rice land (Garrity et al 1986). Flood-prone rice ecosystems include about 11 million ha of shallow rainfed lowlands and about 11.6 million ha of intermediate rainfed lowlands (Huke and Huke 1997, Mackill et al 1996) but the area affected annually varies greatly. Additional constraints arise from the widespread occurrence of soil quality problems. Haefele and Hijmans (2007) estimated that about one-third of rainfed lowland rice is grown on relatively fertile soils, slightly less than one-third grows on soils with low indigenous soil fertility, and slightly more than one-third grows on soils with considerable soil constraints often combined with very low soil fertility (Fig. 4), including problem soils such as acid sulfate soils or saline soils. Rainfed lowland rice in Southeast Asia is much more likely to be constrained by poor soils with various soil constraints than in South and East/northeast Asia.

Management options for water-limited rainfed lowlands

Drought has long been recognized as the primary constraint to rainfed rice production. In the rainfed lowlands of Asia, rice usually grows in bunded fields that are flooded for at least part of the season and the bunds are generally the only available measure of water control. Drought can occur at any time during crop growth and is highly variable in space and in time. However, regional weather patters can cause the dominance of one of the three main types of drought identified (Chang et al 1979, Fukai and Cooper 1995):

1. An early-season drought that occurs during vegetative growth,
2. An intermittent drought that can occur repeatedly between tillering and mid-grain filling, and
3. A late drought that occurs during flowering and grain filling.

An understanding of the water balance of a rice field helps in identifying management interventions to mitigate drought. Water input to a rice field is from rainfall,

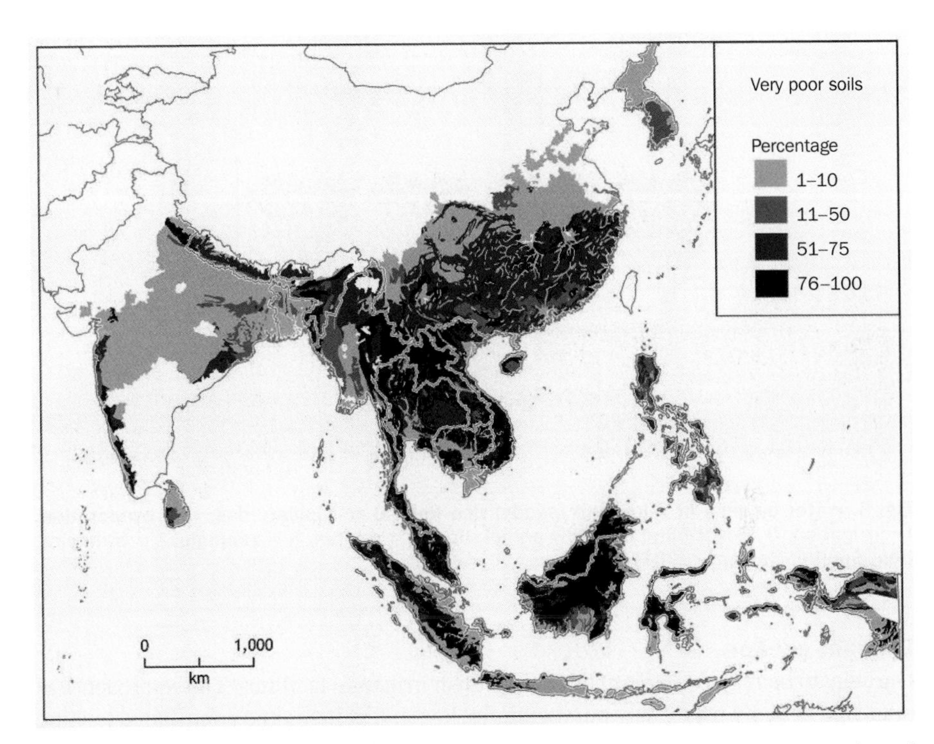

Fig. 4. Distribution of "very poor soils" in areas where rainfed lowland rice occurs based on the Fertility Capability Soil Classification system and the FAO soil map of the world (from Haefele and Hijmans 2007). Crop growth on these soils is potentially limited by combinations of low nutrient reserves, low CEC, Al toxicity, and/or high P fixation. Generally, these are highly weathered soils with very limited indigenous nutrient supplies, low nutrient retention capacity, frequent and often severe P deficiency, acidic to very acidic soil reaction (pH < 5), and Fe/Al toxicities.

capillary rise, and, if available, irrigation. Outputs are by evaporation from the ponded water or the soil, transpiration by the plants, over-bund flow, and lateral seepage and vertical percolation (Fig. 5). The occurrence of drought can loosely be defined as a situation in which there is not enough water in the root zone to allow the plants to transpire according to evaporative demand. Management strategies for drought stress alleviation are either to escape drought (by increased water input or by avoiding dry periods) or to moderate drought by reducing nontranspirational outflows so that more water is left for transpiration. Bouman (2007) and Debaeke and Aboudrare (2004) systematically described generic options to implement these strategies, and here we give specific examples for rainfed lowland rice. However, several management options for drought stress alleviation function based on both strategies and are discussed below under their main mechanism.

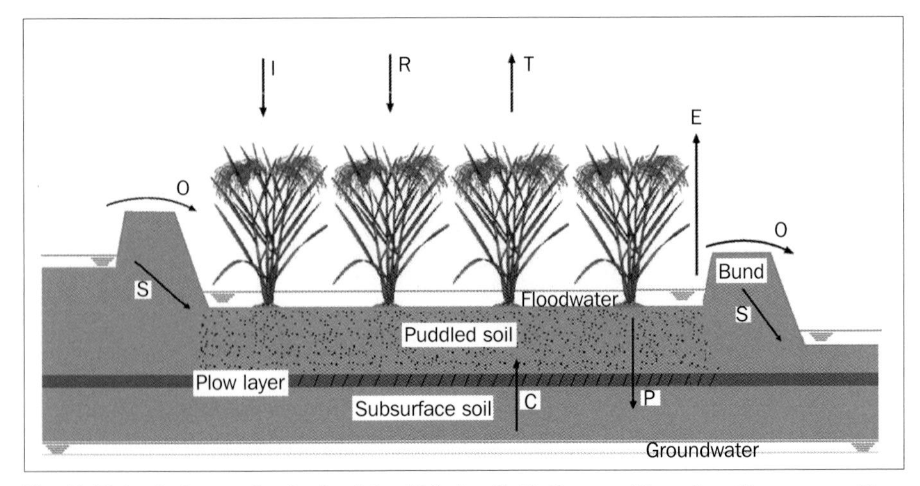

Fig. 5. Water balance of a lowland (paddy) rice field. C = capillary rise, E = evaporation, I = irrigation, O = over-bund flow, P = percolation, R = rainfall, S = seepage, T = transpiration. Source: Bouman (2007).

Drought escape

One way to increase water inputs is to establish irrigation facilities. This approach was practiced in many previously purely rainfed lowland systems and contributed to large productivity increases. Examples are the establishment of large numbers of tube wells in, for example, eastern India and Bangladesh (Singh et al 2003). In northeast Thailand, some regions of eastern India, the Philippines, and Indonesia, considerable numbers of village and farm ponds for rainwater harvesting were established (Patamatamkul 2001, Bhuiyan 1994, Pal and Bhuiyan 1995). Establishing (supplementary) irrigation facilities is a good option to avoid drought in some areas; however, limited groundwater resources and the relatively high costs of rainwater harvesting make it impossible to provide sufficient irrigation water throughout the huge rainfed areas.

If the drought pattern is relatively stable and of the early-season or late-season type, drought avoidance can be achieved by adjusting the cropping season to the time when water availability from rainfall is best or by reducing the length of the cropping season with the help of shorter-duration varieties. Although these options seem obvious, farmers' varietal choice often does not appear to be optimal in this respect. However, even in drought-prone environments, drought stress is often only one of several stresses the crop has to cope with during the season and farmers' variety selection criteria include other characteristics, including yield potential and consumer or market preferences. Therefore, varieties preferred often represent a compromise and several varieties widespread in drought-prone rainfed lowlands are of comparatively long duration (e.g., KDML105, Mahsuri, Swarna). In addition, some rainfed lowlands seem to lack adequate short-duration material, for example, rainfed lowland farmers in large parts of Cambodia use (traditional) varieties with 160 to 180 days' duration. Providing germplasm satisfying farmers' various other demands for varietal characteristics *and* having shorter duration will therefore continue to reduce the effect of drought in many rainfed lowlands.

The last option in this context is to increase access to groundwater resources. In lowland rice fields, the roots of rice are usually restricted to the puddled topsoil and they do not penetrate the plow sole. Breaking the plow sole to allow roots to grow deeper and either tap deeper soil water or the groundwater may be an option (Samson and Wade 1998), but the danger is that deep percolation increases and subsequent rains will no longer be ponded on the surface. Improved access to capillary rise does not seem very promising in most circumstances.

Drought moderation through nutrient management

One important way to reduce unproductive water losses is better nutrient management. The basic principle is that improved nutrition makes the plants a stronger competitor for water (by increasing transpiration) against other outflows (Bouman 2007). Nutrient management is rarely seen as an option to mitigate drought stress although it may alleviate the effects of drought in some circumstances (Biswas et al 1982, Tanguilig and De Datta 1988, Otoo et al 1989, Zaman et al 1990). However, nutrients were repeatedly recognized as a major limiting factor in many rainfed lowlands (Wade et al 1999, Pandey 1998, Garrity et al 1986, Akbar et al 1986, Van Bremen and Pons 1978). If, during periods with good water supply, rice growth is limited by nutrients, this will cause higher unproductive water losses, mainly by evaporation. Assuming ample water supply, evaporation is high at the beginning of the season when the crop cover is sparse and becomes small when the ground is well shaded by the canopy. Inversely, transpiration is low at the beginning of the season and approaches maximum values (about 90% of the potential evapotranspiration) at a leaf area index of about 3 to 4 (Tanner and Sinclair 1983, Ehlers and Goss 2003). Bouman et al (2005) estimated for irrigated rice that about 30% of seasonal evapotranspiration is evaporation and 70% transpiration. However, small changes in that distribution may greatly affect yield. Tuong (1999) gave an example in which the fertilizer-induced decrease in evaporation from 41% to 29% of total evapotranspiration (without any change in total ET) increased rice yield from 2.1 to 4.8 t ha^{-1}. In this case, the high evaporation in the unfertilized treatment was caused solely by a slow and incomplete closure of the crop canopy, a situation that can be regularly observed in many rainfed lowlands. The improved crop growth by decreased evaporation in favor of increased transpiration is also known as "vapor shift," and is one of the main proposed mechanisms to improve yields and crop water productivity in semiarid environments (Falkenmark and Rockström 2004).

Nonetheless, inorganic fertilizer use in most rainfed lowlands is still very low, which is often explained by the high production risk due to abiotic stresses or the low yield response caused by abiotic stresses and/or the use of traditional varieties. To test the second hypothesis, we analyzed a large data set of fertilizer trials conducted in a variety of rainfed environments (Wade et al 1999). Selected data covered 37 different trial sites and three consecutive seasons at most sites: 9 sites in northeast Thailand, 8 sites in the Philippines, 8 sites in Indonesia, 4 sites in Bangladesh, 5 sites in India, and 3 sites in Laos. The term "site" as used here does not necessarily describe far-apart locations but includes close-by sites with clearly different water regimes, including 5 fully irrigated sites. About half of the sites were located on-station whereas the others were situated in farmers' fields. At each site, a set of fertilizer treatments was established in a randomized complete block design with three replications. Only results from two treatments are used in this analysis, namely, the PK treatment (inorganic P and K fertilizer only) and the NPK treatment (inorganic N, P, and K fertilizer applied). Fertilizer rates and application procedures were adjusted to regional recommendations but were identical within each location (for details, see Wade et al 1999). The data were grouped into two sets: one set for all data from northeast Thailand, where only traditional-type varieties were used, and a second set for all other sites, where modern varieties were grown.

To compare the average performance of the fully fertilized NPK treatment between sites and systems, several N-use efficiency indicators were calculated (Table 1). At all sites in northeast Thailand, the applied N rate was homogenous and low according to the regional recommendation (50 kg N ha^{-1}). In contrast, the N rate applied at the other rainfed lowland sites varied considerably (from 60 to 140 kg N ha^{-1}) and appears to be excessively high in the Philippines (140 kg N ha^{-1}). Exclusive use of traditional varieties in northeast Thailand explains the low harvest index, low internal efficiency of N, and low grain yield. The high N rate in the Philippines contributed to the low average agronomic efficiency of N (AEN) and N recovery efficiency for rainfed lowland sites outside of northeast Thailand. Nevertheless, the average AEN in the rainfed systems is only slightly below the average AEN reported for studies on site-specific nutrient management in irrigated systems (Witt 2003), and does not indicate a substantially lower yield response in rainfed systems. Assuming similar input and output prices, this would also suggest comparable economic returns to N fertilizer use (however, partial factor productivity would still be lower in most rainfed systems because grain yield without fertilizer is considerably lower).

To evaluate the effect of drought stress on fertilizer response, grain yields of both treatments (NPK and PK treatment) and for both groups of lowland sites were plotted against the average seasonal field water stress monitored at each site (Fig. 6). The smaller data set from northeast Thailand covers a large range of average field stress levels (Fig. 6A). The exclusive use of traditional varieties caused low yield levels and yield increases were small because of the low N rate applied (50 kg N ha^{-1}). All observations at the lowest water stress level (irrigated conditions) are from one site (Ubon) with very low indigenous soil fertility, explaining the low yield level there. Grain yield differences between the NPK and PK treatments do seem to decrease with increasing water stress but the yield gains due to N application were fairly stable between the water stress levels 1.7 and 2.3. At the highest water stress level (2.7), hardly any yield difference between both treatments was observed and the extremely low harvest index values (data not shown) indicate that drought stress at flowering caused substantial spikelet sterility. The larger data set covered only rainfed lowland sites outside of northeast Thailand where modern varieties were used (Fig. 6B). It shows generally higher yields, and yield gains resulting from N application were larger

Table 1. Mean values for the applied N rate, grain yield (GY), agronomic efficiency of N (AEN), recovery efficiency of applied N (REN), internal efficiency of N (IEN), and harvest index (HI) in the NPK treatment for northeast Thailand, all other rainfed lowland sites, and the respective data from a study in irrigated systems.

System		Rainfed lowland		Irrigated[a]
		Northeast Thailand	All other sites	
		n = 26	n = 87	n = 173
N rate	(kg ha^{-1})	50	102	110
Grain yield	(t ha^{-1})	2.4	4.1	5.6
AEN	(kg grain kg^{-1} Napl)	15.4	14.0	16.3
REN	(kg kg^{-1})	0.33	0.28	0.42
IEN	(kg grain kg^{-1} Nupt)	43	55	57
HI		0.28	0.43	0.48

[a]Data from Witt (2003), where each observation represents the average value from six successive crops. Shown are the results for the treatment with optimized site-specific nutrient management (SSNM).

(average N rate of 102 kg N ha^{-1}). The yield response to N application does not seem to decrease substantially with increasing water stress but the maximum water stress occurring was lower than in northeast Thailand.

It can be concluded that, with the exception of extreme drought or drought around flowering, water stress does not necessarily reduce fertilizer-use efficiency. But traditional varieties and/or limited water resources contribute to lower attainable yields. This causes lower total nutrient requirements to reach the attainable yield and, as a consequence, lower optimal and efficient N rates (Haefele et al 2007). It remains unclear whether fertilizer use can increase drought risk in drought-prone rice environments. In the case of a drought spell in the season, total evapotranspiration should not be affected by the crop biomass as long as the soil surface is wet. However, when the soil surface becomes dry and evaporation very small, a larger canopy will result in higher transpiration. This can lead to a faster decline in remaining plant-available water and higher drought damage, a phenomenon ("haying-off") known from upland crops relying mostly on water stored in the soil (Cantero-Martinez et al 1995). But fertilizer application was also shown to increase access and extraction of soil water, thereby perhaps counteracting higher transpiration needs (Viets 1962). The type of drought and the soil type will further modify the possible "risk" of fertilizer applications in water-limited environments. Given the scarce studies in rice examining this question, the risk related to fertilizer applications can currently not be assessed. Crop growth simulation models might offer an option to better evaluate water-by-nutrient interactions in a variety of typical rice environments (Boling et al 2007).

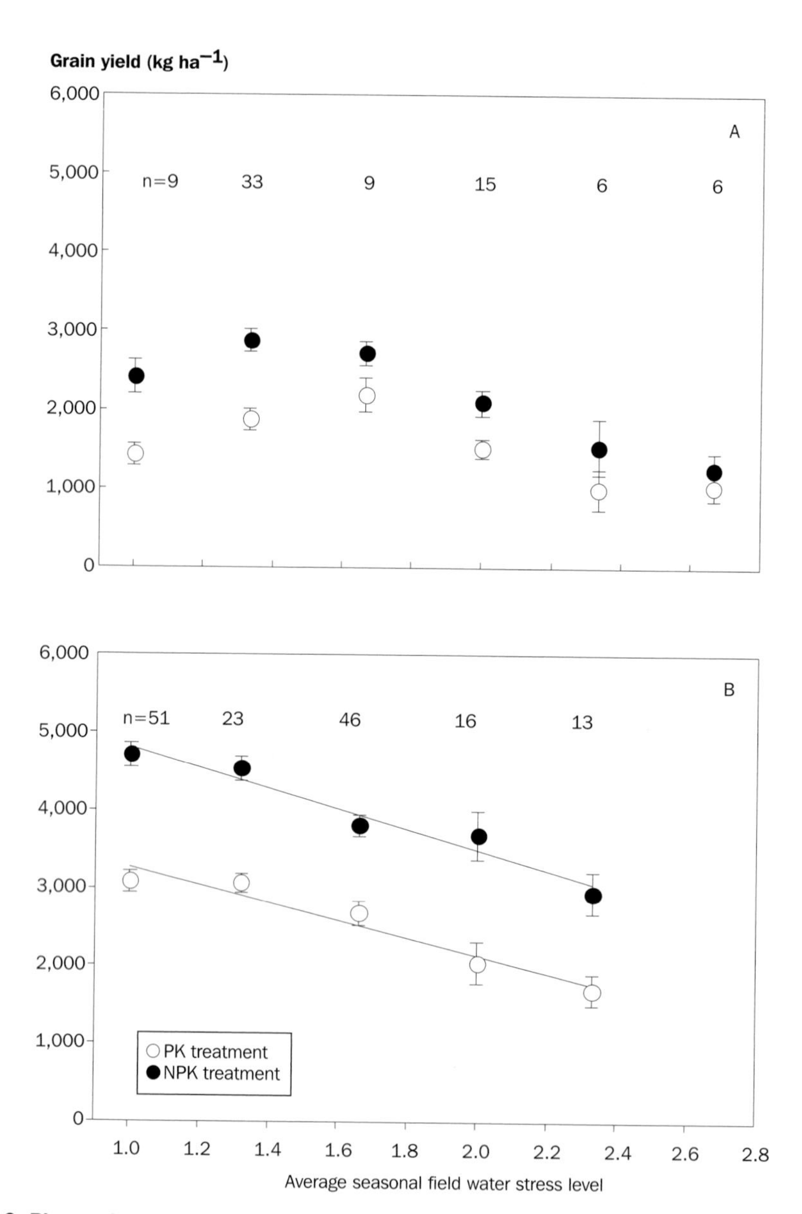

Fig. 6. Rice grain yields for NPK and PK treatments depending on the average field water stress observed during the season in northeast Thailand (A) and all other rainfed lowland sites (B), including sites in India, Philippines, Laos, Bangladesh, and Indonesia. The number of observations included in each data point are given in the upper part of each graph and are identical for NPK and PK treatments. Error bars indicate the standard error. Yield values for each treatment followed by the same letter are not significantly different at $P \leq 0.05$ according to Tukey's Studentized Range Test. Field water stress levels were scored according to 1 = flooded, 2 = wet soil surface, and 3 = dry soil surface.

Table 2. Average time of establishment and corresponding cumulative rainfall in five seasons (1995-2000) depending on rice establishment method in an on-station experiment at Raipur, Chhattisgarh, India. For all establishment methods, rice was grown under rainfed conditions without irrigation.

Establishment method[a]	Event[c]	Day of year	Cumulative rainfall in mm
DDS dry	Sowing	159	0
	Establishment[b]	177	131 ± 51
DDS moist	Sowing	180	134 ± 55
	Establishment[b]	195	261 ± 61
DDS biasi[b]	Sowing	174	113 ± 52
	Establishment[b]	193	255 ± 60
	Biasi-operation	219	532 ± 106
T	Transplanting	219	496 ± 112

[a]Establishment methods were dry direct seeding in lines and dry soil (DDS dry), dry direct seeding in lines and moist soil (DDS moist), dry direct seeding broadcast in moist soil and biasi operation (DDS biasi), and transplanting (T). [b]1998-2000 only. [c]Establishment refers here to a plant size comparable with the seedling size at transplanting.
Source: Rathore and Sahu (2002).

Drought moderation through direct seeding

Another big opportunity to reduce unproductive water losses is direct seeding. Dry direct seeding allows earlier establishment than transplanting, thus reducing deep percolation and evaporation losses from early-season rains. The advantages were analyzed by Rathore et al (n.d.) for four different establishment methods of rainfed rice in eastern India. Dry direct seeding, wet direct seeding, biasi (also known as *beushening* or *beusani*), and transplanting were compared. Biasi is a traditional rice establishment method, popular among farmers in the rainfed lowland rice areas of eastern India (about 12.8 million hectares). In the biasi system, dry rice seed is broadcast after the first rains followed at 20–35 days after emergence and when there is 5–10 cm of water on the fields by wet plowing (biasi) in order to control weeds (Fujisaka et al 1993).

Sowing dry seed in dry soil could be undertaken before the rains start, whereas sowing dry seeds in moist soil, as practiced with the biasi system, requires 112 mm of water on average (Table 2). In comparison, crop establishment by transplanting (including the necessary soil puddling for that purpose) and wet plowing in the biasi system require approximately 500 mm of rainfall. Rice established by DDS on dry soil suffered less from water deficit, resulted in better rainfall-use efficiency, and gave the best yields across the five years of study (Table 3). Further, rice established by DDS in dry soil could be established on average 15 days earlier than DDS rice sown in moist soil and 60 days ahead of transplanted rice (Table 2). This head-start reduced the effect of drought on rice in the moderate and severe drought years of 1999 and 2000

Table 3. Effects of rice establishment method on rice and chickpea grain yield in five seasons with varying rainfall in an on-station experiment at Raipur, Chhattisgarh, India. Rice as well as the postrice crop were grown under rainfed conditions without irrigation.

Crop	Establishment method[a]	Year				
		1995-96[b]	1996-97[b]	1998-99[b]	1999-2000[c]	2000-01[d]
		(Grain yield in t ha^{-1})				
Rice	DDS dry	6.76	5.71	4.61	4.22	3.12
	DDS moist	5.57	3.99	4.21	3.61	0.82
	DDS biasi	–	–	3.55	2.72	0.68
	T	4.54	3.69	3.25	1.69	0.39
Chickpea	DDS dry	0.82	0.92	1.10	0.62	NE
	DDS moist	0.78	0.81	0.96	NE[e]	NE
	DDS biasi	–	–	0.88	NE	NE
	T	0.64	0.68	0.69	NE	NE

[a]Establishment methods were dry direct seeding in lines and dry soil (DDS dry), dry direct seeding in lines and moist soil (DDS moist), dry direct seeding broadcast in moist soil and biasi operation (DDS biasi), and transplanting (T). [b]Good to normal year. [c]Moderate drought year. [d]Severe drought year. [e]NE: no establishment of a postrice crop because of drought.

considerably and increased the chances and performance of a postrice crop. In 1995 to 1998, chickpea after DDS rice in dry soil gave the best yield and, in the moderate drought year of 1999, only DDS in dry soil permitted a second crop (Table 3). No second crop could be established in the severe drought year of 2000, independent of establishment method.

Other advantages of direct seeding include deeper, finer, and more extensive root development and, as a result, consistently better performance under drought conditions (Ingram et al 1994, Singh et al 1995, Castillo et al 1998, Fukai et al 1998). Direct seeding is characterized by much lower labor needs for crop establishment than transplanting or biasi. Increasing labor costs were identified as one major reason for a shift from transplanting to direct seeding in several Asian countries (Pandey and Velasco 2002). In the absence of soil puddling, a disadvantage is that deep percolation rates during the season might be higher. Dry direct seeding in dry soil may also increase the risk of suboptimal crop establishment due to drought spells immediately after seeding. Also, good weed management is much more important for direct-seeded rice because the rice plants have no size advantage over the germinating weeds and there is no flood layer suppressing weeds as is the case in transplanted rice fields. Thus, although direct seeding offers substantial advantages and opportunities for drought-prone systems, direct-seeded systems tend not to be as robust as transplanted systems and management tends to be more critical for successful crop establishment, effective weed control, and high and stable yields.

Other methods tested in rainfed rice to increase the productive use of water and to reduce unproductive water losses include subsoil compaction (Ghildyal 1978, Wickham and Singh 1978, Trébuil et al 1998, Harnpichitvitaya et al 2000), land leveling (Lantican et al 1999), and soil improvement. The effect of subsoil compaction is dependent on site-specific conditions and needs to be carefully evaluated. Also, subsoil compaction requires heavy machinery and considerable investment, which most farmers will not be able to afford. The same applies to land leveling with laser-guided equipment, but often better field leveling can already be achieved with conventional farm equipment. Application and incorporation of organic materials can help to increase soil water retention capacity, especially on coarse-textured soils. This practice is still common in many rainfed lowland systems but is limited by the availability of organic materials, the labor requirements, and the commonly used soil tillage equipment. Mulching, a technique to reduce evaporation in upland crops, is not really an option in periodically flooded rice fields. Minimum tillage or zero-tillage practices might also offer opportunities in some rainfed lowlands but hardly any studies have been conducted for rice.

Aerobic rice

A fundamentally different approach to reduce water outflows from rice fields is to grow the crop like an upland crop, such as wheat or maize. Unlike lowland rice, upland crops are grown in nonpuddled, nonsaturated (i.e., "aerobic") soil without ponded water. The potential water reductions at the field level when rice can be grown as an upland crop are large, especially on soils with high seepage and percolation rates (Bouman 2001). Besides declining seepage and percolation losses, evaporation decreases since there is no ponded water layer, and the large amount of water used for wet land preparation is eliminated altogether.

In Asia, rice is already grown aerobically in the upland environment, but mostly as a low-yielding subsistence crop with minimal inputs (Lafitte et al 2002). Upland rice varieties are often drought tolerant, but have a low yield potential and tend to lodge under high levels of external inputs such as fertilizer and supplemental irrigation. Achieving high yields under irrigated but aerobic soil conditions requires new varieties of "aerobic rice" that combine the drought-adapted characteristics of upland varieties with the high-yielding characteristics of lowland varieties (Lafitte et al 2002, Atlin et al 2006). The development of such aerobic rice varieties for the tropics is of relatively recent origin. De Datta et al (1973) grew the lowland variety IR20 in aerobic soil under furrow irrigation at IRRI in the Philippines. Water savings were 55% compared with flooded conditions, but yield fell from about 8 t ha^{-1} under flooded conditions to 3.4 t ha^{-1} under aerobic conditions. Using improved upland rice varieties, George et al (2002) reported aerobic rice yields of 1.5–7.4 t ha^{-1} in uplands with 2,500 to 4,500 mm of annual rainfall in the Philippines. Yields of 6 t ha^{-1} and more, however, were realized only in the first years of cultivation, and most yields were in the 2–3 t ha^{-1} range. Atlin et al (2006) reported aerobic rice yields of 3–4 t ha^{-1} using recently-developed aerobic rice varieties in farmers' fields in rainfed uplands in the

Philippines. Though the amount of rainfall was not reported, the conditions of the trials were described as "well watered." Bouman et al (2005) and Peng et al (2006) quantified yield and water use of recently-released tropical aerobic rice variety Apo under irrigated aerobic and flooded conditions. In the dry season, yields under aerobic conditions were 4–5.7 t ha^{-1} and, in the wet season, they were 3.5–4.2 t ha^{-1}. On average, the mean yield of all varieties was 32% lower under aerobic conditions than under flooded conditions in the dry season and 22% lower in the wet season. Total water input was 1,240–1,880 mm in flooded fields and 790–1,430 mm in aerobic fields. On average, aerobic fields used 190 mm less water in land preparation and had 250–300 mm less seepage and percolation, 80 mm less evaporation, and 25 mm less transpiration than flooded fields. In some rainfed uplands in Batangas, Philippines, and in the hilly regions of Yunnan Province, southern China, farmers grew rainfed aerobic rice under intensified management, realizing yields of 3–4 t ha^{-1} (Atlin et al 2006). It is suggested that an even distribution of around 600 mm of rainfall would be sufficient to realize 3–4 t ha^{-1} (Bouman et al 2006).

Although considerable research has been done on the development of technologies to maintain crop productivity under water scarcity, little attention has been paid to long-term sustainability. Flooding of rice fields has beneficial effects on soil acidity (pH), soil organic matter buildup, phosphorus, iron, and zinc availability, and biological N fixation that supplies the crop with additional N (Kirk 2004). When fields are not continuously flooded because of water scarcity, these beneficial effects disappear. A change to more aerobic soil conditions will negatively affect the soil pH in some situations and decrease the availability of phosphorus, iron, and zinc. Under fully aerobic conditions, problems with micronutrient deficiencies have been reported by Choudhury et al (2007), Sharma et al (2002), and Singh et al (2002). The introduction of aerobic phases in rice fields may also decrease the soil organic carbon content. There are indications that soil-borne pests and diseases such as nematodes, root aphids, and fungi occur more in nonflooded than in flooded rice systems (Sharma et al 2002, Singh et al 2002, Ventura et al 1981, Prot and Matias 1995). Current experience is that, under fully aerobic soil conditions, rice cannot be grown continuously on the same piece of land each year (which is common practice in flooded rice) without a yield decline (George et al 2002). The mechanisms behind the yield decline are not yet understood, although higher levels of the nematode *Melodoigyne graminicola* can be found in rainfed lowlands and in aerobic rice fields (up to 3,000 counts g^{-1} fresh root) compared with flooded fields (6–400 counts s^{-1} fresh root; unpublished data). The importance of soil-borne pathogens/parasites as a factor limiting rice growth has been often overlooked because they do not seem to be an important constraint in most continuously flooded systems. Yield limitations due to soil-borne pathogens were also indicated by the positive effect of soil solarization in rice-wheat systems of Bangladesh (Banu et al 2005) and in rainfed lowland rice of northeast Thailand (personal communication by Thiess, Padgham et al 2004). Therefore, it is possible that soil-borne pathogens play a much more important role than previously recognized, especially in water-limited rainfed lowlands.

Conclusions

Drought stress is considered the most important limitation to production in rainfed lowlands of Asia, affects huge areas grown to rice, and threatens the livelihood of millions of farmers and their families. Severe and regular droughts affect mainly rainfed lowlands in South and Southeast Asia, but regional weather patterns, topography, and soil characteristics cause considerable drought-risk variations within and beyond these regions. Apart from drought, many of these agroecosystems are simultaneously affected by other abiotic stresses such as submergence, low soil fertility, other soil constraints, and pests. These conditions create a highly diverse and difficult environment for rice cultivation. For any attempt to increase productivity in this environment, improved modern varieties are essential. When they are not available or accepted by farmers, further progress of breeding efforts is required first. However, if such varieties are available and accepted by farmers, their optimal use will require the combination with adequate and, if possible, improved management techniques. Two main management strategies for drought stress alleviation in rice can be distinguished. The first strategy is based on management options that allow escape from drought by either avoiding dry periods (e.g., short-duration varieties, adapted cropping calendars) or by providing access to additional water resources (e.g., irrigation). The second strategy is to moderate drought by reducing unproductive water losses and thereby "saving" water for productive transpiration. Both strategies contain management options that offer considerable scope for improvement of drought-prone rainfed lowlands and the two options discussed in detail are direct seeding and improved nutrient management. A fundamentally different approach to reduce unproductive water losses in rice-based systems is "aerobic rice," in which rice is grown like an upland crop, such as wheat or maize. This relatively new system of rice cultivation is still under development but promises considerable opportunities in specific target environments within the rainfed lowlands of Asia. In addition, research on aerobic rice has revealed that soil-borne pathogens can play a much more important role in water-limited environments than previously recognized. This insight is essential to better understand rice performance in rainfed lowlands and must be considered when developing improved germplasm and management options for this environment.

References

Akbar M, Gunawardena IE, Ponnamperuma FN. 1986. Breeding for soil stresses. In: Progress in rainfed lowland rice. Los Baños (Philippines): International Rice Research Institute. p 263-272.

Atlin GN, Lafitte HR, Tao D, Laza M, Amante M, Courtois B. 2006. Developing rice cultivars for high-fertility upland systems in the Asian tropics. Field Crops Res. 97:43-52.

Banu SP, Shaheed MA, Siddique AA, Nahar MA, Ahmed HU, Devare MH, Duxbury JM, Lauren JG, Abawi GS, Meisner CA. 2005. Soil biological health: a major factor in increasing the productivity of the rice-wheat cropping system. Int. Rice Res. Notes 30(1):5-11.

Bhuiyan SI. 1994. On-farm reservoir systems for rainfed ricelands. Los Baños (Philippines): International Rice Research Institute. 164 p.

Biswas AK Jr, Nayek B, Choudhuri MA. 1982. Effect of calcium on the response of a field-grown rice plant to water stress. Proc. Indian Natl. Sci. Acad. B48:669-705.

Boling A, Bouman BAM, Tuong TP, Murty MVR, Jatmiko SY. 2007. Modelling the effect of groundwater depth on yield-increasing interventions in rainfed lowland rice in Central Java, Indonesia. Agric. Syst. 92:115-139.

Bouman BAM. 2007. A conceptual framework for the improvement of crop water productivity at different spatial scales. Agric. Syst. 93:43-60.

Bouman BAM. 2001. Water-efficient management strategies in rice production. Int. Rice Res. Notes 16(2):17-22.

Bouman BAM, Humphreys E, Tuong TP, Barker R. 2006. Rice and water. Adv. Agron. 92:187-237.

Bouman BAM, Peng S, Castaneda AR, Visperas RM. 2005. Yield and water use of irrigated tropical aerobic rice systems. Agric. Water Manage. 74:87-105.

Cantero-Martinez C, Villar JM, Romagosa I, Fereres E. 1995. Nitrogen fertilization of barley under semi-arid conditions. Mediterranean 7:24-27.

Castillo EG, Tuong TP, Cabangon RC, Boling A, Singh U. 1998. Effects of crop establishment and controlled-release fertilizer on drought stress responses of a lowland rice cultivar. In: Ladha JK, Wade LJ, Dobermann A, Reichardt W, Kirk GJD, Piggin C, editors. Rainfed lowland rice: advances in nutrient management research. Proceedings of the International Workshop on Nutrient Research in Rainfed Lowlands, 12-15 Oct. 1998, Ubon Ratchathani, Thailand. Los Baños (Philippines): International Rice Research Institute. p 201-216.

Chang TT, Somrith B, O'Toole JC. 1979. Potential for improving drought resistance in rainfed lowland rice. In: Rainfed lowland rice: selected papers from the 1978 International Rice Research Conference. Los Baños (Philippines): International Rice Research Institute. p 149-164.

Choudhury BU, Bouman BAM, Singh AK. 2007. Yield and water productivity of rice-wheat on raised beds at New Delhi, India. Field Crops Res. 100:229-239.

Debaeke P, Aboudrare A. 2004. Adaptation of crop management to water-limited environments. Eur. J. Agron. 21:433-446.

De Datta SK, Krupp HK, Alvarez EI, Modgal SC. 1973. Water management in flooded rice. In: Water management in Philippine irrigation systems: research and operations. Los Baños (Philippines): International Rice Research Institute. p 1-18.

Ehlers W, Goss M. 2003. Water dynamics in plant production. Wallingford (UK): CABI.

Falkenmark M, Rockström J. 2004. Balancing water for humans and nature: the new approach in ecohydrology. London (UK): Earthscan. 247 p.

Fujisaka S, Moody K, Ingram KT. 1993. A descriptive study of farming practices for dry seeded rainfed lowland rice in India, Indonesia and Myanmar. Agric. Environ. Ecosyst. 45:115-128.

Fukai S, Cooper M. 1995. Development of drought-resistant cultivars using physio-morphological traits in rice. Field Crops Res. 40:67-86.

Fukai S, Sittisuang P, Chanphengsay M. 1998. Increasing production of rainfed lowland rice in drought-prone environments. Plant Prod. Sci. 1:75-82.

Garrity DP, Oldeman LR, Morris RA, Lenka D, 1986. Rainfed lowland rice ecosystems: characterization and distribution. In: Progress in rainfed lowland rice. Los Baños (Philippines): International Rice Research Institute. p 3-24.

George T, Magbanua R, Garrity DP, Tubaña BS, Quiton J. 2002. Rapid yield loss of rice cropped successively in aerobic soil. Agron. J. 94:981-989.

Ghildyal BP. 1978. Effect of compaction and puddling on soil physical properties and rice growth. In: Soils and rice. Los Baños (Philippines): International Rice Research Institute. p 317-336.

Haefele SM, Hijmans RJ. 2007. Soil quality in rice-based rainfed lowlands of Asia: characterization and distribution. Proceedings of the International Rice Congress 2006, October 9-13 2006, New Delhi, India. p 297-308.

Haefele SM, Konboon Y, Patil S, Mishra VN, Mazid MA, Tuong TP. 2007. Water by nutrient interactions in rainfed lowland rice: mechanisms and implications for improved nutrient management. Paper presented at the CURE Resource Management. Workshop in Dhaka, Bangladesh, March 2006.

Harnpichitvitaya D, Trébuil G, Oberthuer T, Pantuwan G, Craig I, Tuong TP, Wade LJ, Suriya-Arunroj D. 2000. Identifying soil suitability for subsoil compaction to improve water- and nutrient-use efficiency in rainfed lowland rice. In: Tuong TP, Kam SP, Wade LJ, Pandey S, Bouman BAM, Hardy B. editors. Characterizing and understanding rainfed rice environments. Proceedings of the International Workshop on characterizing and understanding rainfed rice environments, 5-9 Dec. 1999, Bali, Indonesia. Los Baños (Philippines): International Rice Research Institute. p 97-110.

Homma K, Horie T, Shiraiwa T, Supapoj N, Matsumoto N, Kabaki N. 2003. Toposequential variation in soil fertility and rice productivity of rainfed lowland paddy fields in a mini-watershed (Nong) in Northeast Thailand. Plant Prod. Sci. 6:147-153.

Huke RE, Huke EH. 1997. Rice area by type of culture: South, Southeast, and East Asia. A revised and updated data base. Los Baños (Philippines): International Rice Research Institute.

Ingram KT, Bueno FD, Namuco OS, Yambao EB, Beyrouty CA. 1994. Rice root traits for drought resistance and their genetic variation. In: Kirk GJD, editor. Rice roots: nutrient and water use. Selected papers from the International Rice Research Conference. Los Baños (Philippines): International Rice Research Institute.

Jearakongman S, Rajatasereekul S, Naklang K, Romyen P, Fukai S, Skulkhu E, Jumpaket B, Nathabutr K. 1995. Growth and grain yield of contrasting rice cultivars under different conditions of water availability. Field Crops Res. 44:139-150.

Kam SP, Dy-Fajardo S, Rala AB, Hossain M, Tuong TP, Bouman BAM, Banik P. 2000. Multi-scale drought risk analysis of rainfed lowland rice environments. Poster presented at the 30th Annual Scientific Conference of the Crop Science Societies of the Philippines, 2-8 May, Batac, Ilocos Norte, Philippines.

Kirk G. 2004. The biochemistry of submerged soils. Chichester (UK): John Wiley and Sons. 291 p.

Lafitte RH, Courtois B, Arraudeau M. 2002. Genetic improvement of rice in aerobic systems: progress from yield to genes. Field Crops Res. 75:171-190.

Lantican MA, Lampayan RM, Bhuiyan SI, Yadav MK. 1999. Determinants of improving productivity of dry-seeded rice in rainfed lowlands. Exp. Agric. 35:127-140.

Ludlow MM. 1989. Strategies of response to water stress. In: Kreeb KH, Ritcher H, Hinckley TM, editors. Structural and functional responses to environmental stresses. The Hague (The Netherlands): SPB Academic Publishing. p 269-281.

Mackill DJ. 1986. Varietal improvement for rainfed lowland rice in South and Southeast Asia: results of a survey. In: Progress in rainfed lowland rice. Los Baños (Philippines): International Rice Research Institute. p 115-144.

Mackill DJ, Coffman WR, Garrity DP. 1996. Rainfed lowland rice improvement. Los Baños (Philippines): International Rice Research Institute. 242 p.

Otoo E, Ishii R, Kumura A. 1989. Interaction of nitrogen supply and soil water stress on photosynthesis and transpiration in rice. Jpn. J. Crop Sci. 58:424-429.

Padgham JL, Duxbury JM, Mazid AM, Abawi GS, Hossain M. 2004. Yield losses caused by *Meloidogyne graminicola* on lowland rainfed rice in Bangladesh. J. Nematol. 36:42-48.

Pal AR, Bhuiyan SI. 1995. Rainwater management for drought alleviation: opportunities and options for sustainable growth in agricultural productivity. In: Fragile lives in fragile ecosystems. Proceedings of the International Rice Research Conference, 13-17 Feb. 1995. Los Baños (Philippines): International Rice Research Institute. p 187-216.

Pandey S. 1998. Nutrient management technologies for rainfed rice in tomorrow's Asia: economic and institutional considerations. In: Ladha JK, Wade L, Dobermann A, Reichhardt W, Kirk GJD, Piggin C, editors. Rainfed lowland rice: advances in nutrient management research. Los Baños (Philippines): International Rice Research Institute. p 3-28.

Pandey S, Velasco L. 2002. Economics of direct seeding in Asia: patterns of adoption and research priorities. In: Pandey S, Mortimer M, Wade L, Tuong TP, Lopez K, Hardy B, editors. Direct seeding: research issues and opportunities. Proceedings of the International Workshop on Direct Seeding in Asian Rice Systems: Strategic Research Issues and Opportunities, 25-28 January 2000, Bangkok, Thailand. Los Baños (Philippines): International Rice Research Institute. p 3-14.

Pane H, Noor ES, Jatmiko SY, Johnson DE, Mortimer M. 2005. Weed communities of gogorancah and walik jerami rice in Indonesia and reflections on management. Proceedings of the BCPC International Congress on Crop Science and Technology, SECC, Glasgow, UK.

Patamatamkul S. 2001. Development and management of water resources in the Korat Basin of northeast Thailand. In: Kam SP, Hoanh CT, Trébuil G, Hardy B, editors. Natural resource management issues in the Korat Basin of northeast Thailand: an overview. Proceedings of the planning workshop on ecoregional approaches to natural resource management in the Korat Basin, northeast Thailand, 26-29 Oct., Khon Kaen, Thailand. Los Baños (Philippines): International Rice Research Institute. p 115-118.

Peng S, Bouman BAM, Visperas RM, Castañeda A, Nie L, Park H-K. 2006. Comparison between aerobic and flooded rice: agronomic performance in a long-term (8-season) experiment. Field Crops Res. 96:252-259.

Prot JC, Matias DM. 1995. Effects of water regime on the distribution of *Meloidogyne graminicola* and other root parasitic nematodes in a rice field toposequence and pathogenicity of *M. graminicola* on rice cultivar UPLR15. Nematologica 41:219-228.

Rathore AL, Romyen P, Mazid AM, Pane H, Haefele SH, Johnson DE. n.d. Challenges and opportunities of direct seeding in rice-based rainfed lowlands of Asia. Proceedings of the CURE Resource Management Workshop in Dhaka, Bangladesh, March 2006. (In press.)

Samson BK, Wade LJ. 1998. Soil physical constraints affecting root growth, water extraction, and nutrient uptake in rainfed lowland rice. In: Ladha JK, Wade LJ, Dobermann A, Reichardt W, Kirk GJD, Piggin C, editors. Rainfed lowland rice: advances in nutrient management research. Proceedings of the International Workshop on Nutrient Research in Rainfed Lowlands, 12-15 Oct. 1998, Ubon Ratchathani, Thailand. Los Baños (Philippines): International Rice Research Institute. p 245-260.

Sharma PK, Bhushan Lav, Ladha JK, Naresh RK, Gupta RK, Balasubramanian BV, Bouman BAM. 2002. Crop-water relations in rice-wheat cropping under different tillage systems and water-management practices in a marginally sodic, medium-textured soil. In: Water-wise rice production. Los Baños (Philippines): International Rice Research Institute. p 223-235.

Singh AK, Choudhury BU, Bouman BAM. 2002. Effects of rice establishment methods on crop performance, water use, and mineral nitrogen. In: Bouman BAM, Hengsdijk H, Hardy B, Bindraban PS, Tuong TP, Ladha JK, editors. Water-wise rice production. Los Baños (Philippines): International Rice Research Institute. p 237-246.

Singh AK, Tuong TP, Wopereis MCS, Boling A, Kropff MJ. 1995. Quantifying lowland rice responses to soil-water deficit. In: Fragile lives in fragile ecosystems. Proceedings of the Internationl Rice Research Conference, 13-17 Feb. 1995. Los Baños (Philippines): International Rice Research Institute. p 507-520.

Singh RK, Hossain M, Thakur R. 2003. Boro rice. International Rice Research Institute - India Office. 234 p.

Tanguilig UC, De Datta SK. 1988. Potassium effects on root growth and yield of upland rice (*Oryza sativa* L.) under drought and compacted soil conditions. Philipp. J. Crop Sci. 13:33.

Tanner CB, Sinclair TR. 1983. Efficient water use in crop production: research or re-search? In: Taylor HM, Jordan WR, Sinclair TR, editors. Limitations to efficient water use in crop production. Madison, Wis. (USA): American Society of Agronomy, Crop Science Society of America, and Soil Science Society of America. p 1-25.

Trébuil G, Harnpichitvitaya D, Tuong TP, Pantuwan G, Wade LJ, Wonprasaid S. 1998. Improved water conservation and nutrient-use efficiency via subsoil compaction and mineral fertilization. In: Ladha JK, Wade LJ, Dobermann A, Reichardt W, Kirk GJD, Piggin C, editors. Rainfed lowland rice: advances in nutrient management research. Proceedings of the International Workshop on Nutrient Research in Rainfed Lowlands, 12-15 Oct. 1998, Ubon Ratchathani, Thailand. Los Baños (Philippines): International Rice Research Institute. p 245-260.

Tuong TP. 1999. Productive water use in rice production: opportunities and limitations. J. Crop Prod. 2:241-264.

Van Bremen N, Pons LJ. 1978. Acid sulfate soils and rice. In: Soils and rice. Los Baños (Philippines): International Rice Research Institute. p 739-762.

Ventura W, Watanabe I, Castillo MB, dela Cruz A. 1981. Involvement of nematodes in the soil sickness of a dryland rice-based cropping system. Soil Sci. Plant Nutr. 27:305-315.

Viets FG Jr. 1962. Fertilizers and the efficient use of water. Adv. Agron. 14:223-264.

Vityakorn P. 1989. Sources of potassium in rainfed agriculture in northeast Thailand. 1989 Annual Report of Farming Systems Research Project. Khon Kaen University, Khon Kaen, Thailand.

Wade LJ, Amarante ST, Olea A, Harnpichitvitaya D, Naklang K, Wihardjaka A, Sengar SS, Mazid MA, Singh G, McLaren CG. 1999. Nutrient requirements in rainfed lowland rice. Field Crops Res. 64:91-107.

Wickham TH, Singh VP. 1978. Water movement through wet soils. In: Soils and rice. Los Baños (Philippines): International Rice Research Institute. p 337-358.

Witt C. 2003. Fertilizer use efficiencies in irrigated rice in Asia. Presented at the IFA Regional Conference for Asia and the Pacific, Cheju Island, Republic of Korea, 6-8 October 2003.

Zaman A, Roy GB, Mallik S, Maiti A. 1990. Response of direct seeded rice to irrigation and nitrogen in drought prone laterite tract of West Bengal. Environ. Ecol. 8:311-314.

Notes

Authors' address: International Rice Research Institute, DAPO Box 7777, Metro Manila, Philippines.

Enhancing rice productivity in water-stressed environments: perspectives for genetic improvement and management

Anil Kumar Singh and Viswanathan Chinnusamy

The worldwide water shortage and uneven distribution of rainfall, made more erratic by global climate change, make the improvement of drought resistance especially important for rice, which is the major cereal crop of monsoon-based agriculture of Asia. The intensity, duration, and occurrence of water stress in relation to various phenological phases differ in the diverse ecosystems in which rice is cultivated. Therefore, the development of specific rice genotypes with enhanced resistance to soil moisture deficit that occurs in a target environment is necessary to enhance water productivity in rice. To maximize production under these conditions, there is a need to have a combination of options that include drought-resistant varieties as well as effective crop management strategies. Important mechanisms of drought resistance include drought avoidance via enhanced water uptake, reduced water loss, and enhanced water-use efficiency (WUE), and drought tolerance via osmotic adjustment, antioxidant capacity, and desiccation tolerance. Depending upon the nature of drought stress, plants may employ some or all of these mechanisms. Most research efforts in rice have been directed toward physiological dissection of complex drought resistance mechanisms into component traits, and mapping of the quantitative trait loci (QTLs) for these physiological traits. QTLs have been shown to contribute from 5% to 50% of the phenotypic variation in a single component trait. This suggests that pyramiding of QTLs by marker-assisted selection (MAS) is necessary for significant improvement of drought resistance. The lack of consistency of QTLs across population limits their immediate application in MAS. The identification of common QTLs across populations under near realistic field stress conditions with standard-stress-assays will enhance the pace of their use in MAS. Genetic engineering efforts are also being made to enhance the drought resistance of rice by overexpressing effector genes or transcriptional regulators. Transcriptome engineering is emerging as an important tool to combat abiotic stresses. Combination of metabolomics with expression QTLs (eQTLs) will increase the pace of our understanding of the molecular basis of drought resistance.

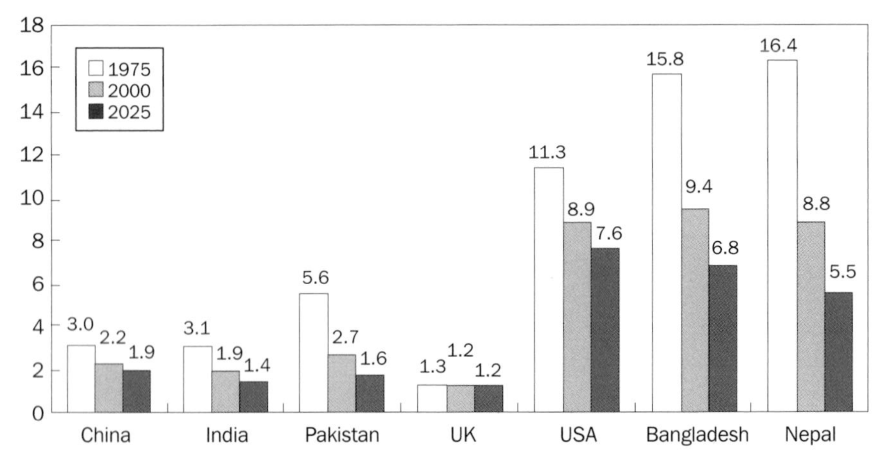

Fig. 1. Per capita water availability in selected countries.

Rice is the staple crop of Asia, with more than 90% of the world's rice being produced and consumed on this continent. It is also the crop which alone consumes 50% of the total irrigation water (Bhuiyan 1992). The proportion of water available for irrigation, however, is becoming increasingly scarce worldwide. In many Asian countries, per capita water availability is expected to decline by a magnitude of 15% to 54% by 2025 compared with 1990 (Guerra et al 1998; Fig. 1). In India, irrigated lowland rice is cultivated in 55% of the total rice-growing area and contributes to 67.5% of production. Irrigated lowland rice has very low water-use efficiency as it consumes anywhere between 3,000 and 5,000 liters of water to produce 1 kilogram of rice (IRRI 2001). The traditional rice production system not only leads to wastage of water but also causes environmental degradation and reduces fertilizer-use efficiency. Hence, the development of alternate production systems such as "aerobic rice," which consumes less water, and the development of rice genotypes suitable for aerobic rice production systems are imperative to enhance the water productivity of rice (Singh and Chinnusamy 2006). Upland rice and rainfed lowland (periodically flooded by rain) rice, which together account for about half of the world's rice, undergo unpredictable periods of drought. Rainfed lowland and upland rice are cultivated in 30% and 15% of the area and contribute to 28% and 4.5% of rice production in India, respectively (Fig. 2). This low productivity in rainfed lowland and upland ecosystems is due to the susceptibility of rice genotypes to water stress. Therefore, the development of rice genotypes with enhanced resistance to various levels of soil moisture deficit in rainfed lowland, upland, and aerobic rice production systems, and the development of crop management practices for these production systems are necessary to enhance the water productivity of rice. During the past decade, significant progress has been made in the understanding of drought resistance mechanisms, with the major emphasis on mapping quantitative trait loci for component traits of drought resistance.

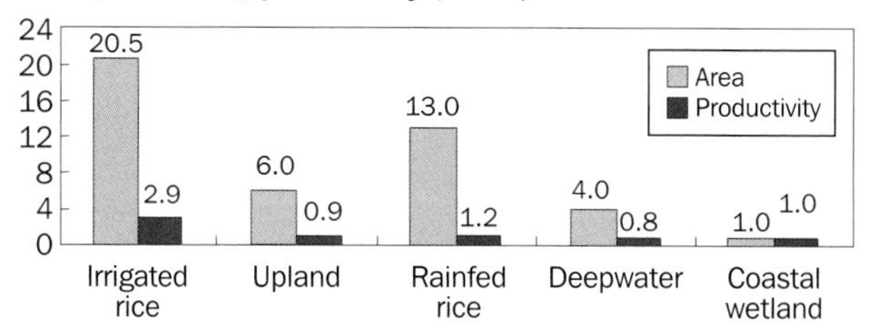

Fig. 2. Area and productivity of rice in different rice ecosystems of India.

Target environments

Agricultural drought can be defined as plant water deficit, caused by a shortage in precipitation and soil water availability (soil water deficit) and excess of evapotranspiration (atmospheric water deficit) that impairs the normal growth and development of plants. Thus, agricultural drought depends upon soil moisture availability and the biology of plants, that is, species, genotype, and sensitivity to water deficit at various growth and development stages of the plant. Because of the varying nature of drought stress and its coincidence with a phenological phase, crop plants may use diverse drought resistance mechanisms. In the unimodal rainfall distribution pattern typical of monsoon Asia, drought stress often occurs during flowering time as cessation of rainfall coincides with this critical stage. The intensity, duration, and occurrence of water stress in relation to various phenological phases differ in rainfed lowland, upland, and aerobic rice production systems; hence, specific genetic improvement and management options need to be focused upon for each ecosystem. In the aerobic rice production system, irrigation water can be applied at a soil moisture tension of around 20 kPa, implying only mild stress conditions, which may not be so detrimental to yield. However, in drought-prone rainfed lowland and upland conditions, the stress intensity and duration are expected to be higher that those of aerobic rice conditions.

Drought resistance

Plants employ many mechanisms to survive water-stress environments and they reproduce under drought conditions. One or more mechanisms may be employed depending upon the stress level (intensity and duration) and its occurrence at different phenological phases of crops (Blum 2005, Tuberosa and Salvi 2006):

1. Maximization of water uptake from the soil
2. Minimizing water loss through plant architecture
3. Enhancing water-use efficiency
4. Cellular tolerance of water deficit
5. Phenotypic and developmental plasticity

In the past two decades, the complex traits of drought resistance mechanisms were dissected into component traits by using physiological, genetic, and molecular approaches, and thus enhanced our understanding of drought resistance mechanisms of plants. Here, we discuss the efforts with QTL mapping and molecular approaches that led to the identification of some of the loci controlling component traits of drought resistance. The objective was not to describe the entire literature on this subject but to highlight some examples for each mechanism of drought resistance in rice. Although the benefits of these approaches have yet to reach farmers' fields in the form of drought-resistant rice genotypes, these approaches have significantly enhanced our understanding of drought resistance.

Maximization of water uptake

Root morphological traits. The roots play a pivotal role in water and nutrient uptake, and act as sensors of water and nutrient status of the soil. Under rainfed unirrigated/inadequately irrigated conditions, the possession of a deep and thick root system and other root morphological traits that allow access to water and nutrients from deeper soil layers will be one of the most promising traits for improving drought resistance and water-use efficiency (WUE). During the past decade, significant progress has been made in mapping quantitative trait loci (QTLs) for root traits such as root length density, root diameter, root volume, root surface area, and root penetration ability in rice (Table 1). Efforts are now being made to introgress QTLs for root traits in marker-assisted selection (MAS) for enhanced drought resistance and WUE. In a marker-assisted backcrossing (MABC) breeding program, efforts were made to transfer four QTLs for deeper roots (on chromosomes 1, 2, 7, and 9) from selected doubled-haploid lines (DHLs) of IR64/Azucena into IR64. Although all the near-isogenic lines carrying target QTLs have not shown significantly improved root traits, some showed significantly improved root traits over IR64 (Shen et al 2001). Another MABC breeding program was conducted to improve the root morphological traits of the Indian upland rice variety Kalinga III from the donor parent Azucena, an upland japonica variety from the Philippines. Pyramids with four root QTLs were obtained in eight generations, completed in 6 years. The target segment on chromosome 9 (RM242-RM201) significantly increased root length under both irrigated and drought-stress treatments, confirming that this root-length QTL from Azucena functions in a novel genetic background (Steele et al 2006).

It is well known that rice establishment methods influence the rate of stress development through their effect on root growth (Boling et al 1998). Significant genotypic differences in drought-induced root growth were found among rice genotypes. As soil moisture tension increased from 0 to 40 kPa, root growth response varied depending upon the rice genotype (Fig. 3; Singh et al, unpublished data). Further, irrigation and nitrogen (the source and quantity) significantly influence root growth in rice (Fig. 4; Singh et al, unpublished data). Increasing soil penetration resistance is also a major factor governing root growth response under drought. Hence, QTL mapping needs to be done in near-realistic field conditions for each ecosystem. A study conducted at IRRI, Philippines, showed that the higher yield of dry-seeded

Table 1. QTLs for various root morphological characters in rice.

Root trait	Population	Environment	Chr. no. and QTL no.; share of major QTLs (%)	Reference
Root thickness, root/shoot ratio, root dry weight per tiller, and deep root weight	CO39 × Moroberekan (RIL)	Grown in soil, greenhouse experiment	12 of the 14 QTLs associated with root morphology and field drought avoidance/tolerance	Champoux et al (1965)
Root number Penetration index	CO39 × Moroberekan (RIL)	Wax-petrolatum layers	19 QTLs (19%) 6 QTLs (13%)	Ray et al (1996)
Total root weight Deep root weight Deep root weight to shoot ratio Deep root weight per tiller Maximum root length Root thickness	IR64 × Azucena (DHL)	Grown in aerobic conditions in well-drained plastic bags	6-(1,5,6,7,9); 5–11 = 43% 5-(1,6,7,9); 4–15 = 23% 9-(1,5-9); 4–22 = 33% 6-(1,2,6-9); 4–20 = 40% 8-(1-3,5-9); 4–21 = 49% 5-(1,2,5,8); 4–10 = 14%	Yadav et al (1997)
Penetrated root no. Total root no. Root penetration index Penetrated root thickness	IR64 × Azucena (DHL)	Wax-petrolatum layers	2-(2,7); 8.4%, 9.0% 2-(1,7,); 8.8%, 14.3% 4-(2,3,7,8); 8.9–13.5% 4-(1,3,4,9); 10.1–16.4%	Zheng et al (2000)
Root traits	IR64 × Azucena (DHL)	Root-related traits at peak vegetative stage and grain yield under both low-moisture stress and nonstress conditions		Venuprasad et al (2002)

Continued on next page

Table 1 continued.

Root trait	Population	Environment	Chr. no. and QTL no.; share of major QTLs (%)	Reference
Maximum root length (28/days)	Azucena × Bala (F_2)	Hydrophonic screen system	4-(3,5,6,11); 5.3–29.8%	Price and Tomos (1997)
Root volume			3-(1,8,12); 6.2–10.2%	
Adventitious root thickness			3-(2,3,5); 7–21.3%	
Root weight	Azucena × Bala (RIL)	Greenhouse; glass-sided chambers; drought at immediately after germination or at 49th day	6	Price et al (2002)
Maximum root length			6	
Root to shoot ratio			11	
Root thickness			14	
Seminal root (SR) length	IR1552 × Azucena (RIL)	Cylindrical-pot experiments under flooding and upland conditions	Chr. 2 QTL common	Zheng et al (2003)
Flooding			2-(2,7); 11.5–12%	
Upland			3-(1,2,9), 11.3–13.4%	
Adventitious root no.				
Flooding			4-(3,4,9); 11.4–20%	
Upland			3-(1,2); 12–18.2%	
Lateral root length on SR				
Flooding			2-(1,6); 11.8%	
Upland			2-(3,5); 13.4–14.4%	
Lateral root no. on SR				
Flooding			1-(6); 13.4%	
Upland			1-(3); 11.7%	
Total root number	IR58821 × IR52561 (RIL)	Wax-petrolatum layers	2-(3,7); 9–12.2 = 15.4%	Ali et al (2000)
Penetrated root number			7-(1,2,3); 7–27.2 = ~43%	
Root penetration index			6-(2,3,10); 7.9–26.2 = ~43%	
Penetrated root thickness			8-(1,2,4,6,7,10); 6.3–13.9 = ~32%	
Penetrated root length			5-(1-3,7,11); 5.8–12.8= ~23%	

Continued on next page

Root trait	Population	Environment	Chr. no. and QTL no.; share of major QTLs (%)	Reference
Total root length, root volume, and total root number per plant and tiller number	IR64 × Azucena (DHL)	Low moisture stress	21 QTLs for all traits under stress	Hemamalini et al (2000)
Deep root mass Deep root ratio Deep root per tiller Rooting depth Root thickness, 0–10 cm Root thickness, 20–25 cm	IR58821 × IR52561 (RIL)	Pot-culture; anaerobic conditions; total QTLs detected in 2 experiments	5-(2-4,9,11); 7.3–21.4% 5-(2-4,9,11); 5.8–27.4% 6-(2,4,6,7); 9.4–21.6% 4-(1,4); 5.7–29.9% 6-(1,3,8,9); 6.2–15.1% 2-(4); 12.4–23.2%	Kamoshita et al (2002)
Penetrated root length Penetrated root thickness Basal root thickness Penetrated root dry weight Total root dry weight Root penetration index Root pulling force	CT9993 × IR62266 (DHL)	Wax-petrolatum layers; root pulling force was measured in the field under rainfed conditions	1-(11); 5.8% 11-(1,2,4,6,7,9,12); 9–31.1 = 53% 6-(2-4,8,9,12); 9.2–37.6 = 52.6% 3-(4,9,12); 11.5–16.8 = 26% 5-(1,2,4,6,10); 8.6–20.2 = 34.1% 4-(3,4,12); 8.3–10.9 = 30% 6-(2-5,11); 9–19.9 = 50.1%	Zhang et al (2001)
Deep root mass Deep root ratio Deep root per tiller Rooting depth Root thickness, 0–10 cm Root thickness, 20–25 cm	CT9993 × IR62266 (DHL)	A cylindrical pot made of polyvinyl chloride; 4 sowing dates	8-(1,2,3,7,11); 9.4–35% 6-(1,2,5,11); 5.6–51.8% 6-(1,2,4,5,11); 4.7–40.4% 9-(1,2,5,7,9,11); 4.8–16.8% 6-(1,2,4,8,11); 6.7–36.4% 3-(2,4); 14.6–21.8%	Kamoshita et al (2002)
Roots	CT9993-5-10-1-M × IR62266-42-6-2 (DHL)	Field-grown, subjected to water stress before anthesis	47 QTLs for phenology and other traits	Babu et al (2003)
Maximum root length Root thickness Root dry weight	IAC165 × Co39 (RIL)	Greenhouse experiment	Most important QTLs found on chromosomes 1, 4, 9, 11, and 12; 5.5–24.5%	Courtois et al (2003)

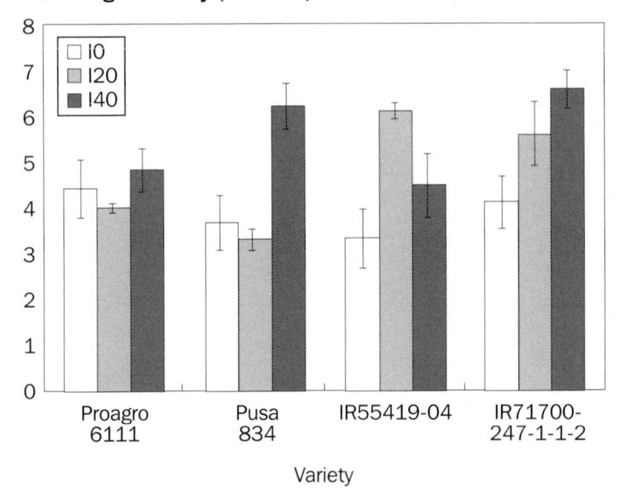

Fig. 3. Genotypic differences in water-deficit stress-induced root growth in aerobic rice (I0, I20, and I40 = irrigation at 0, 20, and 40 kPa soil moisture tension, respectively).

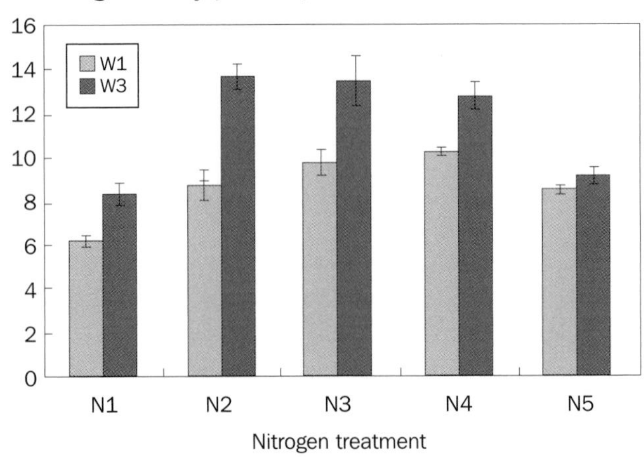

Fig. 4. Effect of irrigation and nitrogen on root growth in transplanted rice variety Pusa Sugandh 3 (recommended fertilizer dose, RFD): 120:60:40 kg NPK ha^{-1}; N1: RFD; N2: RFD (25% N as FYM); N3: RFD (25% N as green manure); N4: RFD (25% N as biofertilizer); N5: organic manure (FYM+ BioF + CropRes.); W1 = submergence; W3 = irrigation after 3-day drainage).

rice under stress was related to better root length density, higher root-shoot ratio, and more uniform root distribution with respect to soil depth (Tuong et al 2002). Cloning of the major genes in the QTL will help in understanding the molecular basis of root growth under drought. Efforts have also been made in this direction. Two genes for cell expansion, expansin (*OsEXP2*) and endo-1,4-β-D-glucanase (*Egase*), were mapped on the intervals carrying the QTLs for seminal root length and lateral root length under upland conditions, respectively (Zheng et al 2003).

Root physiology. In addition to root morphological traits, water uptake by plants also depends upon the resistance (apoplastic and symplastic) of living tissues to radial water flow. Root hydraulic conductance is regulated by a change in root anatomy (suberization, the existence of apoplastic barriers in the exo- and endodermis) and aquaporins, the water channel proteins in membranes that determine water flow through symplast. Transgenic overexpression of an aquaporin gene (*RWC3* driven by the stress-inducible *SWPA2* promoter) from upland rice in lowland rice showed an increase in root osmotic hydraulic conductivity (Lp), leaf water potential, and relative cumulative transpiration at the end of 10 h of PEG treatment in the transgenic plant as compared with control plants (Lian et al 2004). Further studies on the molecular basis of root hydraulic conductance, aquaporins, and root growth under drought stress will help in the genetic enhancement of water uptake by rice under water-deficit conditions.

Osmotic adjustment. Osmotic adjustment (OA) helps in drought avoidance as well as drought tolerance. OA is necessary for the maintenance of cell turgor, and thus maintenance of growth and metabolism, and delays leaf rolling and leaf death under drought. OA lowers root tissue water potential and thus enhances water uptake from soil. Further OA in growing root may help in root growth under drought and thus in extraction of water from deeper layers of soil. Several efforts have been made to map the QTLs for OA in rice (Table 2). OA under drought can also allow better panicle exsertion, which enables higher spikelet fertility, harvest index, and yield stability in rice (Ludlow and Muchow 1990). In wheat, the OA trait has been successfully used to breed a drought-resistant variety, Mulgara (Richards 2006). Heritability of OA is low in rice, and the link between OA and grain yield under drought stress needs to be clearly established before its use in MAS breeding (Serraj and Sinclair 2002).

Minimization of water loss

Water loss from the plant to the atmosphere occurs through transpiration mediated by stomata and direct evaporation of water from the epidermal cell surface. Of these two, water loss through stomata accounts for the majority of water loss. Stomatal pores allow CO_2 influx for photosynthetic carbon fixation and water loss via transpiration to the atmosphere. Thus, the rate of transpiration and photosynthesis depends upon the plant's ability to regulate its stomatal pores. The stress hormone abscisic acid (ABA), synthesized by roots under receding soil water conditions or by leaves when transpiration exceeds water uptake, acts as a signal to control stomatal responses. Genetic analyses of stomatal regulation in *Arabidopsis* have led to the identification of genes that control stomatal response to water deficit (Schroeder et al 2001). Further, the molecular basis of stomatal number and distribution is also well understood in

Table 2. QTLs for osmotic adjustment in rice.

Population	Trait	No. of QTLs	Remarks	Reference
CO39 × Moroberekan (RIL)	Tiller and root number, thickness, dry weight, osmotic adjustment	18		Champoux et al (1995)
CO39 × Moroberekan (RIL)	Osmotic adjustment and dehydration tolerance	7	A major QTL for osmotic adjustment, dehydration tolerance	Lilley et al (1996)
CT9993 × IR62266 (DHL)[a]	Yield, biomass, osmotic adjustment, roots	47	5–59%	Babu et al (2003)
IR62266-42-6-2 × IR60080-46A: BC_3F_3	Osmotic adjustment	14	Together contribute to 58% of phenotypic variability	Robin et al (2003)

[a]CT9993 (an upland japonica type possessing a deep and thick root system and low OA) and IR62266 (an indica type with a shallow root system and high OA).

Arabidopsis. Transpiration depends upon leaf area (normal leaf area, drought-induced rolling and drying) and leaf reflectance characters (wax load and pubescence, leaf angle, leaf rolling). QTLs for traits that minimize water loss through plants such as controlling stomatal regulation, leaf ABA accumulation, and leaf rolling have been identified and tagged with molecular markers in rice (Table 3). Canopy temperature is a surrogate for transpiration rate and, hence, can be easily measured under field conditions but may not be so effective in large populations. Infrared thermal imaging can be applied to measure canopy temperature accurately in breeding populations. High epicuticular wax can help minimize water loss. Rice has low epicuticular wax. Larger genetic variation in cuticle thickness and composition was found within the rice germplasm and this trait exhibits high heritability (H = 0.77) (Haque et al 1992). The relationship of cuticle thickness/wax with grain yield under drought stress and QTLs for this trait need to be established. The molecular basis of water loss minimization under water-deficit stress in rice is poorly understood. Further understanding of the molecular genetic basis of water loss minimization under water-deficit stress will further help to enhance drought resistance of rice.

Water-use efficiency

For agronomists, WUE is the yield of harvested product per unit amount of evapotranspiration. The physiological definition or leaf-level WUE is defined as the ratio of photosynthesis (A) to water loss in transpiration (E) (Condon et al 2004):

Table 3. QTLs for water loss minimization mechanisms in rice.

Population	Trait	No. of QTLs	Reference
Azucena × Bala (F_2)	Stomatal conductance, leaf rolling, and heading date	Slow leaf rolling 1 Stomatal conductance 2 Rate of stomatal closure 2 Days to heading 3	Price et al (1997)
IR20 × 63-83 (F_2)	Leaf size and ABA accumulation	17	Quarrie et al (1997)
IR64 × Azucena (DHL)	Leaf rolling, leaf drying, RWC, growth rate	42 (leaf rolling 11, leaf drying 10, RWC 11, RGR 10 under stress)	Courtois et al (2000)

$$WUE = \frac{A}{B} = \frac{0.6\,(C_a - C_i)}{(W_i - W_a)}$$

where C_a = air CO_2 concentration, C_i = leaf intracellular CO_2 concentration, W_a = air vapor density, and W_i = vapor density inside the stomatal cavity.

Among these parameters, only C_i is determined by the plant, whereas the rest of the parameters are under environmental control. C_i is determined by the balance between stomatal opening and the carboxylating capacity of the Rubisco, and photochemical reactions in the leaf. Direct measurement of WUE can be done only in a limited number of plants. As ^{13}C diffuses slowly and is less used in carboxylation than ^{12}C, ^{13}C discriminates during photosynthesis. The magnitude of discrimination depends on the C_i. Since ^{13}C discrimination and leaf-level WUE both depend on C_i, ^{13}C discrimination can be used as a surrogate for greater agronomic water-use efficiency via greater leaf-level WUE (Farquhar and Richards 1984). This can be easily measured in a large number of plants in breeding populations. High WUE has been found to be associated with low ^{13}C-isotope discrimination in rice (Peng et al 1998). In rice, QTLs for ^{13}C discrimination have been mapped on chromosomes 2, 4, 8, 9, 11, and 12 under flooded conditions. Further, differential expression of QTLs for ^{13}C discrimination was observed (Laza et al 2006). Identification of QTLs under drought and understanding of the molecular genetic basis of leaf-level WUE will help enhance water productivity and yield under water-deficit rice ecosystems (Singh and Chinnusamy 2004).

Cellular tolerance of water deficit

The perception of abiotic stresses and signal transduction to switch on adaptive responses are critical steps in determining the survival and reproduction of plants exposed to adverse environments (Chinnusamy et al 2004). The ability of plant cells to maintain their membrane integrity and ionic-, osmotic-, and metabolic-homeostasis under drought stress determines their tolerance of cellular water deficit.

Cellular water-deficit stress tolerance in plants depends on a modification of metabolism, the production of organic-compatible solutes (proline, sugars, polyols, betaine, etc.), late embryogenesis abundant (LEA) proteins, and antioxidants. The osmoprotectants help in detoxifying radical oxygen species and stabilizing the quaternary structure of proteins. LEA proteins protect membranes and macromolecules under stress. Antioxidant enzymes and metabolites play a crucial role in detoxifying reactive oxygen species produced under cellular water deficit (Chinnusamy et al 2004). Cell membrane stability (CMS) is an effective measure of cellular tolerance of water-deficit stress. In rice, nine QTLs have been mapped for CMS under drought (Tripathy et al 2000). Transgenic rice plants engineered to overproduce organic-compatible solutes such as glycine betaine (Sakamoto et al 1998), proline (Su and Wu 2004), and trehalose (Garg et al 2002) showed enhanced tolerance of abiotic stresses, including drought. Recently, significant progress has been made in cloning and characterizing rice genes involved in cellular tolerance of dehydration stress (Table 4). Transcriptome engineering or overexpression of master switch genes (such as stress sensors, protein kinases, or transcription factors), which regulate several target genes coding for osmolyte biosynthesis enzymes, antioxidants, LEA proteins, and growth regulation, is emerging as an important tool to combat abiotic stresses (Table 4). Hence, molecular genetic dissection of water-deficit stress signal transduction will help in more effective genetic engineering for drought tolerance in rice.

Grain yield and yield components

Phenotypic plasticity and resistance in the development of yield components determine final crop yield. Drought susceptibility index (DSI, Fischer and Maurer 1978) analysis is useful in evaluating the drought resistance of crops as a measure of the stability of yield and its components across various water regimes. DSI analysis revealed that sensitivity of rice genotypes depends upon the intensity of stress (soil moisture tension). Under the aerobic rice system, some rice genotypes were able to maintain yield stability only when irrigated at 20 kPa. Other genotypes maintain yield stability in both 20 and 40 kPa irrigation treatments, whereas some other rice genotypes showed high susceptibility at both irrigation levels under the aerobic rice system (Fig. 5; Singh et al, unpublished data). Further, yield components also showed differential sensitivity to water deficit, namely, 20 and 40 kPa soil moisture tension (Fig. 6). Among the yield components, grains m^{-2} and grains ear^{-1} were more susceptible to aerobic production conditions than grain weight and ears m^{-2}. Grains m^{-2} showed a significantly high and positive correlation with grain yield, and grains per ear correlated positively and significantly with grain per unit area under aerobic cultivation conditions (Fig. 7; Singh et al, unpublished data); hence, improvement of drought resistance of these traits will help in enhancing yield in aerobic rice production systems. Wide variation for yield and yield component stability for water deficit under aerobic rice systems is available among rice genotypes (Table 5; Singh et al, unpublished data), which will be useful in understanding drought resistance at the yield component level. An aerobic rice system

Table 4. Genes for cellular tolerance of water deficit and other stresses.

Gene	Transgenic	Stress phenotype	Reference
OsSNAC1	Constitutive overexpression (P35S) in rice	Transgenic rice showed high spikelet fertility (22–34% higher seed setting than control) in the field under severe drought stress at the reproductive stage Also improves salt tolerance	Hu et al (2006)
OsDREB1A	Rice	Transgenic overexpression—growth retardation under normal growth conditions, improved tolerance of drought, high-salt and low-temperature stresses Higher free proline and various soluble sugars	Ito et al (2006)
		Microarray—target genes identified	Vannini et al (2004)
OsMYB4	*Arabidopsis*	Transgenic overexpression—enhances accumulation of compatible solutes, increases cold and drought tolerance	Mattana et al (2005)
OsISAP1	Tobacco	Transgenic overexpression—enhances tolerance of cold, dehydration, and salt stress	Mukhopadhyay et al (2004)
OsCDPK7	Rice	Transgenic overexpression—enhanced expression of RAB16 Enhanced tolerance of cold, salt, and drought	Saijo et al (2000)
OsMAPK5	Rice	Transgenic overexpression—increased tolerance of drought, salt, and cold stresses Antisense rice/RNAi: defective in drought, salt, and cold tolerance	Xiong and Yang (2003)
OsDREB1A	*Arabidopsis*	Transgenic overexpression—induction of target COR genes Enhanced tolerance of freezing and drought stresses	Dubouzet et al (2003)

with suitable rice genotypes can save up to 50% irrigation water and has great potential to enhance the water productivity of rice (Singh and Chinnusamy 2006).

QTLs for many of these traits for yield under drought stress have been mapped in rice (Table 6). Recently, a major QTL controlling grain yield under drought (explaining 51% of the genetic variance) was identified in random F_3-derived lines from a cross between the upland rice cultivars Vandana and Way Rarem. A larger effect QTL (*qtl12.1*) on chromosome 12 in the 10.2 cM region between SSR markers RM28048 and RM511 was linked to high yield, biomass, harvest index, and plant height and reduced number of days to flowering under reproductive-stage drought stress (Bernier et al 2007).

Heritability of the QTL is important for its use in MAS. Flowering delay and spikelet fertility traits have high correlation with grain yield under flowering-stage drought stress, but these traits have a moderate heritability (Lafitte et al 2003). In a

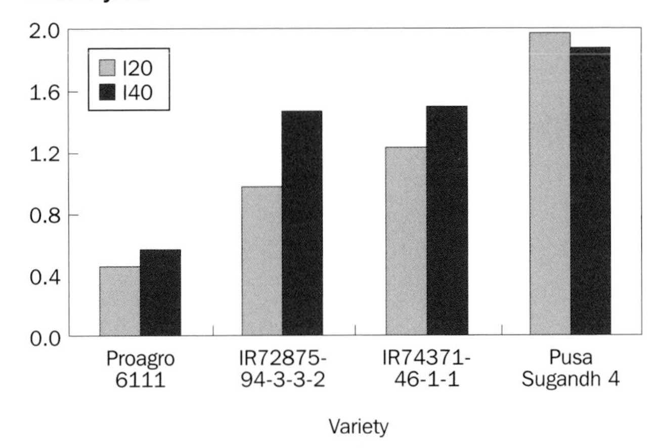

Fig. 5. Drought susceptibility index (DSI) of rice genotypes at two levels of water deficit compared with soil moisture near saturation in an aerobic rice production system (I20 and I40 = irrigation at 20 and 40 kPa soil moisture tension, respectively).

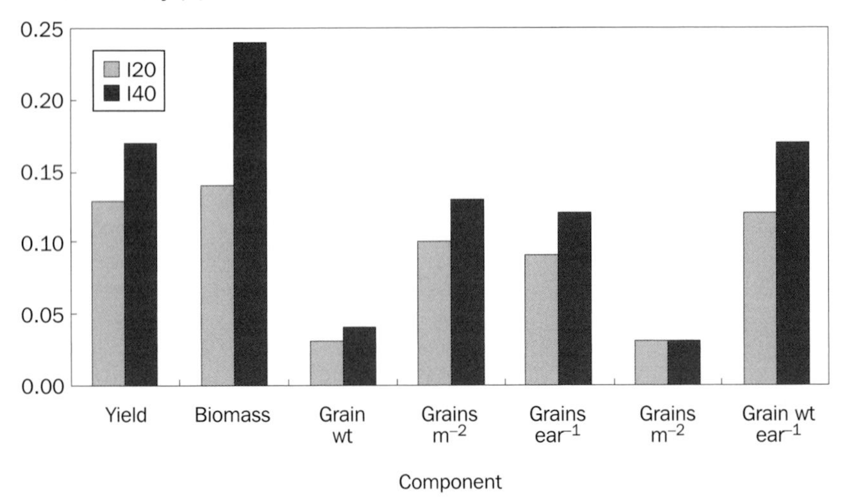

Fig. 6. Differential sensitivity of yield components of rice (measured by drought susceptibility index) at two levels of water deficit compared with soil moisture near saturation in an aerobic rice production system (I20 and I40 = irrigation at 20 and 40 kPa soil moisture tension, respectively).

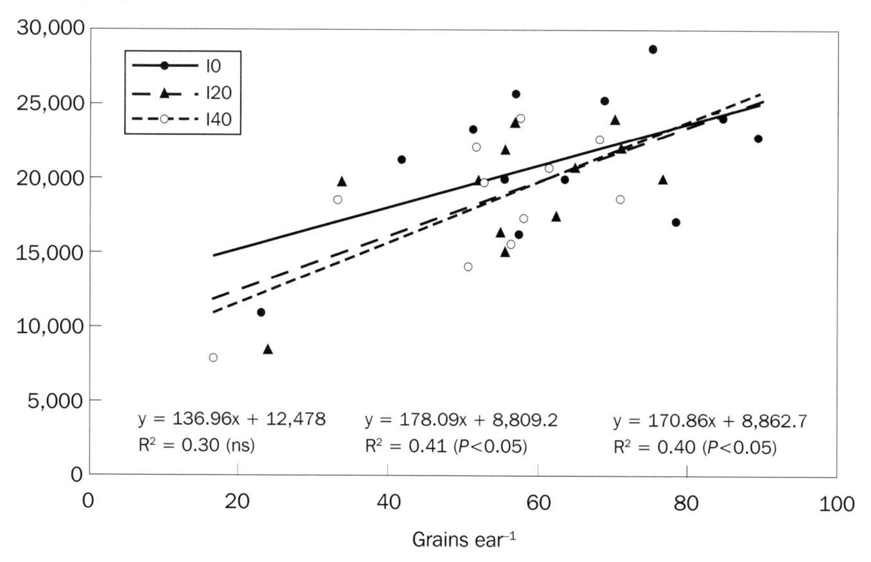

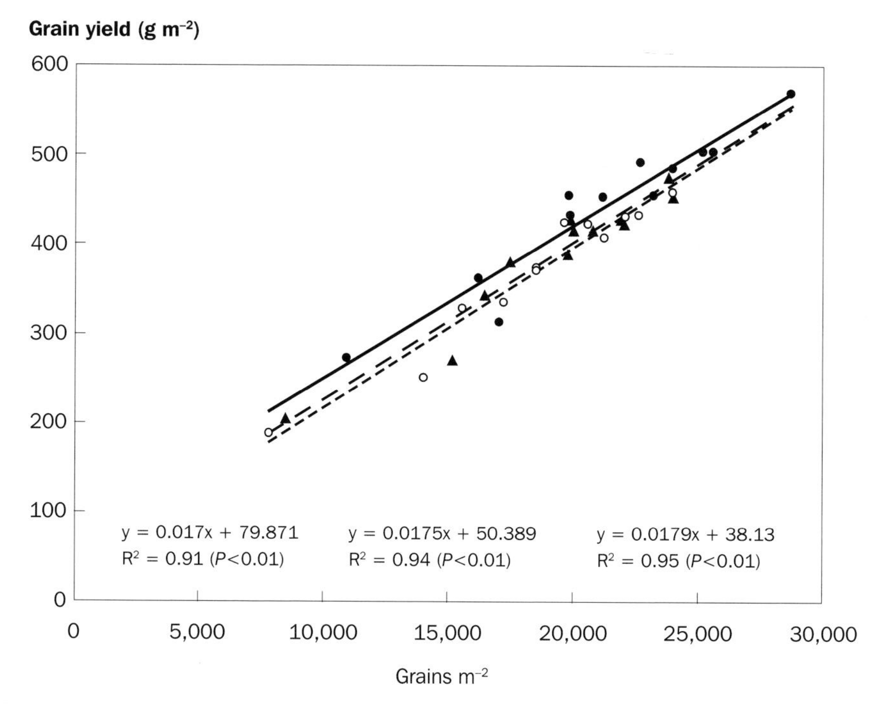

Fig. 7. Correlation between yield and yield components in rice at three levels of water availability in an aerobic rice production system (I0, I20, and I40 = irrigation at 0, 20, and 40 kPa soil moisture tension, respectively).

Table 5. Drought susceptibility index of rice genotypes for yield components at I40 compared with I0 soil moisture tension (DSI>1, susceptible; DSI<1, tolerant).

Genotype	Biomass	Grain weight	Grains m^{-2}	Grains ear^{-1}	Ears m^{-2}	Grain wt ear^{-1}
Proagro 6111	0.70	0.72	0.48	–0.11	2.21	0.11
Pusa 834	0.37	0.04	0.39	0.10	2.08	0.08
Pusa 677	0.71	1.20	0.99	1.53	–1.96	1.27
Pusa 1462-03-14-1	0.60	0.61	1.40	2.99	–7.24	2.10
Pusa RH 10	0.19	1.30	0.06	1.30	–5.37	1.15
IR55419-04	1.19	–0.33	1.10	2.45	–4.82	1.71
IR55423-01	1.35	1.00	1.21	0.07	4.56	0.30
IR71700-247-1-1-2	0.98	2.81	–0.48	–0.20	–2.27	0.69
IR72875-94-3-3-2	1.22	0.69	1.71	–0.08	7.39	0.13
IR74371-54-1-1	1.33	0.87	1.65	0.95	3.40	0.83
Pusa Sugandh 4	1.17	0.95	2.19	1.15	5.42	1.68
IR74371-46-1-1	1.81	1.80	1.42	0.82	2.38	1.48

study to map QTLs for grain yield and its components in 154 DHLs of CT9993-510 × IR62266-42 cross under a line-source sprinkler irrigation system, it was found that broad-sense heritability (H^2) of grain yield and yield components across water regime was relatively low. The H^2 of grain yield and biomass was the lowest at 0.32 and 0.31, respectively, whereas panicle number and plant height had the highest H^2 at 0.61 and 0.73, respectively (Lanceras et al 2004). Understanding and dissecting the complex traits of drought resistance into its component traits with higher heritability by using QTL, eQTL, and molecular approaches will greatly help achieve enhanced drought resistance in rice.

Other production constraints

Nutrient deficiency

Availability of iron (Fe^{2+}) is often limiting in the oxidizing aerobic environment. Iron deficiency leads to a reduction in redox reactions, including photosynthesis and thus biomass production. In addition to iron deficiency, the availability of several other micronutrients such as zinc and manganese may be adversely affected. Phosphorus and zinc availability may also become limiting under aerobic conditions (Singh and Chinnusamy 2006, Choudhury et al 2007). Hence, the development of management options to enhance nutrient availability and genetic enhancement of genotypes for enhanced nutrient uptake and use under water-deficit stress environments are necessary.

Weed management

One of the major advantages of puddling and maintaining standing water continuously is weed control. In aerobic rice systems, puddling and flooding are not practiced, and weeds and rice seeds germinate around the same time. Therefore, weed infestation is a

Table 6. QTLs for yield and yield components in rice.

Trait	Cross and population	QTL no.; share of major QTLs (%)	Reference
Yield components	Tesanai 2 × CB, F_3	Total 44 QTLs; 3 for number of panicles 5 for number of filled grains 6 for total number of spikelets 3 for spikelet fertility 7 for 1,000-grain weight 5 for grain weight per plant 8 for plant height 7 for panicle length	Zhuang et al (1997)
Yield and root traits under limited water	IR64 × Azucena, DHL	A QTL for days to flowering in nonstress negatively influenced root volume and dry weight under stress	Venuprasad et al (2002)
Biomass plant^{-1} Fertility percentage Heading date Harvest index Panicles plant^{-1} Spikelets panicle^{-1} Panicle exsertion Plant height Panicle length 1,000-grain weight Grain yield plant^{-1}	IR64 (an indica variety adapted to irrigated conditions) × Azucena (an upland traditional japonica variety), DHLs	Across 9 locations, 126 QTLs were identified for these 11 traits, of which 34 QTLs were common in more than one environment. 6 QTLs for biomass, 1 for fertility, 7 for heading date, 11 for HI, 6 for panicle no., 6 spikelets panicle^{-1}, 3 for panicle exsertion, 4 for plant height, 8 for panicle length, 10 for 1,000-grain wt., and 3 for grain yield plant^{-1}	Hittalmani et al (2003)
Plant water-stress indicators, phenology, and production traits under control and water-stress conditions in the field	CT9993-5-10-1-M/ IR62266-42-6-2, DHL	DHLs were subjected to drought stress before anthesis in 3 field experiments at 2 locations; a total of 47 QTLs were identified, which individually explained 5% to 59% of the phenotypic variation	Babu et al (2003)
Yield, HI, biomass, days to flowering, % spikelet sterility, total spikelet number, panicle number, plant height	CT9993 × IR62266, DHL	77 QTLs; Y = 7, biomass = 8, HI = 6, DFAIG = 5, TSN = 10, PSS = 7, PN = 23, PH = 11	Lanceras et al (2004)

Continued on next page

Table 6 continued.

Trait	Cross and population	QTL no.; share of major QTLs (%)	Reference
Yield and its components under drought	Bala × Azucena, RILs	Total no. of QTLs 31; for flowering delay 12, panicle length 7, harvest index 5, nodal root thickness 2 A major QTL on Chr 3 for flowering date, panicle number, and biomass production A major QTL for plant height on Chr 1 near the *sd1* gene	Lafitte et al (2004)
Grain yield and its components Spikelet fertility Spikelets per panicle 1,000-grain weight Panicle number	Upland cv. × lowland cv.	32 QTLs 1.29–14.76%	Zou et al (2005)
Grain yield and its components, heading date, plant height	Lemont (japonica) × Teqing (indica), 254 introgression lines	36 QTLs 12 QTLS expressed under drought stress and irrigated conditions 17 QTLs expressed under irrigated conditions only 7 QTLs were induced by drought	Xu et al (2005)
Yield components, root traits	Zhenshan 97 (indica) × IRAT109 (japonica), RILs	27 QTLs for yield, 36 QTLs for root traits under control, 38 QTLs for root traits under drought	Yue et al (2006)
Grain yield under reproductive-stage drought stress	Vandana × Way Rarem, random F_3-derived lines	A larger effect QTL (*qtl12.1*) on chromosome 12 in 10.2 cM region between SSR markers RM28048 and RM511 explains 51% of the genetic variance in yield. This QTL was also linked to high yield, biomass, harvest index, and plant height and reduced number of days to flowering under reproductive-stage drought stress	Bernier et al (2007)

potential threat to successful aerobic rice cultivation. Hence, rice genotypes with high weed competitive ability are required (Singh et al 2002, Zhao et al 2006, Choudhury et al 2007). Development of an effective weed management strategy is therefore essential for enhancing rice production in upland and aerobic production systems.

Pest management

The traditional transplanted lowland rice system is characterized by high humidity within the crop canopy, thus favoring the incidence of insect pests that attack above-ground parts of the plant. In contrast, nematodes appear to be a major threat to aerobic rice cultivation. The rice root-knot nematode (*Meloidogyne graminicola*) is a limiting factor in aerobic rice (Choudhury et al 2007). It might become a yield-reducing factor under successive aerobic rice cultivation.

Conclusions and perspectives

The complex drought resistance mechanism needs to be dissected into highly heritable component traits. QTLs for many component traits of drought resistance have been mapped in rice. However, these QTLs have not been used in marker-assisted selection (MAS) breeding because these genetic studies were largely conducted independent of breeding and because of the unpredictable nature of drought. MAS breeding appears to be successful in producing a promising plant type for enhanced drought resistance (Steele et al 2006). Each trait is controlled by a large number of QTLs. Moreover, for many traits, QTLs for irrigated and drought environments are different. QTLs may also vary from population to population and are influenced by the growing environment. Therefore, the identification of common QTLs across populations under near-realistic field stress (considering target environment) conditions, the combination of metabolomics with eQTLs, map-based cloning, and validation of the master genes of the QTLs involved in drought resistance will be imperative for genetic engineering and molecular breeding to enhance drought resistance and WUE in rice. The results from transgenic plants of rice expressing transcription factors and signaling proteins suggest that transcriptome engineering can be a promising method to engineer drought resistance in rice. Unlike in model plants, in which survival is considered as stress tolerance, in crop plants survival as well as reproduction are equally important. The reproductive phase of crop plants is often more sensitive to drought stress than the vegetative stage. The use of drought-responsive, tissue- and organ-specific promoters thus appears vital for abiotic stress resistance in transgenic crop plants. In most of the reports, transgenic plants have been evaluated for stress resistance under controlled growing conditions for short durations. The effects of stresses in relation to plant ontogeny should be assessed under field conditions.

Defining target environments and accurate phenotyping for the specific target environment (rainfed lowland, upland, and aerobic rice production systems) needs further attention. Nutrient acquisition under soil moisture–deficit conditions is poorly understood. Iron and other-nutrient uptake efficiency of rice under water deficit needs to be studied. As the aerobic rice production system appears to be a promising method to increase the water productivity of rice, understanding of the physiological and molecular basis of rice adaptation to aerobic conditions is pivotal for sustaining rice production. The development of appropriate genotypes with enhanced nutrient uptake (specifically iron) and use, nematode resistance, and weed competitiveness under water-deficit aerobic and upland conditions is the key to increasing grain yield under these environments.

As expected, substantial difference exists in the stress response of rice and *Arabidopsis* at the transcriptome level (Rabbani et al 2003). Thus, knowledge generated in *Arabidopsis* is directly applicable to the manipulation of only some traits of drought resistance of rice. Hence, generation of mutants, accurate phenotyping, map-/eQTL-based cloning, and functional validation of rice genes are imperative to enhance the water productivity and drought resistance of rice. Successful completion of rice genome sequencing (Yu et al 2002, Goff et al 2002, International Rice Genome Sequencing Project 2005) is a major step toward rice crop improvement. The functions for a majority of the genes in the rice genome have not been experimentally studied. Application of forward and reverse genetic approaches and functional genomics tools to understand the function of genes of drought resistance in rice is the need of the hour. TILLING for the homologs of *Arabidopsis* genes that regulate drought resistance and development (e.g., *ERECTA*, Masle et al 2005) will help enhance the pace of our understanding of drought resistance in rice.

The development of an accurate phenotyping protocol and use of uniform protocols will further enhance our understating of stress-responsive mechanisms. Crop management practices such as tillage and nutrient and water management have a significant influence on drought resistance of crops. Hence, studies on the genetic basis of the interaction between crop management practices and drought resistance will be necessary.

Only a few facets of the myriad drought resistance mechanisms of rice have been unraveled thus far by applying molecular tools such as gene disruption and transgenic approaches. The molecular basis of phenotypic and developmental plasticity, and yield component development under drought stress, is poorly understood. Hence, concerted efforts are needed in this direction.

Optimization of crop management practices and the development of drought-resistant and input-responsive genotypes for the aerobic rice production system will greatly enhance the water productivity of rice.

References

Ali ML, Pathan MS, Zhang J, Bai G, Sarkarung S, Nguyen HT. 2000. Mapping QTL for root traits in a recombinant inbred population from two indica ecotypes in rice. Theor. Appl. Genet. 101:756-766.

Babu RC, Nguyen BD, Chamarerk V, Shanmugasundaram P, Chezhian P, Jeyaprakash P, Ganesh SK, Palchamy A, Sadasivam S, Sarkarung S, Wade LJ, Nguyen HT. 2003. Genetic analysis of drought resistance in rice by molecular markers association between secondary traits and field performance. Crop Sci. 43:1457-1469.

Bernier J, Kumar A, Ramaiah V, Spaner D, Atlin G. 2007. A large-effect QTL for grain yield under reproductive-stage drought stress in upland rice. Crop Sci. 47:507-516.
Bhuiyan SI. 1992. Water management in relation to crop production: case study on rice. Outlook Agric. 21:293-299.
Blum A. 2005. Drought resistance, water-use efficiency, and yield potential—are they compatible, dissonant, or mutually exclusive? Aust. J. Agri. Res. 56:1159-1168.
Boling A, Tuong TP, Singh AK, Wopereis MCS. 1998. Comparative root growth and soil water extraction of dry-seeded, wet-seeded and transplanted rice in a greenhouse experiment. Philipp. J. Crop Sci. 23:45-52.
Champoux MC, Wang G, Sarkarung S, Mackill DJ, O'Toole JC, Huang N, McCouch SR. 1995. Locating genes associated with root morphology and drought avoidance in rice via linkage to molecular markers. Theor. Appl. Genet. 90:969-981.
Chinnusamy V, Schumaker K, Zhu JK. 2004. Molecular genetic perspectives on cross-talk and specificity in abiotic stress signalling in plants. J. Exp. Bot. 55:225-236.
Choudhury BU, Bouman BAM, Singh AK. 2007. Yield and water productivity of rice–wheat on raised beds at New Delhi, India. Field Crops Res. 100:229-239.
Condon AG, Richards RA, Rebetzke GJ, Farquhar GD. 2004. Breeding for high water-use efficiency. J. Exp. Bot. 55:2447-2460.
Courtois B, McLaren G, Sinha PK, Prasad K, Yadav R, Shen L. 2000. Mapping QTL associated with drought avoidance in upland rice. Mol. Breed. 6:55-66.
Courtois B, Shen L, Petalcorin W, Carandang S, Mauleon R, Li Z. 2003. Locating QTLs controlling constitutive root traits in the rice population IAC 165 × Co39. Euphytica 134:335-345.
Dubouzet JG, Sakuma Y, Ito Y, Kasuga M, Dubouzet EG, Miura S, Seki M, Shinozaki K, Yamaguchi-Shinozaki K. 2003. *OsDREB* genes in rice, *Oryza sativa* L., encode transcription activators that function in drought-, high-salt- and cold-responsive gene expression. Plant J. 33:751-763.
Farquhar GD, Richards RA. 1984. Isotopic composition of plant carbon correlates with water-use efficiency of wheat genotypes. Aust. J. Plant Physiol. 11:539-552.
Fischer RA, Maurer R. 1978. Drought resistance in spring wheat cultivars. Aust. J. Agric. Res. 29:897-912.
Garg AK, Kim JK, Owens TG, Ranwala AP, Choi YD, Kochian LV, Wu RJ. 2002. Trehalose accumulation in rice plants confers high tolerance levels to different abiotic stresses. Proc. Natl. Acad. Sci. USA 99:15898-15903.
Goff SA, Ricke D, Lan TH, Presting G, Wang R, Dunn M, Glazebrook J, Sessions A, Oeller P, Varma H, Hadley D, Hutchison D, Martin C, Katagiri F, Lange BM, Moughamer T, Xia Y, Budworth P, Zhong J, Miguel T, Paszkowski U, Zhang S, Colbert M, Sun WL, Chen L, Cooper B, Park S, Wood TC, Mao L, Quail P, Wing R, Dean R, Yu Y, Zharkikh A, Shen R, Sahasrabudhe S, Thomas A, Cannings R, Gutin A, Pruss D, Reid J, Tavtigian S, Mitchell J, Eldredge G, Scholl T, Miller RM, Bhatnagar S, Adey N, Rubano T, Tusneem N, Robinson R, Feldhaus J, Macalma T, Oliphant A, Briggs S. 2002. A draft sequence of the rice genome (*Oryza sativa* L. ssp. *japonica*). Science 296:92-100.
Guerra LC, Bhuiyan SI, Tuong TP, Barker R. 1998. Producing more rice with less water from irrigated systems. SWIM Paper No. 5. International Water Management Institute, Colombo, Sri Lanka.
Haque MM, Mackill DJ, Ingram KT. 1992. Inheritance of leaf epicuticular wax content in rice. Crop Sci. 32:865-868.

Hemamalini GS, Shashidar HE, Hittalmani S. 2000. Molecular marker assisted tagging of morphological and physiological traits under two contrasting moisture regimes at peak vegetative stage in rice (*Oryza sativa* L.). Euphytica 112:69-78.

Hittalmani S, Huang N, Courtois B, Venuprasad R, Shashidhar HE, Zhuang JY, Zheng KL, Liu GF, Wang GC, Sidhu JS, Srivantaneeyakul S, Singh VP, Bagali PG, Prasanna HC, McLaren G, Khush G. 2003. Identification of QTL for growth and grain yield-related traits in rice across nine locations in Asia. Theor. Appl. Genet. 107:679-690.

Hu H, Dai M, Yao J, Xiao B, Li X, Zhang Q, Xiong L. 2006. Overexpressing a NAM, ATAF, and CUC (NAC) transcription factor enhances drought resistance and salt tolerance in rice. Proc. Natl. Acad. Sci. USA 103:12987-12992.

International Rice Genome Sequencing Project. 2005. The map-based sequence of the rice genome. Nature 436:793-800.

IRRI. 2001. Annual report 2000-2001: rice research: the way forward. Los Baños (Philippines): International Rice Research Institute. 71 p.

Ito Y, Katsura K, Maruyama K, Taji T, Kobayashi M, Seki M, Shinozaki K, Yamaguchi-Shinozaki K. 2006. Functional analysis of rice DREB1/CBF-type transcription factors involved in cold-responsive gene expression in transgenic rice. Plant Cell Physiol. 47:141-153.

Kamoshita A, Wade LJ, Ali ML, Pathan MS, Zhang J, Sarkarung S, Nguyen HT. 2002b. Mapping QTLs for root morphology of a rice population adapted to rainfed lowland conditions. Theor. Appl. Genet. 104:880-893.

Kamoshita A, Zhang J, Siopongco J, Sarkarung S, Nguyen HT, Wade LJ. 2002a. Effects of phenotyping environment on identification of quantitative trait loci for rice root morphology under anaerobic conditions. Crop Sci. 42:255-265.

Lafitte R, Blum A, Atlin G. 2003. Using secondary traits to help identify drought-tolerant genotypes. In: Fisher KS, Lafitte R, Fukai S, Atlin G, Hardy B, editors. Breeding rice for drought-prone environments. Los Baños (Philippines): International Rice Research Institute. p 37-48.

Lafitte HR, Price AH, Courtois B. 2004. Yield response to water deficit in an upland rice mapping population: associations among traits and genetic markers. Theor. Appl. Genet. 109:1237-1246.

Lanceras JC, Pantuwan G, Jongdee B, Toojinda T. 2004. Quantitative trait loci associated with drought tolerance at reproductive stage in rice. Plant Physiol. 135:384-399.

Laza MR, Kondo M, Ideta O, Barlaan E, Imbe T. 2006. Identification of quantitative trait loci for $\delta^{13}C$ and productivity in irrigated lowland rice. Crop Sci. 46:763-773.

Lian HL, Yu X, Ye Q, Ding X, Kitagawa Y, Kwak SS, Su WA, Tang ZC. 2004. The role of aquaporin RWC3 in drought avoidance in rice. Plant Cell Physiol. 45:481-489.

Lilley JM, Ludlow MM, McCouch SR, O'Toole JC. 1996. Locating QTL for osmotic adjustment and dehydration tolerance in rice. J. Exp. Bot. 47:1427-1436.

Ludlow M, Muchow RC. 1990. A critical evaluation of traits for improving crop yields in water-limited environments. Adv. Agron. 43:107-153.

Masle J, Gilmore SR, Farquhar GD. 2005. The *ERECTA* gene regulates plant transpiration efficiency in *Arabidopsis*. Nature 436:866-870.

Mattana M, Biazzi E, Consonni R, Locatelli F, Vannini C, Provera S, Coraggio I. 2005. Overexpression of *OsMYB4* enhances compatible solute accumulation and increases stress tolerance of *Arabidopsis thaliana*. Physiol. Plant. 125:212-223.

Mukhopadhyay A, Vij S, Tyagi AK. 2004. Overexpression of a zinc-finger protein gene from rice confers tolerance to cold, dehydration, and salt stress in transgenic tobacco. Proc. Natl. Acad. Sci. USA 101:6309-6314.

Peng S, Laza RC, Khush GS, Sanico AL, Visperas RM, Garcia FE. 1998. Transpiration efficiencies of *indica* and improved tropical *japonica* rice grown under irrigated conditions. Euphytica 103:103-108.
Price AH, Steele KA, Moore BJ, Jones RGW. 2002. Upland rice grown in soil-filled chambers and exposed to contrasting water-deficit regimes. II. Mapping quantitative trait loci for root morphology and distribution. Field Crops. Res. 76:25-43.
Price AH, Tomos AD. 1997. Genetic dissection of root growth in rice (*Oryza sativa* L.). II. Mapping quantitative trait loci using molecular markers. Theor. Appl. Genet. 95:143-152.
Price AH, Young EM, Tomos AD. 1997. Quantitative trait loci associated with stomatal conductance, leaf rolling and heading date mapped in upland rice (*Oryza sativa* L.). New Phytol. 137:83-91.
Quarrie SA, Laurie DA, Zhu J, Lebreton C, Semikhodskii A, Steed A, Witsenboer H, Calestani C. 1997. QTL analysis to study the association between leaf size and abscisic acid accumulation in droughted rice leaves and comparisons across cereals. Plant Mol. Biol. 35:155-165.
Rabbani MA, Maruyama K, Abe H, Khan MA, Katsura K, Ito Y, Yoshiwara K, Seki M, Shinozaki K, Yamaguchi-Shinozaki K. 2003. Monitoring expression profiles of rice genes under cold, drought, and high-salinity stresses and abscisic acid application using cDNA microarray and RNA gel-blot analyses. Plant Physiol. 133:1755-1767.
Ray JD, Yu LX, McCouch SR, Champoux MC, Wang G, Nguyen HT. 1996. Mapping quantitative trait loci associated with root penetration ability in rice (*Oryza sativa* L.). Theor. Appl. Genet. 92:627-636.
Richards RA. 2006. Physiological traits used in the breeding of new cultivars for water-scarce environments. Agric. Water Manage. 80:197-211.
Robin S, Pathan MS, Courtois B, Lafitte R, Carandang S, Lanceras S, Amante M, Nguyen HT, Li Z. 2003. Mapping osmotic adjustment in an advanced back-cross inbred population of rice. Theor. Appl. Genet. 107:1288-1296.
Saijo Y, Hata S, Kyozuka J, Shimamoto K, Izui K. 2000. Over-expression of a single Ca^{2+} dependent protein kinase confers both cold and salt/drought tolerance on rice plants. Plant J. 23:319-327.
Sakamoto A, Alia, Murata N. 1998. Metabolic engineering of rice leading to biosynthesis of glycinebetaine and tolerance to salt and cold. Plant Mol. Biol. 38:1011-1019.
Schroeder JI, Kwak JM, Allen GJ. 2001. Guard cell abscisic acid signalling and engineering drought hardiness in plants. Nature 410:327-330.
Serraj R, Sinclair TR. 2002. Osmolyte accumulation: Can it really help increase crop yield under drought conditions? Plant Cell Environ. 25:333-341.
Shen L, Courtois B, McNally KL, Robin S, Li Z. 2001. Evaluation of near-isogenic lines of rice introgressed with QTLs for root depth through marker-aided selection. Theor. Appl. Genet. 103:75-83.
Singh AK, Chinnusamy V. 2004. Biotechnological options for enhancing water use efficiency of rice. Asian Biotechnol. Dev. Rev. 7:109-115.
Singh AK, Chinnusamy V. 2006. Aerobic rice: prospects for enhancing water productivity. Indian Farming 56(7):58-61.
Singh AK, Choudury BU, Bouman BAM. 2002. The effect of rice establishment techniques and water management on crop-water relations. In: Bouman BAM, Hengsdijk H, Hardy B, Bindraban PS, Tuong TP, Lafitte H, Ladha JK, editors. Water-wise rice production. Los Baños (Philippines): International Rice Research Institute. p 237-246.

Steele KA, Price AH, Shashidhar HE, Witcombe JR. 2006. Marker-assisted selection to introgress rice QTLs controlling root traits into an Indian upland rice variety. Theor. Appl. Genet. 112:208-221.

Su J, Wu R. 2004. Stress-inducible synthesis of proline in transgenic rice confers faster growth under stress conditions than that with constitutive synthesis. Plant Sci. 166:941-948.

Tripathy JN, Zhang J, Robin S, Nguyen TT, Nguyen HT. 2000. QTLs for cell-membrane stability mapped in rice (*Oryza sativa* L.) under drought stress. Theor. Appl. Genet. 100:1197-1202.

Tuberosa R, Salvi S. 2006. Genomics-based approaches to improve drought tolerance of crops. Trends Plant Sci. 11:405-412.

Tuong TP, Castillo EG, Cabangon RC, Boling A, Singh U. 2002. The drought response of lowland rice to crop establishment practices and N-fertilizer sources. Field Crops Res. 74:243-257.

Vannini C, Locatelli F, Bracale M, Magnani E, Marsoni M, Osnato M, Mattana M, Baldoni E, Coraggio I. 2004. Overexpression of the rice *OsMYB4* gene increases chilling and freezing tolerance of *Arabidopsis thaliana* plants. Plant J. 37:115-127.

Venuprasad R, Shashidhar HE, Hittalmani S, Hemamalini GS. 2002. Tagging quantitative trait loci associated with grain yield and root morphological traits in rice (*Oryza sativa* L.) under contrasting moisture regimes. Euphytica 128:293-300.

Xiong L, Yang Y. 2003. Disease resistance and abiotic stress tolerance in rice are inversely modulated by an abscisic acid–inducible mitogen-activated protein kinase. Plant Cell 15:745-759.

Xu JL, Lafitte HR, Gao YM, Fu BY, Torres R, Li ZK. 2005. QTLs for drought escape and tolerance identified in a set of random introgression lines of rice. Theor. Appl. Genet. 111:1642-1650.

Yadav R, Courtois B, Huang N, McLaren G. 1997. Mapping genes controlling root morphology and root distribution in a doubled-haploid population of rice. Theor. Appl. Genet. 94:619-632.

Yu J, Hu S, Wang J, Wong GK, Li S, Liu B, Deng Y, Dai L, Zhou Y, Zhang X, Cao M, Liu J, Sun J, Tang J, Chen Y, Huang X, Lin W, Ye C, Tong W, Cong L, Geng J, Han Y, Li L, Li W, Hu G, Huang X, Li W, Li J, Liu Z, Li L, Liu J, Qi Q, Liu J, Li L, Li T, Wang X, Lu H, Wu T, Zhu M, Ni P, Han H, Dong W, Ren X, Feng X, Cui P, Li X, Wang H, Xu X, Zhai W, Xu Z, Zhang J, He S, Zhang J, Xu J, Zhang K, Zheng X, Dong J, Zeng W, Tao L, Ye J, Tan J, Ren X, Chen X, He J, Liu D, Tian W, Tian C, Xia H, Bao Q, Li G, Gao H, Cao T, Wang J, Zhao W, Li P, Chen W, Wang X, Zhang Y, Hu J, Wang J, Liu S, Yang J, Zhang G, Xiong Y, Li Z, Mao L, Zhou C, Zhu Z, Chen R, Hao B, Zheng W, Chen S, Guo W, Li G, Liu S, Tao M, Wang J, Zhu L, Yuan L, Yang H. 2002. A draft sequence of the rice genome (*Oryza sativa* L. ssp. *indica*). Science 296:79-92.

Yue B, Xue W, Xiong L, Yu X, Luo L, Cui K, Jin D, Xing Y, Zhang Q. 2006. Genetic basis of drought resistance at reproductive stage in rice: separation of drought tolerance from drought avoidance. Genetics 172:1213-1228.

Zhang J, Zheng HG, Aarti A, Pantuwan G, Nguyen TT, Tripathy JN, Sarial AK, Robin S, Babu RC, Nguyen BD, Sarkarung S, Blum A, Nguyen HT. 2001. Locating genomic regions associated with components of drought resistance in rice: comparative mapping within and across species. Theor. Appl. Genet. 103:19-29.

Zhao DL, Atlin GN, Bastiaans L, Spiertza JHJ. 2006. Developing selection protocols for weed competitiveness in aerobic rice. Field Crops Res. 97:272-285.

Zheng BS, Yang L, Zhang WP, Mao CZ, Wu YR, Yi KK, Liu FY, Wu P. 2003. Mapping QTLs and candidate genes for rice root traits under different water-supply conditions and comparative analysis across three populations. Theor. Appl. Genet. 107:1505-1515.
Zheng H, Babu R, Pathan M, Ali L, Huang N, Courtois B, Nguyen HT. 2000. Quantitative trait loci for root-penetration ability and root thickness in rice: comparison of genetic backgrounds. Genome 43:53-61.
Zhuang JY, Lin HX, Lu J, Qian HR, Hittalmani S, Huang N, Zheng KL. 1997. Analysis of QTL × environment interaction for yield components and plant height in rice. Theor. Appl. Genet. 95:799-808.
Zou GH, Mei HW, Liu HY, Liu GL, Hu SP, Yu XQ, Li MS, Wu JH, Luo LJ. 2005. Grain yield responses to moisture regimes in a rice population: association among traits and genetic markers. Theor. Appl. Genet. 112:106-113.

Notes

Authors' address: Water Technology Centre, Indian Agricultural Research Institute, New Delhi, India.
Acknowledgments: Work done by us on aerobic rice was supported by CPWF-funded project on "Developing a System of Temperate and Tropical Aerobic Rice (STAR) in Asia."

Effects of irrigation treatment on rice growth and development: comparing a study of rice farming between nonflooding and flooding cultivation

Longxing Tao, Xi Wang, Huijuan Tan, and Shihua Cheng

Interspecific hybrid rice cultivars Xeiyou 9308 and Liangyoupeijiu were used as test materials and three levels of soil water content were applied during grain filling in irrigated fields to observe their effects on translocation and allocation of carbohydrates. The results showed that, in conventional flooding or nonflooding cultivation, the export rates of stored carbohydrates from stems and carbon assimilates from leaves were 60% and 90%, respectively. The export rate of carbohydrates decreased significantly in nonflooding cultivation. The filling grains were the major sink for carbohydrate storage during the grain-filling stage. Grains received nearly 50% of leaf-sheath-stored carbohydrates and 80% of leaf-stored carbon assimilates. In nonflooding conditions, the C assimilates import rates of grains decreased significantly by 10% and 20% from leaf sheath and leaf blades, respectively. Drought stress caused a large decrease in absorbing ability of inferior grains. A two-time water saturation during the grain-filling stage slowed the decline in root respiration and root exudates. The effects caused by drought stress during the grain-filling stage were mostly associated with inferior spikelets, resulting in a lower seed-setting rate and lower grain weight. This may be the main cause for lower yield under water stress during grain filling. We showed in this experiment that there was less effect of water stress during grain filling in Xieyou 9308 than in Liangyoupeijiu. Therefore, it seemed that Xieyou 9308 showed higher drought resistance than Liangyoupeijiu.

China is one of the most water-deficient countries in the world, as more than 72% of the total land area suffers from water stress. The average water resource per unit cropland is only 67% of the average world level, and the average water volume per unit irrigation area is only 19% of the world level (Qu 1998). Almost half of the fresh water is used for rice irrigation, resulting in very low water-use efficiency (the so-called "half kg of rice with 1 ton of water") and high cost of rice production. It is time to limit irrigation water in rice production by using the nonflooding rice farming technique (NFRF). This technique has been studied for many years, but the grain yield of NFRF was not always as high and stable as that of traditional flooding rice farming (FRF) (Luo and Zhang 2001). The main cause may be premature senescence of NFRF during grain filling. It was reported that phenomena such as leaf yellowing, stem softening, and early lodging were usually associated with NFRF during the grain-maturing stage (Zhang and Liu 1988, Huang et al 1999, Yang et al 2002). Some field trials conducted by Zhang and Liu (1988) in Hunan, Anhui, and Jiangsu provinces showed similar results. Wang et al (2004) compared the root system and root activity between NFRF and FRF, and pointed out that NFRF usually showed a larger root system but lower root physiological activity and earlier stem senescence than FRF. With different irrigation treatments, different soil water contents were applied to study their effects on transportation and distribution of both the stored carbohydrates in leaf sheaths and carbon assimilates in leaves. This paper tries to obtain some useful physiological data to support and improve irrigation techniques and to increase grain-filling enrichment using the nonflooding rice farming technique.

Materials and methods

Plant materials

Three-line intersubspecific hybrid rice Xeiyou 9308 (released from the China National Rice Research Institute) and two-line intersubspecific hybrid rice Liangyoupeijiu (released by the Academy of Agricultural Sciences of Jiangsu Province) were used as the test materials. Those two hybrids showed a very high yield potential in large-scale farmers' fields. Xieyou9308 was typical for having large panicles and low tillering ability, and Liangyoupeijiu was typical for more panicles and high tillering ability.

^{3}H-glucose used in this experiment was D-[6-^{3}H]glucose, produced by the China Institute of Atomic Energy, with an irradiation intensity of 18 ci mM^{-1} and irradiation concentration of 3.7×10^7Bq mL^{-1}.

Experimental methods

Treatments. The experiments were conducted at the experiment station of the China National Rice Research Institute. Hybrid rice was planted in cement slots at $2 \times 20 \times 0.7$ m and a bottom sealed with cement. A shield (made of transparent plastic film) was built to prevent rainwater from entering. Water irrigation was controlled and monitored by a water meter. The seedlings were single planted in the slots, and plant density, fertilizer, irrigation, and agro-chemical application were the same as in the main field management (Wang et al 2004). At initial heading stage (5–10% heading), three irrigation treatments were used: flooding irrigation (treatment A), dry cultivation (no watering after initial heading, treatment B), and alternate drying and wetting (after the B treatment, but giving two irrigations at 15 days and 30 days after initial heading stage, soil was saturated, and excess water was allowed to drain off, treatment C), with three replications. Soil samples were taken at a fixed time from treatment A to monitor soil water content under normal irrigation conditions (gravimetric soil water content of the flooded treatment was 63.5% in this soil). Treatment B was dry cultivation, with soil water content higher than 48% (that is, 75% of the saturated SWC) at initial heading stage, and was decreasing to less than 45% for the rice plant to suffer a water deficit stress (a water deficit stress point at SWC less than 70% of saturated SWC). The water stress was becoming severe during maturation. Treatment C was a dry-watering alternation; there was no standing water in the field, but the soil water content was always higher than 70% of the saturated SWC.

Sampling for superior and inferior grains. Sampling of superior and inferior grains followed the method of Wang et al (2001, 2003), that is, grains from the top three primary branches of one panicle are defined as superior grains; usually, most of those spikelets finish their flowering on the first or second day during the heading stage, and very few of them (less than 5%) flower on the third day during the heading stage. Those grains from secondary branches of three base primary branches are defined as inferior grains; usually, those spikelets start flowering at 5 to 9 days after the heading stage. Each sample contained 50 panicles.

Isotope trace experiments. Isotope trace materials were sampled from treatments A, B, and C. Five uniform shoots were labeled on each hill for 20 plants at the initial heading stage, and the other shoots of the labeled plant were cut. Using 5% Tween-20 as a dilute liquid, 20 µL of ^{3}H-glucose was dabbed on each leaf sheath for the labeled shoots for each hill at the pollen cell maturity stage. The same dabbing was done for each flag leaf of the labeled shoots for each hill. ^{3}H-glucose dabbed on the leaf sheaths was considered as stored carbohydrates and ^{3}H-glucose dabbed on the flag leaf was considered as synthesizing carbon assimilates. The whole plant (above the ground) was sampled near maturity stage.

The method of radioactive sample preparation and measurement followed Yu et al (1995). Some 50 mg of dried and smashed sample was oxidized and measured by an LKB-1217 Liquid Scintillation Counting (LSC) instrument. The scintillation liquid was toluene:trition.X-100 = 2:1(v/v), containing 6 g L^{-1} PPO.

Drying of the samples. First, the samples were quickly killed at 105 °C for 15 min, and then were dried at 80 °C for 24 h.

Root respiration measurement. One whole plant with 50 cm of uninjured root system was dug from the field and cut, and we used a portion from 6 to 15 cm measuring downward. The roots were carefully cleaned with water. Root respiration rate was determined using a Warburg Manometer (Wang et al 2003).

Root exudate intensity. Ten uniform plants were selected for each treatment and cut 10 cm upward on the stem. Each stem was covered with a small plastic bag containing absorbent cotton soon after cutting in the late afternoon. Bags were collected 12 h after cutting and root exudates were squeezed out for hormone analysis (Wang et al 2003).

Zeatin measurement. Zeatin analysis was conducted using HPLC (HP1100 Chemical Station) with an HP-ODS C-18 column, 12.5 mm × 4.6 mm, and a DAD detector at 270 nm (Wang et al 2003, Ding and Shen 1985).

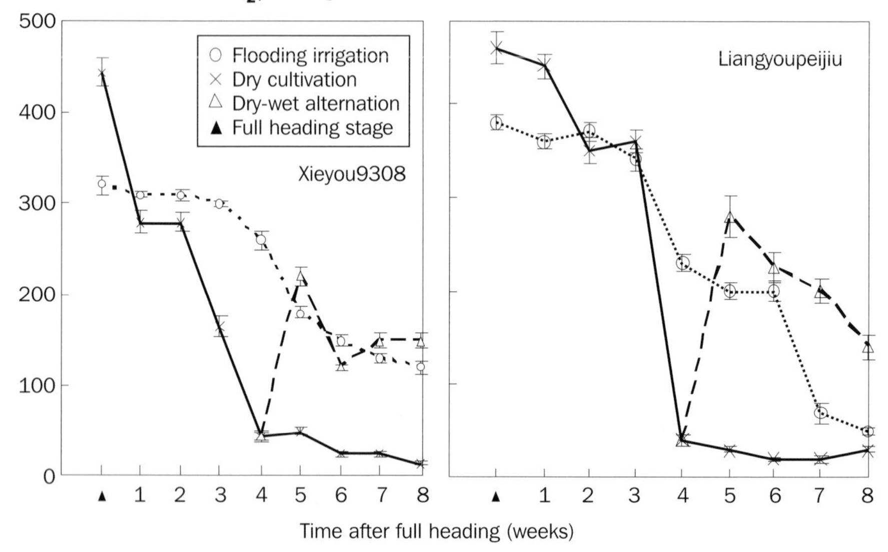

Fig. 1. Effect of different irrigation methods on root respiration rate during the grain-filling stage.

Results and analysis

Effects of soil water content on root respiration

Root respiration intensity decreased after the initial heading stage (Fig. 1). Respiration intensity dropped from 320 O_2 µL h^{-1} g^{-1} fresh weight (FW) at initial heading to 130 O_2 µL h^{-1} g^{-1} FW within 8 weeks in treatment A for Xieyou9308, and an even larger decrease occurred with hybrid Liangyoupeijiu.

At initial heading, root respiration rate was higher in dry cultivation (treatment B) than in treatment A, but it was decreasing faster than in treatment A during grain filling for Xieyou9308. The root respiration intensity of Liangyoupeijiu tended to decrease faster than for Xieyou9308 in treatment B. A two-time pass-through irrigation at 3 weeks after heading (treatment C) showed a significantly better effect in increasing root respiration intensity (recovery from a decrease) than treatment B: (1) root respiration intensity of treatment C for Xieyou9308 recovered to the level of treatment A after a pass-through irrigation; (2) root respiration intensity of treatment C for Liangyoupeijiu was higher than the level of treatment A after a pass-through irrigation.

Effects of soil water content on root exudate volume and zeatin content

The volume of root exudates was declining during maturation for both hybrids, and the rate decreased faster in treatment B than in treatment A (Fig. 2), whereas treatment C was in between treatments A and B.

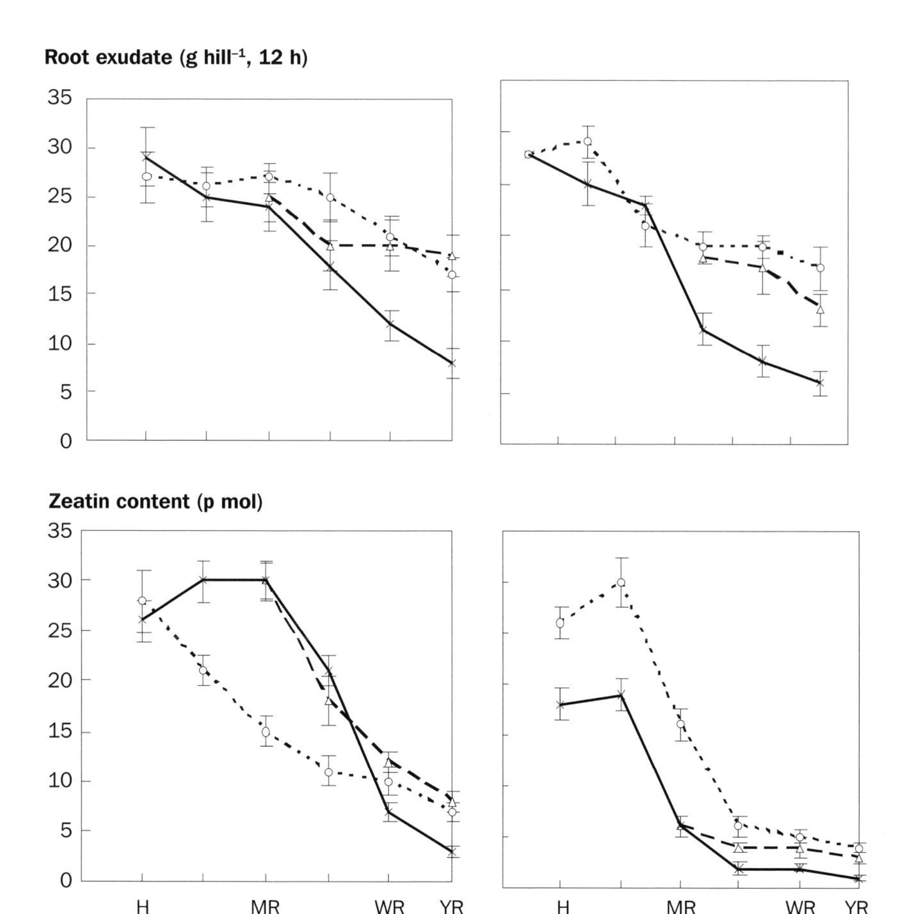

Fig. 2. Effect of irrigation treatment on root exudates and zeatin content. H = heading stage, YR = yellow maturity stage, o = flooding irrigation, MR = milk ripening stage, × = dry cultivation, WR = wax ripening stage, and Δ = dry-wet alternation.

Zeatin content was also declining during maturation, but there was some difference among treatments. In treatment A, the zeatin content in Liangyoupeijiu was higher than in Xieyou9308 before milk ripening stage (MR); then, zeatin content in Liangyoupeijiu declined to a similar value as Xieyou9308 after MR until maturity. The zeatin content in treatment B was higher than in treatment A for Xieyou9308 and lower than in treatment A for Liangyoupeijiu, but the zeatin content for both hybrids was lower in treatment B than in treatment A near maturity. Treatment C was able to maintain the zeatin content for both hybrids.

Table 1. Effects of irrigation treatments on the transportation of ^{3}H-glucose during the grain-filling stage (Bq × 10^4).

Combination	Treatment[a]	Stem and sheath[b]		Leaf		Stalk		Grain	
		Bq	% of total	Bq	% of total	Bq	% of total	Bq	% of total
Xieyou9308	A	3,145	43.6	346	4.8	42	0.6	3,676	51.0
	B	3,306	48.7**[c]	481	7.1*	27	0.4	2,974	43.8*
	C	3,115	39.6	589	7.5*	47	0.6	4,113	52.3
Liangyoupeijiu	A	2,723	42.5	512	8.0	50	0.8	3,115	48.7
	B	3,519	50.2**	527	7.5	49	0.7	2,908	41.5**
	C	3,619	45.3*	584	7.3	44	0.6	3,744	46.9*

[a]A = water irrigation, B = dry cultivation, C = dry-wet alternation. [b]Labeling part. [c]*, ** significantly different from the control (treatment A) at 0.05 and 0.01 level, respectively.

Transportation and allocation of labeled carbohydrates stored in leaf sheath

^{3}H-glucose was fed to the leaf sheath at pollen maturity. This ^{3}H-glucose represented stored carbohydrate. About 40–50% was allocated to grains and less than 10% was translocated to the leaf and panicle branch. This distribution was affected by irrigation treatment (Table 1). Of the two tested hybrids in this experiment, 60% of the stored carbohydrate was translocated into grain in the flooding irrigation treatment (treatment A), and less stored carbohydrate was translocated into grain in dry cultivation (treatment B).

With dry cultivation, the export of stored carbohydrates decreased by 31% (P<0.01) and 18% (P<0.05) for Liangyoupeijiu and Xieyou9308, respectively, when compared with flooding irrigation (treatment A). There were different effects of alternate drying and wetting (treatment C) on the transportation rate of stored carbohydrates for the two hybrids. For Liangyoupeijiu, the export of stored carbohydrates was an 18% (P<0.05) decrease compared with that of flooding irrigation, and there was no significant effect for that of Xieyou9308. The transport efficiency of stored carbohydrates to grain, which was also considered as a physiological activity of spikelets (Lu et al 1988), was affected by the irrigation treatment. Import rate of the stored carbohydrates to grain was 17% (P<0.05) and 23% (P<0.01) less for Xieyou9308 and Liangyoupeijiu, respectively, by dry cultivation (treatment B) compared with treatment A. In alternate drying and wetting (treatment C), the import rate of stored carbohydrates to grain was 13% (P<0.05) less for Liangyoupeijiu when compared with that of treatment A, and no significant effects on Xieyou9308.

Table 2 shows the translocation rate for stored carbohydrates: in the flooding irrigation treatment (A), the translocation rate for Xieyou9308 was higher than that of Liangyoupeijiu. Alternate drying and wetting (C) had no effect on carbohydrate translocation to panicles for both hybrids. The dry-cultivation treatment (B) showed a decrease in translocation rate for carbohydrates of 1.3 Bq × 10^4 grain^{-1}

Table 2. Effects of irrigation treatment on allocation of ^{3}H-glucose from sheath to grains during the grain-filling stage (Bq × 10^4).

Hybrid	Treatment	Whole panicle			Superior grains			Inferior grains			Other grains		
		Panicle		Bq grain^{-1}	Panicle		Bq grain^{-1}	Panicle		Bq grain^{-1}	Panicle		Bq grain^{-1}
		(Bq)	(%)		(Bq)	(%)		(Bq)	(%)		(Bq)	(%)	
Xieyou9308	A	3,676	100	14.5	1,875	51.0	16.7	526	14.3	12.4	1,076	34.7	14.3
	B	2,974	100	13.2*[a]	1,454	48.9	16.0	289	9.7	7.2**	1,338	44.9	13.6
	C	4,113	100	14.8	2,081	50.6	16.4	505	12.3	13.0	1,526	37.1	14.0
Liangyoupeijiu	A	3,115	100	15.3	1,464	47.6	17.2	258	8.3	13.1	1,115	35.8	15.0
	B	2,980	100	11.3**	1,287	43.2	17.0	186	6.4	6.3**	1,465	50.4	13.6*
	C	3,744	100	15.4	1,760	47.0	16.7	262	7.0	12.0*	1,722	46.0	14.2

[a]*, ** = significantly different from the control (treatment A) at 0.05 and 0.01 level, respectively.

(8.9% decrease, $P<0.05$) and 4.0 Bq × 10^4grain^{-1} (26% decrease, $P<0.01$) for Xieyou9308 and Liangyoupeijiu, respectively.

Translocation of stored carbohydrates into grains was greater in superior spikelets than in inferior spikelets, regardless of irrigation treatment. There were significant effects of irrigation treatment on the translocation of stored carbohydrate to inferior spikelets. In the dry-cultivation treatment (B), the stored carbohydrate translocation rate of inferior spikelets decreased to 5.2 Bq × 10^4 grain^{-1} (41.9%, $P<0.01$) and 6.8 Bq × 10^4 grain^{-1} (51.9%, $P<0.01$) for Xieyou9308 and Liangyoupeijiu, respectively, when compared with the flooding irrigation treatment. Similar results were also obtained with alternate drying and wetting, and a 1.1 Bq × 10^4 grain^{-1} (8.4%, $P<0.05$) decrease was observed in hybrid rice Liangyoupeijiu compared with treatment A.

Transportation and allocation of labeled carbon assimilates from flag leaf

The data in Table 3 were based on data collected from the aboveground part of the rice plant. Some 80–90% of the labeled carbon assimilates were translocated out from the labeled leaf and the two hybrids showed no significant differences. However, the irrigation treatment affected the export rate of the labeled carbon assimilates. The carbon assimilate export rate decreased by 62% and 61% for Xieyou9308 and Liangyoupeijiu, respectively, in dry cultivation (treatment B) when compared with flooding irrigation (treatment A).

Almost 80% of the carbon assimilates were allocated to grain for both hybrids, but this was affected by irrigation treatment. There was a 9.2% and 11.5% ($P<0.01$) decrease for Xieyou9308 and Liangyoupeijiu, respectively, caused by dry cultivation compared with flooding irrigation. The export rate from the flag leaf to other parts except grain was not significantly affected by irrigation treatment.

Translocation of carbon assimilates per single grain was affected by irrigation treatment (Table 4) more in Liangyoupeijiu than in Xieyou9308, and Liangyoupeijiu showed a 6.9 Bq × 10^4 grain^{-1} (20.2%, $P<0.01$) decrease in dry cultivation (treatment B) compared with treatment A. There was no significant effect of treatment C on translocation of carbon assimilates per single grain for the two hybrids. Translocation of carbon assimilates per single superior grain was not significantly affected by irrigation treatment for Xieyou9308, but a 6.8 Bq × 10^4 grain^{-1} (18.6%, $P<0.01$) decrease for Liangyoupeijiu was caused by dry cultivation compared with flooding irrigation. Translocation of carbon assimilates per single inferior grain was negatively affected by the dry-cultivation treatment: there was a 6.5 Bq × 10^4 grain^{-1} (decrease of 36.7%, $P<0.01$) and 7.6 Bq × 10^4 grain^{-1} (decrease of 41.7%, $P<0.01$) decrease for Xieyou9308 and Liangyoupeijiu, respectively, when compared with that of the flooding irrigation treatment.

Grain weight and seed setting rate of superior and inferior grains

The seed setting rate of superior grains for both hybrids decreased in the dry-cultivation treatment, and dry cultivation had more negative effects in Liangyoupeijiu than in Xieyou9308. There were no significant differences in seed setting rate in the dry-

Table 3. Effects of irrigation treatments on translocation and allocation of ^3H-glucose from flag leaf to organs during the grain-filling stage (Bq × 10^4).

Combination	Treatment	Flag leaf[a,b]		Other leaves		Stem and sheath		Stalk		Grains	
		Bq	%	Bq	%	Bq	%	Bq	%	Bq	%
Xieyou9308	A	766	12.0	165	2.6	185	2.9	48	0.7	5,243	81.8
	B	1,235	18.0*	179	2.6	212	3.1	51	0.7	5,176	75.5**
	C	726	10.5	163	2.4	172	2.5	47	0.7	5,796	84.0
Liangyoupeijiu	A	804	12.2	145	2.2	167	2.5	47	0.7	5,432	82.4
	B	1,300	19.7*	190	2.9	253	3.8	50	0.8	4,812	72.9**
	C	763	12.3	160	2.6	171	2.8	49	0.8	5,072	81.6

[a]Labeling part. [b]*, ** = significantly different from the control (treatment A) at 0.05 and 0.01 level, respectively.

Table 4. Effects of irrigation treatment on allocation of ^{3}H-glucose from sheath to grains during the grain-filling stage (Bq × 10^4).

Combination	Treatment	Total grains per panicle			Superior grains			Inferior grains			Other grains		
		Panicle		Bq grain^{-1}	Panicle		Bq grain^{-1}	Panicle		Bq grain^{-1}	Panicle		Bq grain^{-1}
		(Bq)	(%)		(Bq)	(%)		(Bq)	(%)		(Bq)	(%)	
Xieyou9308	A	5,743	100	31.6	3,704	64.5	45.3	834	14.6	17.7	1,201	20.9	31.6
	B	5,176	100	30.6	3,395	65.6	43.9	352	6.8	11.0**	1,428	27.6	30.6
	C	5,796	100	32.4	3,709	64.0	45.3	725	12.5	16.8	1,363	23.5	33.0
Liangyoupeijiu	A	5,432	100	34.2	2,889	53.2	46.4	769	14.0	18.2	1,781	32.8	35.2
	B	4,812	100	27.3*[a]	2,540	52.8	39.6**	269	5.6	10.6**	2,001	41.6	39.8*
	C	5,072	100	31.6	3,719	53.6	44.8	598	11.8	16.9	1,755	34.6	30.7

[a]*, ** = significantly different from the control (treatment A) at 0.05 and 0.01 level, respectively.

watering alternation treatment (treatment C) compared with that of treatment B. A decrease in grain weight was also caused by the dry-cultivation treatment for the two hybrids, and Liangyoupeijiu had a more significant grain weight decrease ($P<0.05$) than Xieyou9308.

Inferior grains suffered more under dry cultivation than superior grains for seed setting rate and grain weight, and seed setting rate and grain weight decreased significantly for both Xieyou9308 ($P<0.05$) and Liangyoupeijiu ($P<0.01$). Treatment C was significantly better for seed setting rate and grain weight of superior and inferior grains.

Discussion

Scientists conducted many studies on hybrid rice, especially on intersubspecific hybrid rice. They cut the leaves, thinned spikelets, or carried out isotope tracing experiments or dry matter accumulation measurement of organs to explore the physiological reasons for lower seed setting rate and premature plant senescence during the maturing stage. It was reported that phenomena such as early leaf senescence, a decline in photosynthesis or biomass accumulation rate, and lower seed setting rate were always associated with hybrid rice, especially for intersubspecific hybrid rice with a large sink and source (Lu et al 1988, Wang et al 1966). Regarding the effects of export of stored carbohydrates in the leaf sheath on seed setting rate, it was previously assumed that it was not a major cause of the lower seed setting rate of hybrid rice, including intersubspecific hybrid rice. However, other scientists had the opposite opinion. We had reported that the drawing potential (or physiological activity) of spikelets for carbon assimilates (or all kinds of carbohydrates) was important to seed setting rate (Wang et al 2000). There were few reports on the physiological effects of water stress on seed setting of hybrid rice. We showed a physiological mechanism of drought stress during the grain-filling stage on seed setting. Under normal flooding irrigation (treatment A), the translocation of stored carbohydrate from the leaf sheath (radioactive isotope labeled) and carbon assimilates (radioactive isotope labeled) of the flag leaf was 60% and 90%, respectively, of which 50% and 80%, respectively, was distributed to grains. In dry cultivation (treatment B), however, the translocation of stored carbohydrates from leaf sheath to grains was 16.7% ($P<0.05$) and 23.0% ($P<0.01$), which decreased, respectively, for Xieyou9308 and Liangyoupeijiu. In addition, the translocation of carbon assimilates from the flag leaf to grains was 9.3% ($P<0.05$) and 11.5% ($P<0.01$), a decrease for both Xieyou9308 and Liangyoupeijiu, respectively. Therefore, it was considered that, under dry-cultivation conditions, translocation of carbohydrates from source to sink was not rapid, leading to a higher percentage of carbon assimilates nontranslocated. We also consider in this paper that, under dry cultivation, the physiological activity of spikelets decreased (Yang et al 2001, Beltrano et al 1998, Parthier 1990); therefore, the translocation of carbon assimilates from the source of filling grains declined (Wang et al 2000).

In this experiment, the translocation of carbon assimilates of superior spikelets was relatively stable for two hybrid combinations. There was a small effect of drought stress on translocation, only a slight decrease. However, there was a significant decrease in translocation caused by drought stress for inferior spikelets ($P<0.01$). This was an example of "intergrain apical superiority" (Wang et al 2001). Under dry-cultivation conditions, the physiological activity of superior spikelets was stronger than that of inferior spikelets, as was the translocation of carbon assimilates. Usually, the filling duration of superior spikelets was 10–20 days shorter than that of inferior spikelets; in other words, superior spikelets suffered from a shorter time of drought stress than inferior spikelets. Therefore, blight effects caused by drought stress during the grain-filling stage were mostly associated with inferior spikelets, resulting in a lower seed setting rate and lower grain weight. This may be the main cause for lower yield under water stress.

We showed in this experiment that there was less effect of water stress during grain filling in Xieyou9308 than in Liangyoupeijiu. It seemed that Xieyou9308 showed higher drought resistance than Liangyoupeijiu. Further experimentation is needed to understand the physiological mechanisms.

References

Beltrano J, Ronco MG, Montalaldi ER. 1998. Carbon senesence of flag leaves in ears of wheat hastened by methyl jasmonate. J. Plant Growth Regul. 17:53-57.

Ding J, Shen ZD. 1985. Cytokinin substances in cotton exudate. Acta Phytophysiol. Sin. 11(3):249-259.

Huang YD, Zhang ZL, Wei FZ. 1999. Ecophysiological effect of dry-cultivated and plastic film-mulched rice planting. Chinese J. Appl. Ecol. 10(3):305-308.

Lu DZ, Pan YC, Ma YF, et al. 1988. Physiological and biochemical studies on leaf senescence at heading and grain formation stage in hybrid rice. Sci. Agric. Sin. 21(3):21-26.

Luo LJ, Zhang QF. 2001. The status and strategy on drought resistance of rice (*Oryza sativa* L.). Chinese J. Rice Sci. 15(3):209-214.

Parthier BJ. 1990. Hormonal regulators or stress factors in leaf senescence. J. Plant Growth Regul. 9:57-63.

Qu GP. 1998. Some problems of water environmental management in China. China Environ. Sci. 8(3):1-4.

Wang HX, Zhang ZL, Ren GG, et al. 1966. Study on the distribution of photosynthetic products and plumpness of two-line intersubspecific hybrid rice. Acta Agric. Naclwatae Sin. 10(3):166-172.

Wang X, Tao LX, Huang XL, et al. 2003. Seed setting characteristics and physiological bases of subspecific hybrid rice Xieyou 9308. Acta Agron. Sin. 29(4):530-533.

Wang X, Tao LX, Tian SL, et al. 2000. The absorption for carbohydrate between vigorous and weak spikelets in hybrid rice and its restorers. Acta Agric. Naclwatae Sin. 14(2):76-84.

Wang X, Tao LX, Xu RS, et al. 2001. Apical-grain superiority in hybrid rice. Acta Agron. Sin. 27(6):980-985.

Wang Xi, Tao Longxing, Huang Xiaolin, et al. 2004. Study on non-flooding farming technique in paddy field: technique specification and formation of yield components. Sci. Agric. Sin. 37(4):502-509.

Yang JC, Zhang WH, Wang ZQ, et al. 2001. Source-sink characteristics and the translocation of assimilates in new plant type and japonica/indica hybrid rice. Sci. Agric. Sin. 34(5):511-518.

Yang JC, Wang ZQ, Liu LJ. 2002. Growth and development characteristics and yield formation of dry-cultivated rice. Acta Agron. Sin. 28(1):11-17.
Yu MY, Wang X, Tao LX, et al. 1995. Effect of S-07 on $^{14}CO_2$-assimilation and distribution of assimilates during ripening stage of wheat. Acta Agric. Naclwatae Sin. 9(2):102-106.
Zhang RK, Liu BK. 1988. Dry-cultivated rice production and cultivation techniques. Hunan J. Agric. Sci. 2:30-36.

Notes

Authors' address: State Key Laboratory of Rice Biology/China National Rice Research Institute, Hangzhou 310006.

Genes and genomics for drought-resistant rice

Gene expression analysis and data mining from microarray analysis applied to drought stress in rice

Kouji Satoh, Koji Doi, Toshifumi Nagata, Aeni Hosaka, Kohji Suzuki, Xumei Ji, Muturajan Raveendran, Hei Leung, John Bennett, and Shoshi Kikuchi

To elucidate global responses to drought stress in rice, we used a 60-mer oligomer microarray covering 22K of unique genes based on the sequences of full-length cDNA clones to profile gene expression changes in shoots at the seedling stage and in peduncles at the heading stage. Cluster analysis of genes up- and down-regulated by drought stress at these two different growth stages revealed stage-specific and stress treatment–specific gene expression profiles. Among 503 differentially expressed transcription factor–encoding genes, all the paralogous members of PHD and SNF2 were up-regulated, and those of Jumonji and TCP were down-regulated, by four drought-stress treatments. AP2-EREBP, AUX/IAA, bZIP, C2C2-GATA, C3H, CPP, HB, HMG, HSF, MYB-related, NAC, SBP, SNF2, and Trihelix were commonly up-regulated, and Alfin-like, AUX/IAA, BES1, bHLH, bZIP, MYB, NAC, WRKY, and ZIM were commonly down-regulated, by the four drought-stress treatments. The promoter regions (1 kb upstream from the start site of transcription) of the genes clustered after microarray experiments were examined by using information from the PLACE database. The *cis*-elements known to be localized in the promoter regions of drought stress–responsive genes in *Arabidopsis* were also found to be localized in the promoter regions of the rice genes. Data mining using gene annotation data (e.g., Rice Annotation Project database, Osa1 from TIGR Gene Ontology term), pathway data, and genome-mapping data suggested the existence of transcription networks of drought stress–responsive genes.

Drought stress is a serious problem that threatens the sustainability of agricultural crops worldwide (Widawsky and O'Toole 1996). It is highly important that we breed or genetically engineer crops with improved stress tolerance, but one problem that remains to be solved is the identification of key genes for breeding or engineering of transgenic crops with improved drought tolerance. Drought induces many biochemical and physiological responses in plants attempting to survive such environmental challenges. Unraveling the molecular mechanism underlying stress tolerance may allow us to improve stress tolerance in plants. In response to drought stress, the expression of various genes is up-regulated or down-regulated. These changes in expression can detoxify the effect of stress through adjustments in the cellular environment and plant tolerance.

Gene expression studies have identified a number of the genes involved in drought response (Kreps et al 2002, Ozturk et al 2002, Seki et al 2002, Bray 2004, Kawaguchi et al 2004, Talamè et al 2007). Many of these genes are regulated by common transcription factors such as those of the CBFs/DREBs (AP2/EREBP family), MYB, MYC, and bZIP families. Stress-responsive transcription factors have been among the foci of studies of drought stress tolerance. Major stress-responsive transcription factors have been extensively analyzed in *Arabidopsis thaliana* (Stockinger et al 1997, Zhang et al 2004, Yamaguchi-Shinozaki and Shinozaki 2006, Yamasaki et al 2008). One subset of stress-responsive genes is under the control of the CBF/DREBs (C-repeat binding factor/dehydration responsive element binding protein) family of transcription factors (Stockinger et al 1997, Liu et al 1998). Ectopic expression of the normally stress-induced CBF/DREB genes in transgenic plants has led to constitutive expression of downstream stress-responsive genes and improved tolerance of drought, salinity, and freezing stress (Jaglo-Ottosen et al 1998, Kasuga et al 1999, Gilmour et al 2000).

Recently, an *Arabidopsis* NFYB (B subunit of nuclear factor Y) family transcription factor was reported to provide drought tolerance in *Arabidopsis* and maize (*Zea mays*), implying the existence of a previously unknown drought-tolerance mechanism (Nelson et al 2007). In addition, overexpression of a rice (*Oryza sativa*) NAC-type transcription factor endows drought tolerance in rice (Hu et al 2006).

Two aspects of reproductive development are highly drought-sensitive in rice: spikelet fertility (Ekanayake et al 1989) and panicle exsertion (O'Toole and Namuco 1983). Exsertion of the panicle is driven by elongation of the peduncle, the uppermost internode of the stem. In most cereal crops, panicle or head exsertion is completed several days before anthesis begins in the florets, but, in rice, panicle exsertion overlaps in time with anthesis of the spikelets, with any given spikelet emerging from the flag-leaf sheath only about a day before it undergoes anthesis. Depending on its severity, drought stress either slows or stops both exsertion and anthesis (O'Toole and Namuco 1983). Re-watering may allow both processes to resume, but usually the peduncle fails to achieve full length after stress, leaving a significant fraction of the panicle trapped within the flag-leaf sheath. The unexserted spikelets are sterile (Mackill et al 1996).

The completion of rice genome sequence analysis by the International Rice Genome Sequencing Project (IRGSP 2005), the Beijing Genomics Institute (Yu et al 2002), and Syngenta (Goff et al 2002) has yielded many kinds of rice functional genomics tools, including whole-genome sequences from the japonica cultivar Nipponbare and the indica cultivar 93-11, collections of rice full-length cDNA clones and their complete and partial end sequences (Kikuchi et al 2003, Satoh et al 2007), microarray systems based on full-length cDNA sequences, expressed sequence tags (ESTs) and predicted genes in the genome sequences, and many kinds of mutants with insertion of *Tos17*, *Ac-Ds*, and T-DNAs (Hirochika et al 2004). Our research has focused on transcriptome analysis in rice and has contributed to the collection of full-length cDNA clones and the establishment of a microarray system. To date,

37,133 full-length cDNA clones have been collected from more than 20 rice tissues; their complete sequences were presented at the KOME site (Kikuchi et al 2003, Satoh et al 2007).

In addition to the 37K completely sequenced full-length cDNA clones, we also have 530K single pass–sequenced cDNA clones. Mapping of the sequences of these clones has revealed 29,800 transcription units (TU) on the rice genome pseudomolecules (Satoh et al 2007). Using the sequence information for ESTs and full-length cDNA clones, we have developed a rice microarray system for gene expression analysis. We first established 1,265 and 8,987 cDNA-based microarray systems (Yazaki et al 2000, 2003). Probed cDNA clones were originated in the first phase of the rice genome project (1991-96) (Yamamoto and Sasaki 1997). Second, a 60-mer oligoarray system covering 22K genes was established and commercialized by Agilent Technologies; the probe sequences are based on 29,100 rice full-length cDNA sequences (Yazaki et al 2004). Many experiments have been conducted with these microarray systems, and the gene expression data have been deposited in the Rice Expression Database. In this study, using the 22K oligo-microarray system based on the sequences of full-length cDNA clones, we analyzed global gene expression in response to drought-stress treatments. Rice plants in two developmental stages—10-day-old seedlings and the panicle exsertion stage—were targeted. We isolated the genes specifically and nonspecifically responsive to drought by comparison of responses to drought and other stress treatments. Analyses and data mining revealed gene sets that were up-regulated by certain transcription factors in response to drought-stress treatment and others that were down-regulated. Among the genes encoding proteins involved in many metabolic pathways, those responsible for drought stress, especially in relation to energy formation processes, such as carbohydrate-, lipid-, and energy-related pathways, are listed.

Materials and methods

Plant materials and RNA preparation

Laboratory-scale stress treatments of young seedlings at NIAS. Rice seeds (*Oryza sativa* L. subsp. *japonica* Nipponbare) were supplied by Dr. M. Yano of the National Institute of Agrobiological Sciences (NIAS). Seedlings were grown under hydroponic conditions at 28 °C under a 16:8 h light:dark cycle for 10 days after water absorption.

For stress treatment, 10-day-old seedlings were exposed to one of the following: cold stress (incubation at 10 °C for 24, 48, or 72 h); salt stress (addition of 150 mM sodium chloride); osmotic stress (addition of 260 mM mannitol, 24 h); flood stress (submergence for 1, 6, 24, or 72 h); or PEG stress (a treatment mimicking drought stress in the laboratory: addition of 25% polyethylene glycol 6000 for 1, 9, or 24 h). Control seedlings were grown under standard hydroponic conditions for 11 days. Controls for all microarray experiments were 10-day-old seedlings before each stress treatment.

Phenotype was monitored during treatment and plant tissue samples were harvested after the indicated hour treatment duration and stored at –80 °C. After 11 days of growth, control seedlings were harvested and stored at –80 °C. To test the reproducibility of the microarray experiments, three biological repeats were made by separating the samples into three parts and thus extracting the RNA three times.

Total RNAs were prepared from each sample by using an RNeasy Midi Kit (Qiagen, Tokyo, Japan). mRNAs were purified with an Oligotex-dT30 Kit (Takara, Shiga, Japan). The quality of the preparations was assessed by using an RNA 6000 Nano Lab-on-a-Chip Kit with a 2100-Bioanalyzer System (Agilent Technologies, Tokyo, Japan), and the quantity was assessed with a NanoDrop microscale spectrophotometer (NanoDrop Technologies Inc., Wilmington, DE).

Drought-stress treatment of two indica cultivars, Apo and IR64. IR64 (a high-yielding, drought-susceptible, lowland indica variety; Lafitte et al 2001) and Apo (a moderately drought-tolerant, upland indica variety; Atlin et al 2004) were grown in pots under greenhouse conditions at the International Rice Research Institute, Los Baños, Philippines. Plants were grown in pots (three per pot) filled with 3.5 kg of oven-dried soil (upland soil + coconut husk in a 3:1 v/v ratio) and the recommended basal fertilizer. Plants were watered twice daily. Drought stress was induced at 33 days after seeding (DAS) by withholding water from one set of plants and by maintaining corresponding well-watered control plants. Physiological parameters—namely, stomatal conductance, leaf transpiration rate, relative water content, and leaf water potential—were recorded in both control and stressed plants when the available soil moisture level reached around 25% in the stressed pots. The top two leaf tissues after 24 h stress were collected from both control and stressed plants and frozen in liquid N for expression analysis. To test the reproducibility of the microarray, two biological repeats were made by separating the samples into two parts and thus extracting the RNA twice.

Total RNA was extracted from both control and drought-stressed leaves of IR64 and Apo by the TRIZOL reagent method as per the manufacturer's instructions (Invitrogen Inc., Carlsbad, CA). The quality of the RNA extracted was checked by both electrophoresis and a bioanalyzer. RNA was quantified by spectrophotometric reading.

Drought-stress treatment of IR64 during peduncle elongation. IR64 was grown under well-watered conditions in pots. Starting 3 days before heading, water was withheld to induce mild drought stress and delay heading. RNA was purified from whole peduncles (uppermost internodes) excised from well-watered plants at heading and from drought-stressed plants at the end of the 4-day stress period. For the reproducibility check of the microarray experiments, two biological repeats were made by separating the samples into two parts and thus extracting the RNA twice.

Total RNA was extracted from rice whole peduncles by the TRIZOL protocol, in accordance with the manufacturer's instructions (Invitrogen). The quality of the preparations was assessed by using an RNA 6000 Nano Lab-on-a-Chip Kit with a 2100-Bioanalyzer System (Agilent Technologies, Tokyo, Japan), and the quantity was assessed with a NanoDrop microscale spectrophotometer (NanoDrop Technologies).

Gene expression analysis using a rice oligoarray system

Target cRNA was amplified and labeled with cyanine-3 CTP (Cy3) and cyanine-5 CTP (Cy5) from 200 to 500 ng mRNA by using a Fluorescent Linear Amplification Kit (Agilent Technologies). To check the technical reproducibility of the experiments, we attempted dye-swap labeling and hybridization. Among the 22K spots, the number shown in violation of the change of dye was always less than 10. The quality and concentration of the targets were again determined by the 2100-Bioanalyzer and NanoDrop, respectively. Some 1,000 ng of Cy3- and Cy5-labeled cRNAs were mixed and fragmented to an average size of 100 to 200 bases with an In Situ Hybridization Kit Plus (Agilent Technologies) by incubation at 60 °C for 30 min. Fragmented cRNAs were added to hybridization buffer, applied to the 22K rice oligo array (NCBI-GEO platform ID = GPL892; the Agilent Technologies G4138A), and hybridized for 17 h at 60 °C. The array slides were washed with SSC (10 min in 6× SSC and 0.005% Triton X-102 at room temperature, then 5 min in 0.1× SSC and 0.005% Triton X-102 at 4 °C). Slides were dried and scanned on a G2565BA Microarray Scanner System (Agilent Technologies).

Data analysis

Scanned images were processed by Feature Extraction ver. 8.1 (Agilent Technologies). The ratio between Cy3 and Cy5 signal intensities was normalized by the LOWESS method and significant differences (*P* values) were calculated. To select genes differentially expressed under each condition, genes whose *P* values were less than 0.01 were defined as significantly differentially expressed.

The genes on the rice 22K rice oligo array were mapped using TIGR Pseudomolecules rel. 3, and genome position and functional information were derived from OSA1 and other public databases (RiceCyc, Rice TF DB) on the basis of the TIGR annotation. Gene annotation was also collected from RAP-DB on the basis of rice full-length cDNA, and each gene annotation was manually defined. *Cis*-element information based on information from PLACE-DB was obtained from KOME. For small-scale stress treatments at NIAS (cold, salt, osmotic, PEG, and flood), many time points were obtained and the pooled data from each stress treatment were used for the data-mining analysis.

Results and discussion

Commonly and specifically up- and down-regulated genes among eight microarray experiments

We determined the numbers of genes universally up-regulated and down-regulated during the eight stress experiments (PEG drought-mimicking, cold, salt, flood, and osmotic treatments in Nipponbare seedlings; drought-stress treatments in IR64 and Apo seedlings; and drought-stress treatment of IR64 during the peduncle elongation

Table 1. Number of commonly up- and down-regulated genes among eight stress treatments. APO: drought-stress treatment for seedlings of Apo. IR64: drought-stress treatment for seedlings of IR64. Ped: drought-stress treatment for the peduncle elongation stage of IR64. PEG, Cold, Salt, Osmotic, and Flood: PEG, cold, salt, osmotic, and flood treatment for seedlings of Nipponbare. Details of the treatments and physiological conditions are described in the text.

	APO	IR64	Ped	PEG	Cold	Salt	Osmotic	Flood
Up-regulation								
APO	2,466							
IR64	1,340	1,871						
Ped	922	745	2,459					
PEG	1,285	895	1,128	2,834				
Cold	1,005	708	968	1,320	3,737			
Salt	256	224	226	369	244	467		
Osmotic	306	283	298	456	293	297	551	
Flood	798	553	737	1,268	766	208	228	2,628
Down-regulation								
APO	2,707							
IR64	1,551	1,864						
Ped	692	451	2,451					
PEG	1,140	891	1,272	2,682				
Cold	1,039	798	1,015	1,700	3,295			
Salt	174	140	284	421	417	542		
Osmotic	156	148	247	415	397	288	492	
Flood	994	814	970	1,676	1,327	265	304	2,581

stage) (Table 1). The numbers of genes universally up- and down-regulated during the salt and osmotic stress treatments were lower than with the other stress treatments. This result suggests that salt and osmotic stresses are different and their responses may be affected by different types of signal transduction pathways. Among the four drought-stress experiments, about half of the genes up-regulated were also up-regulated by other stress treatments.

We also determined the numbers of genes up- and down-regulated only by specific treatments (Table 2). Among the eight stress treatments, cold stress induced the up-regulation of a particularly large number of treatment-specific genes. As reported in many organisms, salt, osmotic, and drought-stress treatments led to cross-talk of signal transduction networks, but cold-stress treatment gave responses common to these three stress treatments and cold-specific responses (Seki et al 2002). Thirty-three genes were universally up-regulated and 17 were universally down-regulated by all eight stress treatments (Table 3). Genes that were universally up-regulated were typically those encoding stress-induced proteins such as low temperature–induced stress protein (lt101–2), temperature stress–induced lipocalin, proteins of the universal stress–induced protein (Usp) family, and thaumatin-like protein. Among the commonly down-regulated genes were those typically encoding proteins within organelles, such as

Table 2. Number of genes specifically up- and down-regulated among eight stress treatments. The percentage of the number of specifically changed genes in the total number of changed genes is also shown.

Treatment profile	APO	IR64	Ped	PEG	Cold	Salt	Osmotic	Flood
Specific-up	251	94	346	177	870	1	6	339
Total-up	2,466	1,871	2,459	2,834	3,737	467	551	2,628
%	10.2	5.0	14.1	6.2	23.3	0.2	1.1	12.9
Specific-up	260	94	375	58	374	8	3	109
Total-down	2,707	1,864	2,451	2,682	3,295	542	492	2,581
%	9.6	5.0	15.3	2.2	11.4	1.5	0.6	4.2

Table 3. List of genes commonly up- or down-regulated by eight stress treatments.

Up or down	Gene name (acc. number of probe)	RAP description
Up	AK058555	DnaJ protein homolog ANJ1
Up	AK060423	Alanine: glyoxylate aminotransferase-like protein
Up	AK060757	Aldehyde dehydrogenase family 7 member A1 (EC 1.2.1.3) (Antiquitin 1) (matured fruit 60-kDa protein) (MF-60)
Up	AK061438	Ribonuclease T2 family protein
Up	AK062784	Conserved hypothetical protein
Up	AK063334	Protein phosphatase 2C (EC 3.1.3.16) (PP2C)
Up	AK063685	Short, highly repeated, interspersed DNA (fragment)
Up	AK063835	Twin-arginine translocation pathway signal domain-containing protein
Up	AK063896	Hypothetical protein
Up	AK063923	No annotation
Up	AK065206	Phytepsin precursor (EC 3.4.23.40) (aspartic proteinase)
Up	AK068233	Importin alpha 2
Up	AK068727	ERD1 protein, chloroplast precursor
Up	AK069748	Amino acid/polyamine transporter II family protein
Up	AK070268	Gibberellin-regulated protein family protein
Up	AK070556	No annotation
Up	AK070872	Low temperature–induced protein 1t101.2
Up	AK070914	Universal stress protein (Usp) family protein
Up	AK070973	Hypothetical protein
Up	AK071108	Temperature stress–induced lipocalin
Up	AK071205	ChaC-like protein family protein
Up	AK072280	Universal stress protein (Usp) family protein
Up	AK099336	Importin alpha 2
Up	AK099618	Dormancy auxin-associated family protein
Up	AK100465	Cys/Met metabolism pyridoxal-phosphate-dependent enzyme family protein
Up	AK101209	Myb, DNA binding-domain-containing protein

Continued on next page

Table 3 continued.

Up or down	Gene name (acc. number of probe)	RAP description
Up	AK101609	Single-stranded nucleic acid-binding R3H domain-containing protein
Up	AK101837	Thaumatin-like protein
Up	AK102352	Conserved hypothetical protein
Up	AK103194	Acetoacetyl-coenzyme A thiolase (EC 2.3.1.9)
Up	AK105896	Electron transfer flavoprotein alpha-subunit, mitochondrial precursor (Alpha-ETF)
Up	AK111578	YT521-B-like protein family protein
Down	AK058436	No annotation
Down	AK058756	Hypothetical protein
Down	AK058858	Conserved hypothetical protein
Down	AK059151	Magnesium-protoporphyrin O-methyltransferase (EC 2.1.1.11) (magnesium-protoporphyrin IX)
Down	AK061654	Ferredoxin I, chloroplast precursor (anti-disease protein)
Down	AK061690	Chloroplast 50S ribosomal protein L27 (fragment)
Down	AK062322	Conserved hypothetical protein
Down	AK062627	Hypothetical protein
Down	AK063148	Plant disease resistance response protein family protein
Down	AK064950	Lipase, class 3 family protein
Down	AK068586	UDP-glucuronosyl/UDP-glucosyltransferase family protein
Down	AK068620	Allene oxide synthase (EC 4.2.1.92)
Down	AK068710	Isocitrate lyase and phosphorylmutase family protein
Down	AK072286	Hypothetical protein
Down	AK099409	Mitochondrial substrate carrier family protein
Down	AK099444	Nitrilase-associated protein-like
Down	AK105813	Photosystem II protein PsbX family protein

ferredoxin I, isocitrate lyase and phosphorylmutase family protein, and mitochondrial substrate carrier family protein.

Genes commonly and specifically up- and down-regulated among four drought-stress experiments

The total number of genes responsive to the four drought stresses was 9,707. Among these, 4,284 were up-regulated and 4,306 were down-regulated. These data suggest that about half of the genes on the microarray are drought-responsive. The remaining 1,117 genes could be either up- or down-regulated by drought stress, depending on the treatment. The profiles of some drought-responsive genes are given in Table 4A. We classified the drought-responsive genes into three groups according to their profiles: Dr_up, Dr_down, and Other. Genes classified as Other were up-regulated by some stress treatments and down-regulated by other stress treatments. The numbers of genes commonly up- and down-regulated among the four drought-stress treatments are shown in Table 4B. The commonality of gene expression changes among samples from seedlings (IR64, Apo, and PEG) was greater than that between samples from

Table 4. (A) Example of profiles of some drought-responsive genes. According to these profiles, responsive genes are classified into three groups: Dr_up, Dr_down, and Other. (B) Number of genes commonly up- and down-regulated by four drought-stress treatments. (C) Number of genes classified into Other group is shown. For the Apo treatment, 281 genes are up-regulated; among these, 1, 194, and 135 are down-regulated in IR64, Ped, and PEG treatments, respectively. As described in the text, PEG data contain two time points; therefore, the PEG-PEG position is not zero.

(A)

Gene name (acc. number of probe)	APO		IR64		Ped		PEG		Profile	Group
	Up	Down	Up	Down	Up	Down	Up	Down		
AK058252	1	0	0	0	1	0	1	0		Dr_up
AK058261	0	0	1	0	1	0	1	0		Dr_up
AK058268	1	0	0	0	0	0	0	0	APO_u	Dr_up
AK058272	0	0	0	0	1	0	0	0	Ped_u	Dr_up
AK058336	1	0	1	0	1	0	1	0	Dr_comm_u	Dr_up
AK058258	0	–1	0	–1	0	–1	0	0		Dr_down
AK058259	0	0	0	0	0	–1	0	–1		Dr_down
AK058260	0	0	0	0	0	0	0	–1	PEG_d	Dr_down
AK058263	0	0	0	0	0	–1	0	0	Ped_d	Dr_down
AK058271	0	–1	0	–1	0	–1	0	–1	Dr_comm_d	Dr_down
AK058224	0	0	0	–1	1	0	0	–1		Other
AK058235	0	–1	0	0	1	0	0	0		Other
AK058236	0	0	1	0	0	–1	0	0		Other
AK058239	0	0	1	0	0	0	0	–1		Other

(B)

	APO	IR64	Ped	PEG
Up-regulation				
APO	2,185			
IR64	1,182	1,543		
Ped	903	723	2,091	
PEG	1,202	833	1,046	2,338
Down-regulation				
APO	2,228			
IR64	1,294	1,543		
Ped	659	423	2,032	
PEG	1,041	805	1,142	2,262

Continued on next page

Table 4 continued.

(C)

Treatment	Number of genes	Down-regulation			
		APO	IR64	Ped	PEG
		(no. of genes)			
Up-regulation	Total	479	321	419	420
APO	281	0	1	194	135
IR64	328	3	0	215	192
Ped	368	273	170	0	129
PEG	496	266	192	190	99

seedlings and peduncle samples. This may have been because of the difference in growth stages. A similar pattern was observed in the numbers of genes classified as Other (Table 4C). Ninety-nine genes showed up- and down-regulation in response to PEG treatment. This is because the PEG data included data from two time points. A relatively high number of directions of gene expression change were observed between PEG and the other treatments (Table 4). In contrast, the direction of change between Apo and IR64 was extremely small. This might have occurred because of stage-specific gene expression in the reproductive stage.

Combined analysis of genes encoding transcription factor proteins and *cis*-elements found in the promoter regions of possible downstream genes

Examination of the gene annotations revealed that some 1,125 genes on the 22K microarray may encode transcription factor proteins. The expression of 503 of these genes was up- or down-regulated by the drought-stress treatments. The genes encoding transcription factors can be classified into 36 families, and their up- and down-regulation profiles are summarized in Table 5. All the PHD and SNF2 family members were up-regulated, whereas all the members of Jumonji and TCP were down-regulated. Many C3H, HSF, TUB, SBP, and Trihelix family members were up-regulated, whereas many bHLH, GRAS, and ABI3VP1 family members were down-regulated. Transcription factor genes commonly up-regulated among the four treatments were AP2-EREBP, AUX/IAA, bZIP, C2C2-GATA, C3H, HB, HMG, HSF, MYB-related, NAC, SBP, SNF2, and Trihelix, whereas transcription factor genes commonly down-regulated were Alfin-like, AUX/IAA, bHLH, bZIP, MYB, NAC, WRKY, and ZIM (Table 6).

Clustered genes with similar expression profiles after microarray analysis could be downstream target genes of the same transcription factor. To check for this possibility, a search of common regulatory or binding sequence *cis*-elements in the promoter region, 500 to 1,000 bp upstream from the transcription start site, is very important. Using the PLACE database, we searched *cis*-elements in the 1-kb upstream sequences of 2,427 genes up-regulated by the four drought-stress treatments. The

Table 5. Genes encoding transcription factors and their expression profile by drought-stress treatments. Some 1,125 genes encoding transcription factors are classified into gene families by the structural information of DNA binding domains. Their expression profiles are also classified according to the classification of Dr_up and Dr_down.

Family	Total in 221K	Dr_up	Ratio_Rd_up	Dr_down (%)	Ratio_Dr_down (%)	Comment
AP2-EREBP	80	19	23.8	14	17.5	
bHLH	72	11	15.3	25	34.7	
NAC	64	20	31.3	16	25.0	
MYB	60	15	25.0	16	26.7	
HB	58	16	27.6	10	17.2	
bZIP	54	15	27.8	11	20.4	
C3H	54	20	37.0	5	9.3	
C2H2	52	9	17.3	6	11.5	
WRKY	52	14	26.9	10	19.2	
MYB-related	47	14	29.8	7	14.9	
PHD	43	16	37.2	0	0.0	Dr_up_only
AUX/IAA	27	7	25.9	12	44.4	
SNF2	27	8	29.6	0	0.0	Dr_up_only
G2-like	26	5	19.2	6	23.1	
GRAS	26	4	15.4	10	38.5	
ABI3VP1	23	1	4.3	3	13.0	
MADS	23	2	8.7	4	17.4	
SET	23	2	8.7	4	17.4	
ARF	21	5	23.8	6	28.6	
C2C2-GATA	18	2	11.1	4	22.2	
HSF	16	9	56.3	2	12.5	
TUB	15	5	33.3	2	13.3	
ZIM	15	3	20.0	6	40.0	
C2C2-Dof	14	5	35.7	3	21.4	
SBP	14	4	28.6	1	7.1	
Jumonji	11	0	0.0	3	27.3	Dr_down_only
TCP	11	0	0.0	5	45.5	Dr_down_only
Alfin-like	10	3	30.0	3	30.0	
CCAAT_HAP5	10	2	20.0	3	30.0	
C2C2-CO-like	9	2	22.2	2	22.2	
Trihelix	9	3	33.3	1	11.1	
CCAAT-HAP2	8	3	37.5	2	25.0	
HMG	8	3	37.5	3	37.5	
ARID	7	1	14.3	1	14.3	
EIL	7	2	28.6	1	14.3	
LUG	7	2	28.6	1	14.3	

Table 6. List of commonly up- and down-regulated genes encoding transcription factors by four drought-stress treatments.

Profile	Family	Gene name (acc. number of probe)	RAP_locus_description_Description
Dr_comm_U	AP2-EREBP	AK058349	Pathogenesis-related transcriptional factor and ERF domain-containing protein
Dr_comm_U	AP2-EREBP	AK060543	Ethylene responsive element-binding factor 5 (AtERF5)
Dr_comm_U	AP2-EREBP	AK064252	AP2 domain-containing protein RAP2.2 (fragment)
Dr_comm_U	AP2-EREBP	AK066670	Ethylene responsive element-binding factor 5 (AtERF5)
Dr_comm_U	AP2-EREBP	AK067313	DRE binding protein 2
Dr_comm_U	AP2-EREBP	AK107775	Dehydration responsive element-binding protein 1D (DREB1D) protein (C-repeat binding factor 4) (C-repeat/dehydration) responsive element-binding factor 4) (CRT/DRE binding factor)
Dr_comm_U	AUX/IAA	AK073365	AUX/IAA protein family
Dr_comm_U	bZIP	AK063889	Nonprotein coding transcript, unclassified transcript
Dr_comm_U	bZIP	AK067919	OSE2-like protein (fragment)
Dr_comm_U	bZIP	AK073142	Conserved hypothetical protein
Dr_comm_U	C2C2-GATA	AK068931	Zn-finger, GATA-type domain-containing protein
Dr_comm_U	C3H	AK063896	Hypothetical protein
Dr_comm_U	C3H	AK070857	Zn-finger, C-x8-C-x5-C-x3-H-type domain-containing protein
Dr_comm_U	C3H	AK106392	Zn-finger, C-x8-C-x5-C-x3-H-type domain-containing protein
Dr_comm_U	CPP	AK108664	Tesmin/TSO1-like, CXC domain-containing protein
Dr_comm_U	HB	AK063685	Short, highly repeated, interspersed DNA (fragment)
Dr_comm_U	HMG	AK102592	SSRPI protein
Dr_comm_U	HSF	AK106488	Heat shock transcription factor 29 (fragment)
Dr_comm_U	HSF	AK106545	Heat shock factor protein 3 (HSF 3) (heat shock transcription factor 3) (HSTF 3)
Dr_comm_U	MYB-related	AK065594	Transcription factor MYBS2
Dr_comm_U	MYB-related	AK101209	Myb, DNA binding-domain-containing protein
Dr_comm_U	NAC	AK108080	No apical meristem (NAM) protein domain-containing protein
Dr_comm_U	SBP	AK062581	SBP domain-containing protein
Dr_comm_U	SNF2	AK065390	Chromatin remodeling factor CHD3 (GYMNOS/PICKLE)
Dr_comm_U	Trihelix	AK059666	Myb, DNA binding-domain-containing protein
Dr_comm_D	Alfin-like	AK109447	Zn-finger-like, PHD finger domain-containing protein

Continued on next page

Table 6 continued.

Profile	Family	Gene name (acc. number of probe)	RAP_locus_description_Description
Dr_comm_D	AUX/IAA	AK059838	AUX/IAA protein family
Dr_comm_D	AUX/IAA	AK066518	AUX/IAA protein family
Dr_comm_D	AUX/IAA	AK072001	Auxin-responsive protein (Aux/IAA) (fragment)
Dr_comm_D	BES1	AK106748	Plant protein of unknown function, DUF822 family protein
Dr_comm_D	bHLH	AK064946	Transcription factor ICE1 (inducer of CBF expression 1) (basic helix-loop-helix protein 116) (bHLH116) (AtbHLH116)
Dr_comm_D	bHLH	AK101063	TA1 protein (fragment)
Dr_comm_D	bZIP	AK071639	Eukaryotic transcription factor, DNA binding-domain-containing protein
Dr_comm_D	MYB	AK111803	MYB transcription factor
Dr_comm_D	NAC	AK073667	OsNAC3 protein
Dr_comm_D	WRKY	AK066255	WRKY transcription factor 45
Dr_comm_D	ZIM	AK065170	ZIM domain-containing protein

frequency of occurrence of each *cis*-element in the 1-kb upstream sequences of the clustered gene sets after microarray analysis and in all 22K genes is shown in Table 7. The sequence ABRELATERD1 (ACGTG) is the binding sequence of the bZIP (ABRE) transcription factor. It appeared often (43.5%) in the promoter region of the "Common" gene set up-regulated by the four drought-stress treatments. Abundantly accumulated *cis*-elements in the clustered gene sets are indicated in Table 7 in bold. Typical drought-responsive transcription factor binding sites such as bZIP(ABRE) and AP2/EREBP are also listed. The frequencies of localization of the *cis*-elements in the upstream 500-bp DNA sequences are plotted in Figure 1. The ABRELATERD1 element was most frequently located between the –80 to –160 region of the cluster of the commonly up-regulated gene set, whereas the SEF3MOTIFGM element was most frequently located in the –80 and –240 region of the genes up-regulated in IR64 seedlings.

The combination of a change in the expression of transcription factor–encoding genes and *cis*-elements in the clustered genes predicts the presence of gene regulatory networks for drought-stress response. Further confirmation of the presence of these networks should be substantiated by chromosomal-immunoprecipitation promoter chip analysis.

Metabolic pathway analysis

By referring to the relationship between an enzyme-coding gene and its metabolic pathway, it is possible to determine what kind of metabolic pathways are drastically affected by drought stress. We used this information from the Rice Cyc database, which currently lists 302 metabolic pathways. Many of the genes involved in the pathways

Table 7. List of *cis*-elements appearing in the 1-kb upstream sequence of drought-up-regulated (Dr_up) genes. According to information of the PLACE database, the name of the element (SIGNAL_NAME), ID of the element (SIGNAL_ID), and its sequence are listed. The percentage of appearance in 22K genes is shown in the Whole column. The percentage of appearance in Dr_up genes according to the classification in Table 4A is shown in the Dr_up column. Percentages of appearance in up-regulated genes in Apo, IR64, Ped, and PEG treatments are shown in Apo, IR64, Ped, and PEG columns, respectively. The percentage of appearance of commonly up-regulated genes in four drought-stress treatments is shown in the Common column. A possible transcription factor binding to the element is shown in the TF column.

SIGNAL_NAME	SIGNAL_ID	SIGNAL_SEQ	Ratio							TF
			Whole (%)	Dr_up	APO	IR64	Ped	PEG	Common	
ABRELATERD1	S000414	ACGTG	31.6	35.8	33.6	33.0	28.3	37.4	**43.5**	vZIP (ABRE)
SEF3MOTIFGM	S000115	AACCCA	15.2	15.8	18.5	20.9	14.4	15.1	16.2	–
BOXIINTPATPB	S000296	ATAGAA	15.0	15.5	14.4	20.9	15.5	15.6	**20.2**	–
DRECRTCOREAT	S000418	RCCGAC	15.0	16.7	17.0	13.2	16.2	15.6	**18.4**	AP2/EREBP (DREB)
CANBNNAPA	S000148	CNAACAC	13.5	14.2	13.2	17.0	10.2	14.3	**24.7**	bZip (ABRE)
CACGTGMOTIF	S000042	CACGTG	13.3	16.5	15.1	13.7	12.8	15.2	13.2	bZip (ABRE)
TATCCAOSAMY	S000403	TATCCA	13.2	13.8	13.9	18.1	13.4	16.1	12.2	Myb
CAREOSREP1	S000421	CAACTC	12.6	12.9	11.3	18.7	15.0	12.7	12.2	Myb
RAV1BAT	S000315	CACCTG	10.8	11.9	13.2	11.5	9.6	10.3	**15.1**	AP2/EREBP (DREB)
IBOX	S000124	GATAAG	10.1	11.0	9.4	15.4	12.8	9.7	10.8	–
HEXMOTIFTAH3H4	S000053	ACGTCA	9.7	11.3	12.9	8.2	11.2	10.3	**13.4**	Myb
TATABOX3	S000110	TATTAAT	9.3	9.6	9.4	11.0	8.5	11.5	**12.9**	–
S1FBOXSORPS1L21	S000223	ATGGTA	9.2	9.8	10.6	12.1	11.1	6.9	9.2	Trihelix (GT-factor)
GT1CORE	S000125	GGTTAA	8.9	9.0	7.7	13.7	8.7	8.6	8.2	Myb
ACGTABREMOTIFA2OSEM	S000394	ACGTGKC	7.6	10.0	7.2	6.0	6.6	10.3	**17.6**	bZip (ABRE)
DRE2COREZMRAB17	S000402	ACCGAC	7.5	8.8	9.1	4.4	8.3	8.3	**10.4**	AP2/EREBP (DREB)
TATCCAYMOTIFOSRAMY3	S000256	TATCCAY	7.0	7.5	7.2	9.3	8.2	7.6	7.8	Myb
TGACGTVMAMY	S000377	TGACGT	7.0	7.1	9.6	6.6	4.5	7.0	8.9	–
ACGTTBOX	S000132	AACGTT	6.7	6.8	9.6	6.0	6.9	7.5	7.3	bzip?
ABREOSRAB21	S000012	ACGTSSC	6.2	8.0	8.2	6.6	6.4	6.1	**10.1**	bZip (ABRE)
CDTDREHVCBF2	S000411	GTCGAC	5.9	6.0	4.6	9.3	5.4	6.7	4.7	AP2/EREBP (DREB)
IBOXCORENT	S000424	GATAAGR	5.7	6.2	5.3	6.6	8.2	4.8	6.4	Myb
SV40COREENHAN	S000123	GTGGWWHG	5.5	5.8	4.6	4.4	7.3	4.2	7.1	–
REBETALGLHCB21	S000363	CGGATA	5.5	6.4	6.7	4.4	3.8	7.7	7.1	–
SP8BFIBSP8BIB	S000184	TACTATT	5.4	5.7	5.8	6.0	5.4	4.6	7.3	WRKY
BS1EGCCR	S000352	AGCGGG	5.2	5.5	5.5	3.3	5.4	4.6	7.3	–
WBBOXPCWRKY1	S000310	TTTGACT	5.1	5.1	5.0	4.4	4.1	7.0	4.0	WRKY
BOXCPSAS1	S000226	CTCCCAC	5.1	4.9	4.8	3.3	5.1	3.9	8.0	

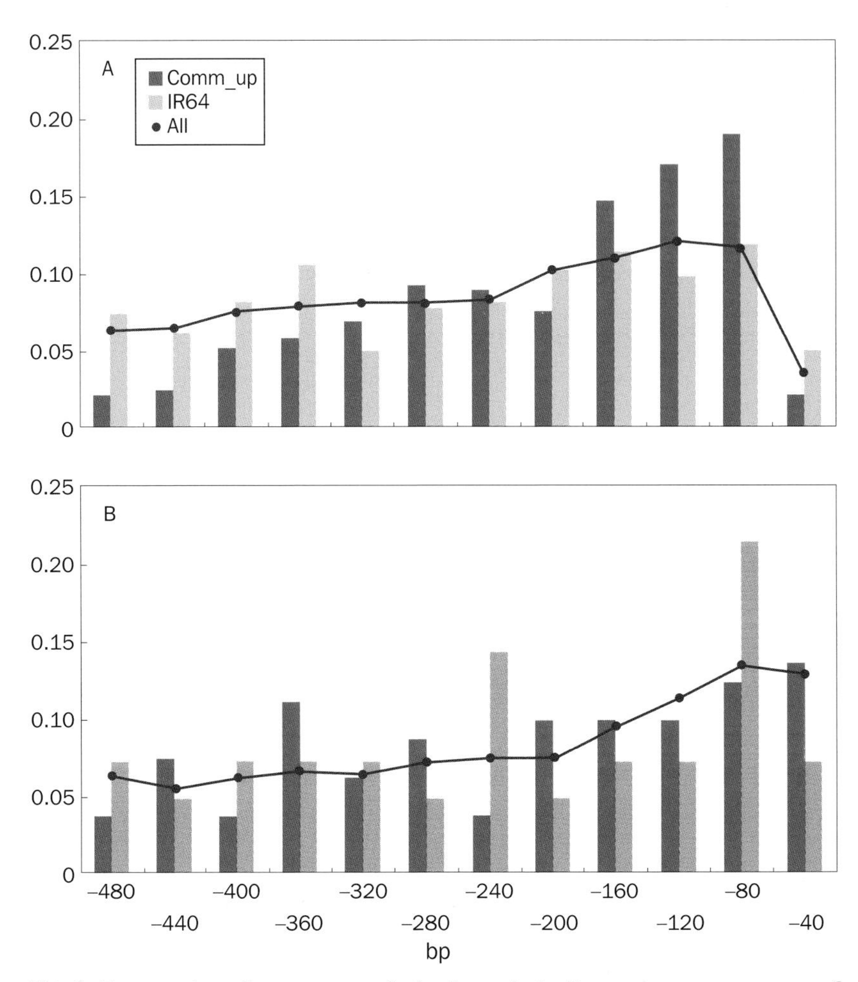

Fig. 1. Frequencies of occurrence of *cis*-elements in the upstream sequences of up-regulated genes. (A) Occurrence of the ABRELATERD1 element in 1 to –500 bp of commonly up-regulated genes (Comm_up) and genes up-regulated under Ped treatment (PED_up). (B) Occurrence of the SEF3MOTIFGM element in 1 to –500 bp of the commonly up-regulated genes and genes up-regulated under drought stress in cultivar IR64 (IR64). For comparison, the average occurrence rate in all the genes examined (All) is shown as a blue curve.

showed changes in expression in response to the four drought-stress treatments; only 15 pathways were not affected. Some 150 metabolic pathways involved enzyme-coding genes commonly up- or down-regulated by the four drought-stress treatments. Sixty-nine of them involved only genes that were up-regulated and 47 involved only genes that were down-regulated, whereas 34 involved genes that were up- or down-regulated. The 302 pathways were grouped into 17 classes (Table 8). Thirty-two amino acid–related pathways contained only up-regulated genes, 10 contained only

Table 8. Classification of 302 metabolic pathways in rice affected by drought-stress treatments and number of member pathways with member genes encoding enzymes in up- and down-regulation.

Metabolic pathway	Up-regulation	Down-regulation	Up- and down-regulation
Amine	4	–	2
Amino acid	32	10	6
Aromatic compounds	1	4	–
C1 compounds	–	–	4
Carbohydrate	9	5	6
Cell structure	1	2	1
Electron	–	2	–
Energy	7	11	5
Lipid	4	3	4
Nitrogen	1	–	–
Nucleotide	5	–	–
Other	7	1	1
Photosynthesis	–	3	–
Plant hormone	2	–	4
Respiration	–	–	1
Sulfur assimilation	–	4	1
Vitamin	2	5	–

down-regulated genes, and six contained both up- and down-regulated genes. As for carbohydrate-related metabolism, nine contained only up-regulated genes and five contained only down-regulated genes, and six contained both up- and down-regulated genes. In the case of energy-related metabolism, seven contained up-regulated genes, 11 contained down-regulated genes, and five contained both. Lipid-related pathways are also affected. These data suggest that foundamental pathways for energy formation are affected by the drought-stress treatments.

Drought-responsive genes and their chromosomal positions and relationships to known drought-related QTLs

With chromosome 1 as an example, the number of genes encoding a transcription factor protein on a 22K array was plotted every 1 Mb. Among the genes, those up- or down-regulated by drought-stress treatments were also plotted. These profiles were compared among the four stress treatments (Fig. 2). We focused on two chromosomal regions: the short-arm region (A) and the long-arm region (B). In the case of region A, the profiles of Apo and IR64 were similar and those of PEG and Peduncle were also similar, but those of Apo and PEG or IR64 and Peduncle were different (Fig. 2A). For region B, the profiles of Apo and IR64 were similar, but those of PEG and Peduncle were different (Fig. 2B). The transcription factors encoded in these two regions are listed in Table 9. bZIP, SBP, and Orphans-type transcription factors in region B were universally up-regulated by the four stress treatments, whereas the AUX/IAA-type transcription factor in region A was commonly down-regulated.

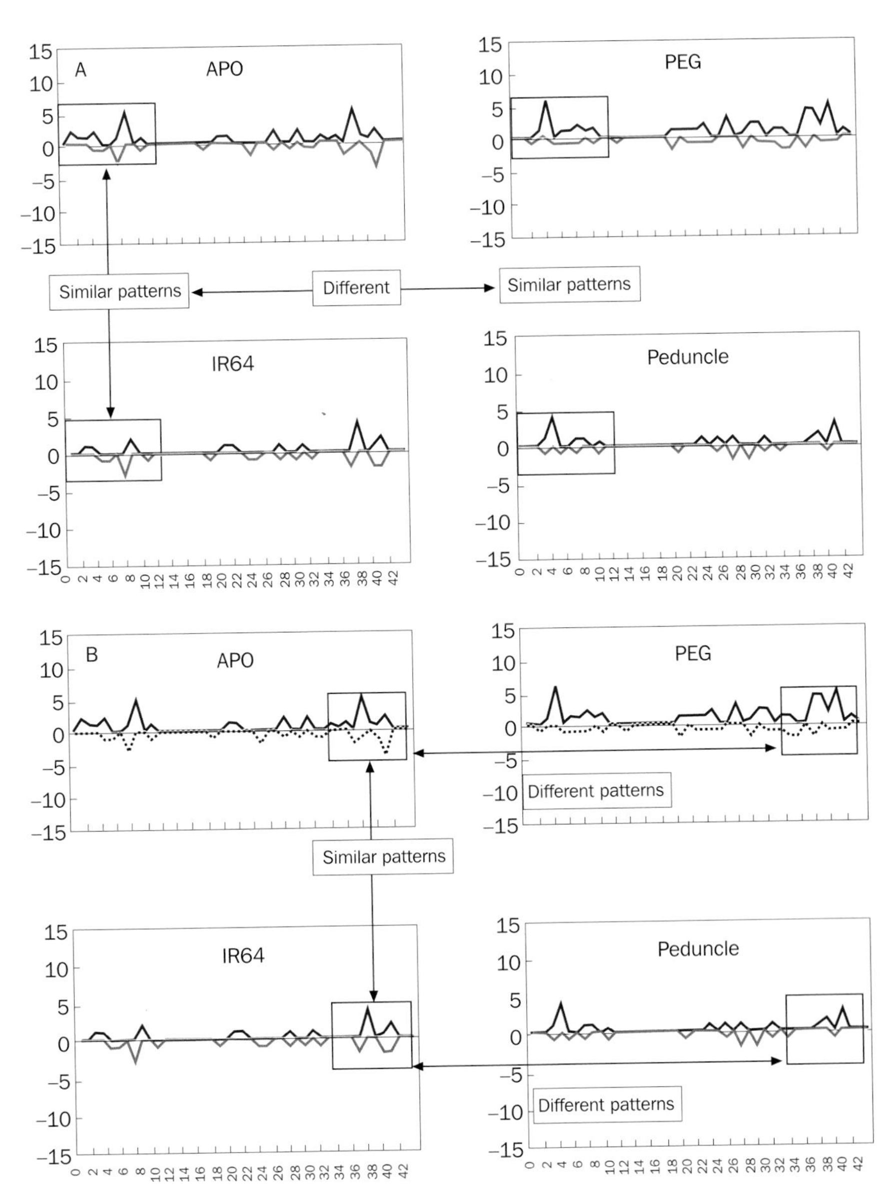

Fig. 2. Comparison of expression profiles of transcription factor–encoding genes on the 22K array, located on chromosome 1. Y axis indicates the relative unit of gene expression after microarray analysis. (A) Comparison in region A among the four drought-stress treatments. (B) Comparison in region B among the four drought-stress treatments. Blue curve indicates up-regulated transcription factor–encoding genes and red curve indicates down-regulated transcription factor–encoding genes.

Table 9. Expression profiles of genes encoding transcription factors in regions A (A) and B (B) of chromosome 1. Genes located in the highly up-regulated region are enclosed in a rectangle.

A

Family	Start position in	APO		IR64		Ped		PEG	
		Up	Down	Up	Down	Up	Down	Up	Down
WRKY	4343952	0	0	0	0	0	0	0	0
HMG	4494520	1	0	0	0	1	0	1	0
WRKY	4569249	0	0	0	0	0	0	0	0
MYB-related	4711275	0	0	0	–1	1	0	1	0
AUX/IAA	4813630	0	–1	0	0	1	0	1	0
NAC	4876010	0	0	0	0	0	0	0	0
MYB	4914685	0	0	0	0	0	0	1	0
C3H	4945933	1	0	0	0	1	0	1	0
MYB-related	4967972	0	0	0	0	0	0	1	0
MYB-related	5025090	0	0	0	0	0	0	0	0
C2H2	5096334	0	0	0	0	0	0	0	0
bHLH	5198663	0	0	0	0	0	0	0	0
AP2-EREBP	5451335	0	–1	0	–1	0	0	0	–1
BES1	5665770	0	0	0	0	0	–1	0	0
bZIP	6091473	0	0	0	0	0	0	0	–1
bHLH	6459518	0	0	0	0	0	0	1	0
AP2-EREBP	6810495	0	0	0	0	0	0	0	0
AUX/IAA	7246542	0	–1	0	–1	0	–1	0	–1
ABI3VP1	7414084	0	0	0	0	0	0	0	0
bHLH	7491902	1	0	0	0	1	0	1	0
ARF	7543666	0	–1	0	–1	0	0	0	0
RWP-RK	7568017	0	–1	0	–1	0	0	0	0
G2-like	7683681	0	0	0	0	0	0	0	0
C2H2	7844271	0	0	0	0	0	0	0	0
RWP-RK	8061641	0	0	0	0	0	0	0	0
WRKY	8080965	1	0	0	0	0	0	1	0
GeBP	8231406	0	0	0	0	0	0	0	0
C3H	8331300	0	0	0	0	0	0	0	0
C3H	8557404	0	0	0	0	0	0	0	0
C3H	8590573	1	0	1	0	0	0	0	0
C3H	8656161	1	0	1	0	0	0	1	0
NAC	8789428	1	0	0	0	0	0	0	0
C2C2-Dof	8946118	1	0	0	0	1	0	0	–1

Continued on next page

Table 9 continued.

B

Family	Start position in	APO		IR64		Ped		PEG	
		Up	Down	Up	Down	Up	Down	Up	Down
C2H2	37126581	1	0	1	0	0	0	1	0
bZIP	37154211	0	0	0	0	0	0	1	0
bZIP	37161137	0	0	0	0	0	0	0	0
NAC	37337869	0	0	0	0	0	0	0	–1
MYB-related	37337869	1	0	1	0	0	0	1	0
TUB	37515610	1	0	0	0	0	0	0	0
bZIP	37536348	1	0	1	0	1	0	1	0
C2H2	37753658	0	–1	0	0	0	0	0	–1
MYB	37911922	1	0	1	0	0	0	0	0
PHD	38061272	0	0	0	0	0	0	0	0
GRAS	38241514	0	0	0	0	1	0	1	0
MADS	38301006	0	0	0	0	0	0	0	0
NAC	38378582	1	0	0	0	0	0	1	0
MADS	38480544	0	0	0	0	0	0	0	0
Alfin-like	38545811	0	0	0	0	1	0	1	0
NAC	38591328	0	0	0	0	0	0	0	0
TAZ	38823389	0	0	0	0	0	0	1	0
bHLH	39190971	0	0	0	0	0	0	1	0
G2-like	39366894	0	0	0	0	0	0	0	0
Jumonji	39472837	0	0	0	0	0	0	0	0
C2H2	39595861	1	0	1	0	0	0	1	0
ABI3VP1	39702781	0	0	0	–1	0	0	0	0
MADS	39806458	0	0	0	0	0	0	0	0
bHLH	39874306	0	–1	0	–1	0	–1	0	–1
C3H	39985731	0	0	0	0	0	0	0	0
SBP	40310690	1	0	1	0	1	0	1	0
MADS	40324036	0	0	0	–1	0	0	1	0
Orphans	40376979	0	0	0	0	0	0	0	0
CAMTA	40376979	0	0	0	0	0	0	0	0
Orphans	40383379	1	0	1	0	1	0	1	0
TCP	40456890	0	–1	0	0	0	0	0	0
NAC	40552486	0	–1	0	0	0	0	1	0
SET	40633852	0	0	0	0	0	0	0	0
ARF	40675306	0	–1	0	0	1	0	1	0
bHLH	40693876	0	–1	0	–1	0	0	0	–1
HB	40964100	0	0	0	0	0	0	0	0
C2H2	40990970	0	0	0	0	0	0	0	0

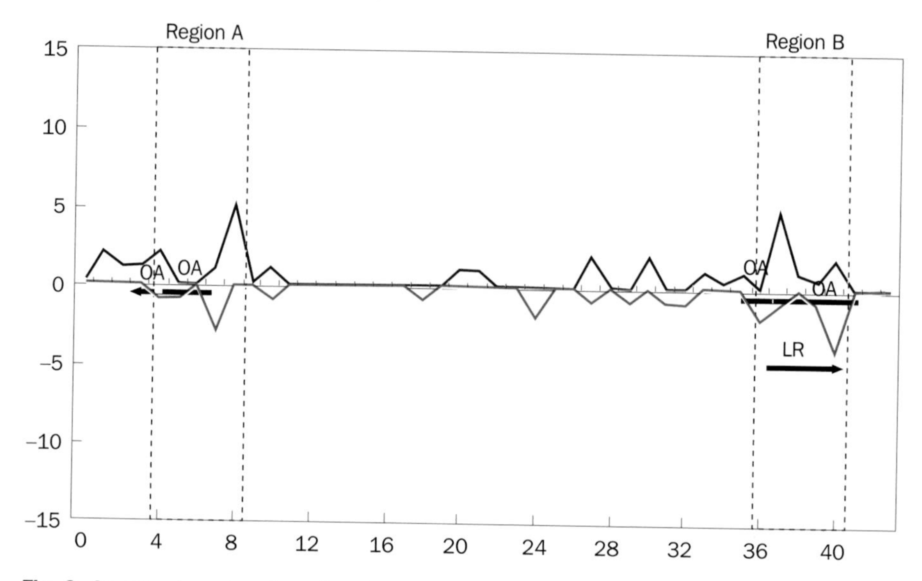

Fig. 3. Overlapping of regions A and B with known QTL regions. OA = osmotic adjustment QTL. LR = leaf-rolling QTL. Y axis indicates the relative unit of gene expression after microarray analysis.

We also surveyed known drought QTLs overlapping with these regions; one leaf-rolling QTL and the osmotic adjustment QTL overlapped with region B (Fig. 3). The bZIP, SBP, and Orphans-type transcription factors might be associated with this drought-related QTL. Though those are insufficient and rough analyses of the genomic locus and gene expression, drought QTLs in this chromosomal region and localization of TF show a good correlation.

Conclusions

We used a microarray system to perform gene expression analyses under four different drought-stress treatments and other abiotic-stress treatments in rice. A comparison of gene expression data under the various stress treatments revealed that salt and osmotic stresses appeared more specific than other stresses, and that many genes were differentially expressed under cold stress. Analysis of genes specifically and commonly responsive to various types of drought-stress treatments revealed that tissue-specific differentially expressed genes were present between seedling and peduncle elongation stages. Analysis of differentially expressed transcription factors, together with a *cis*-element search in the promoter regions of the clustered genes after the microarray analysis, revealed a possible combination of transcription factors and their downstream genes, such as DREB-type transcription factor genes and their *cis*-elements in the promoter region. Metabolic pathway analysis revealed that, among the genes affected from many metabolic pathways, energy-related pathway member genes showed a marked change in gene expression. Genomic localization of the differentially expressed transcription factors revealed the co-localization of the abundance of differentially expressed transcription factors with some known drought-related QTLs.

References

Atlin GN, Laza M, Amante M, Lafitte HR. 2004. Agronomic performance of tropical aerobic, irrigated and traditional upland rice varieties in three hydrological environments at IRRI: new directions for a diverse planet. In: Fischer T et al, editors. Proceedings of the 4th International Crop Science Congress, Brisbane, Australia (Sept. 2004).

Bray EA. 2004. Genes commonly regulated by water-deficit stress in *Arabidopsis thaliana*. J. Exp. Bot. 55:2331-2341.

Ekanayake IJ, De Datta SK, Steponkus PL. 1989. Spikelet sterility and flowering response of rice to water stress at anthesis. Ann. Bot. 63:257-264.

Goff SA, Ricke D, Lan TH, Presting G, Wang R, Dunn M, Glazebrook J, Sessions A, Oeller P, Varma H, Hadley D, Hutchison D, Martin C, Katagiri F, Lange BM, Moughamer T, Xia Y, Budworth P, Zhong J, Miguel T, Paszkowski U, Zhang S, Colbert M, Sun WL, Chen L, Cooper B, Park S, Wood TC, Mao L, Quail P, Wing R, Dean R, Yu Y, Zharkikh A, Shen R, Sahasrabudhe S, Thomas A, Cannings R, Gutin A, Pruss D, Reid J, Tavtigian S, Mitchell J, Eldredge G, Scholl T, Miller RM, Bhatnagar S, Adey N, Rubano T, Tusneem N, Robinson R, Feldhaus J, Macalma T, Oliphant A, Briggs S. 2002. A draft sequence of the rice genome (*Oryza sativa* L. ssp. *japonica*). Science 296:92-100.

Gilmour SJ, Sebolt AM, Salazar MP, Everard JD, Thomashow MF. 2000. Overexpression of the *Arabidopsis* CBF3 transcriptional activator mimics multiple biochemical changes associated with cold acclimation. Plant Physiol. 124:1854-1865.

Hirochika H, Guiderdoni E, An G, Hsing YI, Eun MY, Han CD, Upadhyaya N, Ramachandran S, Zhang Q, Pereira A, Sundaresan V, Leung H. 2004. Rice mutant resources for gene discovery. 1. Plant Mol. Biol. 54:325-334.

Hu H, Dai M, Yao J, Xiao B, Li X, Zhang Q, Xiong L. 2006. Overexpressing a NAM, ATAF, and CUC (NAC) transcription factor enhances drought resistance and salt tolerance in rice. Proc. Natl. Acad. Sci. USA 103:12987-12992.

IRGSP (International Rice Genome Sequencing Project). 2005. The map-based sequence of the rice genome. Nature 436:793-800.

Jaglo-Ottosen KR, Gilmour SJ, Zarka DG, Schabenberger O, Thomashow MF. 1998. *Arabidopsis* CBF1 overexpression induces COR genes and enhances freezing tolerance. Science 280:104-106.

Ji XM, Raveendran M, Oane R, Ismail A, Lafitte R, Bruskiewich R, Cheng SH, Bennett J. 2005. Tissue-specific expression and drought responsiveness of cell-wall invertase genes of rice at flowering. Plant Mol. Biol. 59:945-964.

Kasuga M, Liu Q, Miura S, Yamaguchi-Shinozaki K, Shinozaki K. 1999. Improving plant drought, salt, and freezing tolerance by gene transfer of a single stress-inducible transcription factor. Nat. Biotechnol. 17:287-291.

Kawaguchi R, Girke T, Bray EA, Bailey-Serres J. 2004. Differential mRNA translation contributes to gene regulation under nonstress and dehydration stress conditions in *Arabidopsis thaliana*. Plant J. 38:823-839.

Kikuchi S, Satoh K, Nagata T, Kawagashira N, Doi K, Kishimoto N, Yazaki J, Ishikawa M, Yamada H, Ooka H, Hotta I, Kojima K, Namiki T, Ohneda E, Yahagi W, Suzuki K, Li CJ, Ohtsuki K, Shishiki T, Otomo Y, Murakami K, Iida Y, Sugano S, Fujimura T, Suzuki Y, Tsunoda Y, Kurosaki T, Kodama T, Masuda H, Kobayashi M, Xie Q, Lu M, Narikawa R, Sugiyama A, Mizuno K, Yokomizo S, Niikura J, Ikeda R, Ishibiki J, Kawamata M, Yoshimura A, Miura J, Kusumegi T, Oka M, Ryu R, Ueda M, Matsubara K, Kawai J, Carninci P, Adachi J, Aizawa K, Arakawa T, Fukuda S, Hara A, Hashizume W, Hayatsu N, Imotani K, Ishii Y, Itoh M, Kagawa I, Kondo S, Konno H, Miyazaki A, Osato N, Ota Y, Saito R, Sasaki D, Sato K, Shibata K, Shinagawa A, Shiraki T, Yoshino M, Hayashizaki Y, Yasunishi A. 2003. Collection, mapping, and annotation of over 28,000 cDNA clones from japonica rice. Science 301:376-379.

Kreps JA, Wu Y, Chang HS, Zhu T, Wang X, Harper J. 2002. Transcriptome changes for *Arabidopsis* in response to salt, osmotic, and cold stress. Plant Physiol. 130:2129-2141.

Lafitte HR, Champoux MC, McLaren G, O'Toole JC. 2001. Rice root morphological traits are related to isozyme group and adaptation. Field Crops Res. 71:57-70.

Liu Q, Kasuga M, Sakuma Y, Abe H, Miura S, Yamaguchi-Shinozaki K, Shinozaki K. 1998. Two transcription factors, DREB1 and DREB2, with an EREBP/AP2 DNA binding domain separate two cellular signal transduction pathways in drought- and low-temperature-responsive gene expression, respectively, in *Arabidopsis*. Plant Cell 10:1391-1406.

Mackill DJ, Coffman WR, Garrity DP. 1996. Varietal improvement for rainfed lowland rice in South and Southeast Asia: results of a survey. In: Rainfed lowland rice improvement. Manila (Philippines): International Rice Research Institute. 242 p.

Nelson DE, Repetti PP, Adams TR, Creelman RA, Wu J, Warner DC, Anstrom DC, Bensen RJ, Castiglioni PP, Donnarummo MG, Hinchey BS, Kumimoto RW, Maszle DR, Canales RD, Krolikowski KA, Dotson SB, Gutterson N, Ratcliffe OJ, Heard JE. 2007. Plant nuclear factor Y (NF-Y) B subunits confer drought tolerance and lead to improved corn yields on water-limited acres. Proc. Natl. Acad. Sci. USA 104:16450-16455.

O'Toole JC, Namuco OS. 1983. Role of panicle exsertion in water stress induced sterility. Crop Sci. 23:1093-1097.

Ozturk ZN, Talamè V, Deyholos M, Michalowski CB, Galbraith DW, Gozukirmizi N, Tuberosa R, Bohnert HJ. 2002. Monitoring large-scale changes in transcript abundance in drought- and salt-stressed barley. Plant Physiol. 48:551-573.

Satoh K, Doi K, Nagata T, Kishimoto N, Suzuki K, et al. 2007 Gene organization in rice revealed by full-length cDNA mapping and gene expression analysis through microarray. PLoS ONE 2(11):e1235.

Seki M, Narusaka M, Ishida J, Nanjo T, Fujita M, Oono Y, Kamiya A, Nakajima M, Enju A, Sakurai T, Satou M, Akiyama K, Taji T, Yamaguchi-Shinozaki K, Carninci P, Kawai J, Hayashizaki Y, Shinozaki K. 2002. Monitoring the expression profiles of 7000 *Arabidopsis* genes under drought, cold and high-salinity stresses using a full-length cDNA microarray. Plant J. 31:279-292.

Stockinger EJ, Gilmour SJ, Thomashow MF. 1997. *Arabidopsis thaliana* CBF1 encodes an AP2 domain-containing transcriptional activator that binds to the C-repeat/DRE, a *cis*-acting DNA regulatory element that stimulates transcription in response to low temperature and water deficit. Proc. Natl. Acad. Sci. USA 94:1035-1040.

Talamè V, Ozturk NZ, Bohnert HJ, Tuberosa R. 2007. Barley transcript profiles under dehydration shock and drought stress treatments: a comparative analysis. J. Exp. Bot. 58:229-240.

Widawsky DA, O'Toole JC. 1996. Prioritizing the rice research agenda for eastern India. In: Evension RE, Herdt RW, Hossain M, editors. Rice research in Asia: progress and priorities. Wallingford (UK): CAB International. p 109-130.

Yamaguchi-Shinozaki K, Shinozaki K. 2006. Transcriptional regulatory networks in cellular responses and tolerance to dehydration and cold stresses. Annu. Rev. Plant. Biol. 57:781-803.

Yamamoto K, Sasaki T. 1997. Large-scale EST sequencing in rice. Plant Mol. Biol. 35:135-144.

Yamasaki K, Kigawa T, Inoue M, Watanabe S, Tateno M, Seki M, Shinozaki K, Yokoyama S. 2008. Structures and evolutionary origins of plant-specific transcription factor DNA-binding domains. Plant Physiol. Biochem. 46:394-401.

Yazaki J, Kishimoto N, Nagata Y, Ishikawa M, Fujii F, Hashimoto A, Shimbo K, Shimatani Z, Kojima K, Suzuki K, Yamamoto M, Honda S, Endo A, Yoshida Y, Sato Y, Takeuchi K, Toyoshima K, Miyamoto C, Wu J, Sasaki T, Sakata K, Yamamoto K, Iba K, Oda T, Otomo Y, Murakami K, Matsubara K, Kawai J, Carninci P, Hayashizaki Y, Kikuchi S. 2003. Genomics approach to abscisic acid- and gibberellin-responsive genes in rice. DNA Res. 10:249-261.

Yazaki J, Kishimoto N, Nakamura K, Fujii F, Shimbo K, Otsuka Y, Wu J, Yamamoto K, Sakata K, Sasaki T, Kikuchi S. 2000. Embarking on rice functional genomics via cDNA microarray: use of 3′ UTR probes for specific gene expression analysis. DNA Res. 7:367-370.

Yazaki J, Shimatani Z, Hashimoto A, Nagata Y, Fujii F, Kojima K, Suzuki K, Taya T, Tonouchi M, Nelson C, Nakagawa A, Otomo Y, Murakami K, Matsubara K, Kawai J, Carninci P, Hayashizaki Y, Kikuchi S. 2004. Transcriptional profiling of genes responsive to abscisic acid and gibberellin in rice: phenotyping and comparative analysis between rice and *Arabidopsis*. Physiol. Genomics 17:87-100.

Yu J, Hu S, Wang J, Wong GK, Li S, Liu B, Deng Y, Dai L, Zhou Y, Zhang X, Cao M, Liu J, Sun J, Tang J, Chen Y, Huang X, Lin W, Ye C, Tong W, Cong L, Geng J, Han Y, Li L, Li W, Hu G, Huang X, Li W, Li J, Liu Z, Li L, Liu J, Qi Q, Liu J, Li L, Li T, Wang X, Lu H, Wu T, Zhu M, Ni P, Han H, Dong W, Ren X, Feng X, Cui P, Li X, Wang H, Xu X, Zhai W, Xu Z, Zhang J, He S, Zhang J, Xu J, Zhang K, Zheng X, Dong J, Zeng W, Tao L, Ye J, Tan J, Ren X, Chen X, He J, Liu D, Tian W, Tian C, Xia H, Bao Q, Li G, Gao H, Cao T, Wang J, Zhao W, Li P, Chen W, Wang X, Zhang Y, Hu J, Wang J, Liu S, Yang J, Zhang G, Xiong Y, Li Z, Mao L, Zhou C, Zhu Z, Chen R, Hao B, Zheng W, Chen S, Guo W, Li G, Liu S, Tao M, Wang J, Zhu L, Yuan L, Yang H. 2002. A draft sequence of the rice genome (*Oryza sativa* L. ssp. *indica*). Science 296:79-92.

Zhang JZ, Creelman RA, Zhu JK. 2004. From laboratory to field: using information from *Arabidopsis* to engineer salt, cold, and drought tolerance in crops. Plant Physiol. 135:615-621.

WEB site information

Rice genome sequence
IRGSP Releases Build 4.0 Pseudomolecules of the Rice Genome: http://rgp.dna.affrc.go.jp/E/IRGSP/Build4/build4.html
TIGR Release Ver. 4 Rice Genome Pseudomolecules: www.tigr.org/tdb/e2k1/osa1/pseudomolecules/info.shtml

Gene annotation
Rice Annotation Project Database (RAP): http://rapdb.dna.affrc.go.jp/
TIGR Osa1: www.tigr.org/tdb/e2k1/osa1/

Full-length cDNA clones
Knowledge-based Oryza Molecular biological Encyclopedia (KOME): http://cdna01.dna.affrc.go.jp/cDNA/

Microarrays
Rice Expression Database (RED): http://red.dna.affrc.go.jp/RED/

Metabolic pathways
Rice Cyc: www.gramene.org/pathway/

Promoter and *cis*-elements
PLACE A Database of Plant *Cis*-acting Regulatory DNA Elements: www.dna.affrc.go.jp/PLACE/

Transcription factors
Rice Transcription Factor Database: http://ricetfdb.bio.uni-potsdam.de/v2.1/

Notes

Authors' addresses: K. Satoh, K. Doi, T. Nagata, A. Hosaka, and S. Kikuchi, Plant Genome Research Unit, Division of Genome and Biodiversity Research, National Institute of Agrobiological Sciences, 2-1-2 Kannon-dai, Tsukuba, Ibaraki 305-8602, Japan; K. Suzuki, Hitachi Software Engineering Co., Ltd., 6-8-1 Onoe-cho, Naka-ku, Yokohama Kanagawa 31-0015, Japan; X. Ji, M. Raveendran, H. Leung, and J. Bennett, Plant Breeding, Genetics, and Biotechnology Division, International Rice Research Institute, DAPO Box 7777, Metro Manila, Philippines

Gene discovery for improving drought resistance of irrigated rice by systematic genetic and functional genomics approaches

Lizhong Xiong

Drought resistance is a very complex trait with distinct molecular and physiological mechanisms in different plant species. Irrigated rice has been domesticated in an ecosystem with full irrigation and it is extremely sensitive to drought. With a long-term goal of improving drought resistance in rice, we have adopted a strategy of integrating approaches, including germplasm exploitation and using genetics and functional genomics to identify loci/genes effective for improving drought resistance in rice. In this paper, we describe the approaches and the major progress made to discover genes for improving drought resistance. On the basis of a genetic dissection of drought resistance of rice, more than 30 QTLs have been targeted for the construction of near-isogenic lines and marker-assisted molecular breeding. Several drought resistance–associated genes were identified through drought screening of T-DNA insertion mutants of rice. Hundreds of candidate genes were identified for drought resistance through comparative expression profiling analysis. More than 50 drought-responsive candidate genes were transferred into rice for drought resistance testing, and four genes (*SNAC1, OsCIPK12, OsLEA3-1,* and *OCPI1*) showed a significant effect in improving drought resistance in transgenic rice. Finally, the problems and perspectives of improving drought resistance in rice are discussed.

Drought stress causes major economic losses in crop production throughout the world. For example, global losses of the two major cereal crops, rice and maize, to drought are estimated to be more than US$5 billion annually. Drought has been, and continues to be, the single most devastating factor that threatens food production and food security, especially in areas with inadequate water resources for agriculture. On the other hand, rice production consumes a huge amount of water. For example, water consumption of rice production in China is estimated to be 70% of the water consumed in agriculture. With the global shortage of water, reducing water consumption in crop production has now been generally recognized as an essential strategy for sustainable agriculture.

In recent years, the idea of developing drought-resistant crops has been well recognized as the most promising and effective strategy for alleviating food insecurity caused by drought and water shortage. For that purpose, several groups have conducted research focusing on the genetic basis of drought resistance in rice, using molecular marker-based genetic studies to identify and use quantitative trait loci (QTLs) for drought resistance in rice breeding programs. Most studies have concentrated on root characteristics since the importance of the root system in drought resistance in crops is commonly accepted (O'Toole and Bland 1987, Lynch 1995, Blum 1996). Many QTLs have been identified for traits such as root penetration ability in the soil (Ray et al 1996, Price et al 2000, Ali et al 2000) and root mass and root depth (Champoux et al 1995, Yadav et al 1997, Price et al 2000, Zhang et al 2001, Kamoshita et al 2002, Zheng et al 2003). In addition, the molecular marker-assisted approach has also been applied to map QTLs for physiological parameters such as osmotic adjustment (Lilley et al 1996, Robin et al 2003) and membrane stability (Tripathy et al 2000). However, very few QTLs (Chandra Babu et al 2003) have been identified that directly control grain yield and biomass production under drought stress. Although more than 100 QTLs have been reported for a wide spectrum of traits that are directly or indirectly related to drought resistance, few of them have been fine-mapped and most QTLs could not be repeatedly detected in different populations or even in the same populations, which has largely limited the use of QTLs.

Drought resistance is a complex trait that involves numerous aspects of developmental, physiological, biochemical, and molecular adjustments (Blum 2002). These include, for example, root architecture modification, guard cell regulation, osmotic adjustment, alterations in photosynthesis, and synthesis of protective proteins and antioxidants. The regulatory pathways leading to these adjustments are not well understood and remain focal points of research. Recently, several genes have been demonstrated to be important for drought resistance in model plant species such as *Arabidopsis*. These genes are encoding proteins or enzymes that are involved in a wide spectrum of cellular functions such as ABA synthesis (such as NCED and LOS5), transcriptional regulation (such as CBF3, DREB, and ZAT10), production of compatible osmolytes (such as TPS and CodA), detoxification (such as NPK1 and SOD), and ion homeostasis (NHX1) (reviewed by Zhu 2002). Genetic engineering using some of these genes has shown promise in improving plant drought resistance in laboratory tests with *Arabidopsis* (Thompson et al 2000, Qin and Zeevaart 2002, Kasuga et al 1999, Apse et al 1999, Iuchi et al 2001). In rice, some efforts have also been made in the isolation of genes for drought resistance. Expression of a fused bacterial gene TPS-TPP in rice significantly increased the level of trehalose and drought resistance (Garg et el 2002, Jang et al 2003). Overexpression of a rice MAPK can significantly increase the tolerance of rice of drought, salinity, and cold (Xiong and Yang 2003). Although these results were from greenhouse tests, they suggest that genetic engineering is a promising strategy for exploring a wide range of genes that have potential for improving drought resistance in economically important crops.

Toward a long-term goal of developing drought-resistant rice cultivars or hybrids, we have begun a series of studies on drought resistance in rice, including

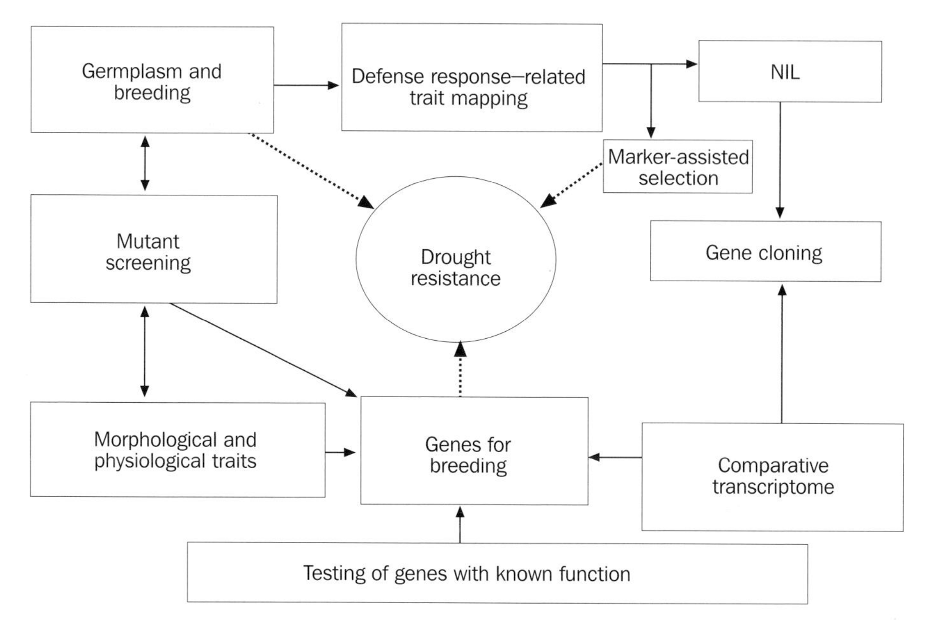

Fig. 1. Schematic strategy for the identification of genes for drought resistance in rice.

mapping and cloning of drought resistance–related QTLs, mutant screening, expression profiling, and genetic transformation and testing of candidate genes. In this paper, the approaches and progress from our group are reviewed and the problems and strategies for improving drought resistance in rice are discussed.

General description of approaches for improving drought resistance in rice

Drought resistance is a very complex quantitative trait. We have adopted a strategy by combining conventional breeding and functional genomics and genetic engineering approaches (Fig. 1) to improve the drought resistance of irrigated rice. In this strategy, the identification of drought-resistant germplasm is a very important component because such germplasm is a key material for both conventional breeding and gene discovery by genetic and functional genomics approaches. Backcrossing (with or without marker-assisted selection) is the major approach to introduce a drought-resistance trait from donors (drought-resistant germplasm) to elite irrigated rice. So far, characterization of drought-resistance genes is still the most challenging task for improving drought resistance through the genetic engineering approach because very few genes have been shown to have a significant effect in improving drought resistance in field conditions. We adopted four integrated approaches to identify critical drought resistance–related genes: (1) QTL mapping and cloning, the most difficult approach but very important for cloning critical genes; (2) mutant screening; (3) expression profiling; and (4) testing of candidate genes (or genes with a known function).

Major progress made in the discovery of genes for drought resistance

QTL mapping and near-isogenic line construction

To initiate genetic analysis and breeding of drought resistance in rice, thousands of rice germplasm accessions from around the world were screened for drought resistance by using relative yield as a major evaluation criterion. Resistant germplasm from this screening has been used as donors for generating introgression lines or backcrossing breeding.

A recombinant inbred line (RIL) population was derived from a cross between irrigated rice Zhenshan 97 (drought sensitive) and upland rice IRAT109 (drought resistant) and the population was tested for two years under fully irrigated and drought conditions, respectively. A genetic map containing 228 simple sequence repeat (SSR) markers was constructed and used to map QTLs controlling drought-related traits (Yue et al 2005). More than 20 QTLs were detected for various drought tolerance–related traits, including drought-resistance index, leaf rolling, days to leaf rolling, canopy temperature, yield, and yield components (Yue et al 2006). Meanwhile, 17 QTLs were also detected for various root traits (maximum root depth, deep root rate, drought-induced deep-root depth, drought-induced root volume, etc.) under normal growth and drought-stressed conditions (Yue et al 2006). Most QTLs have a minor genetic effect (LOD value from 2.5 to 5) and no match to the reported QTLs in other populations. However, the positions of three QTLs for root traits (two QTLs on chromosome 4 and one on chromosome 9) match well with the positions reported in other populations (Price et al 2000, Kamoshita et al 2002). Interestingly, the position of a QTL for relative yield under drought stress (located between markers RM284 and RM531 on chromosome 8) matches that of a QTL for osmotic adjustment detected by other groups (Zhang et al 2001, Robin et al 2003) using different populations.

Based on QTL mapping of morphological (such as root and leaf traits), physiological (such as osmotic adjustment), and yield stability traits related to drought resistance (Yue et al 2005, 2006), a total of 38 QTL regions have been targeted for constructing near-isogenic lines (NILs) by using the elite parental rice Zhenshan 97 as a recurrent parent (Fig. 1). To accelerate progress, the RILs containing targeted regions with maximum genetic background (generally more than 70%) of the recurrent parent were selected from the RIL population as donor parents for backcrossing. To date, more than 20 NILs (at BC_4F_2 or BC_5F_1 generation) have been generated. The NILs are valuable materials for fine mapping and cloning of the QTLs for drought-resistance genes. Field testing has been performed under drought-stressed conditions to identify drought-resistant lines. A few lines with improved drought resistance have been identified. In the process of gene cloning, mapping of differentially expressed genes obtained from the gene chip approach (as described in the following section) will be very helpful in determining the candidate genes for the QTLs.

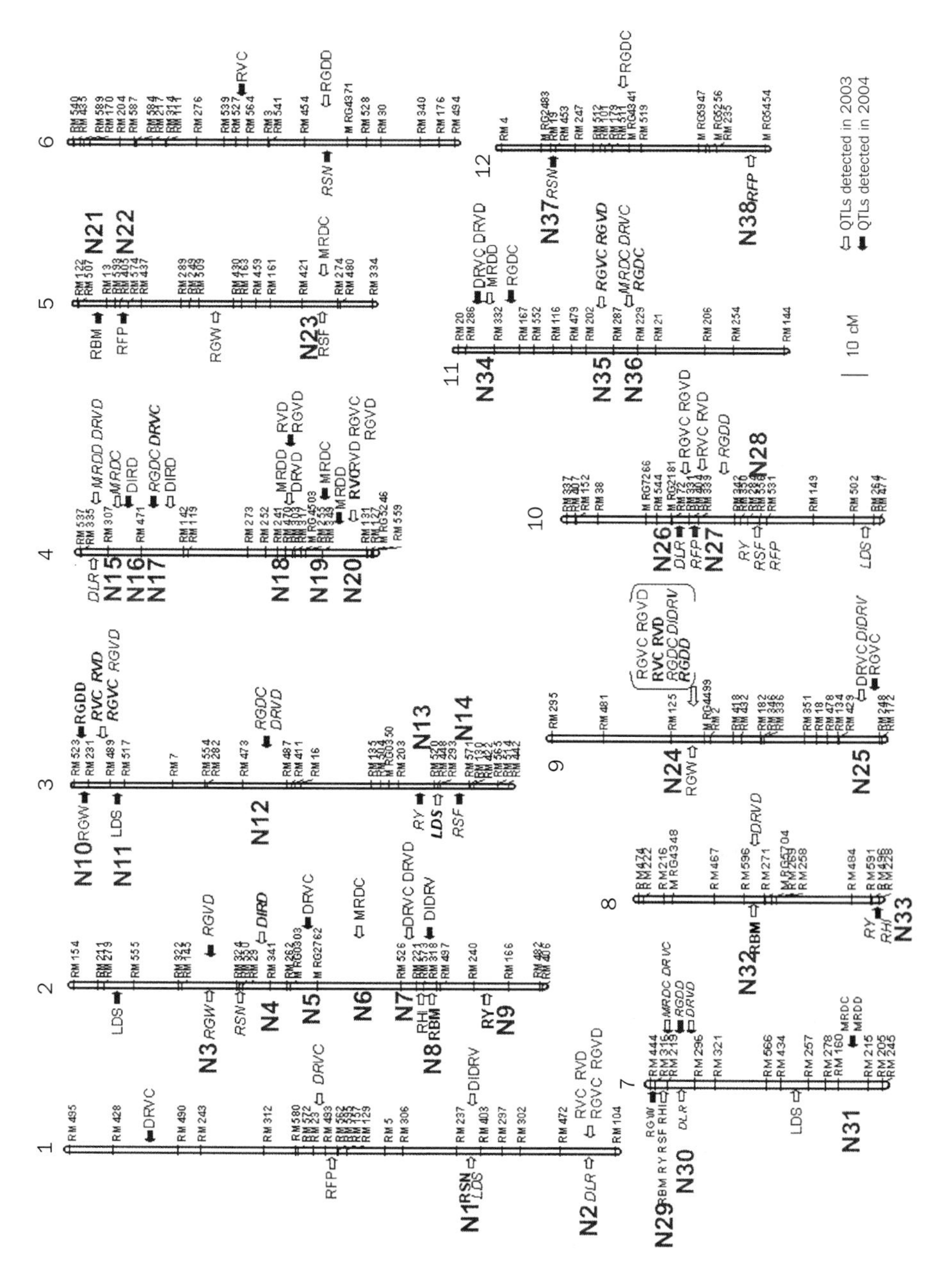

Fig. 2. Selected QTLs for constructing near-isogenic lines (designated N1 to N38) based on the molecular marker linkage map based on the RIL population from a cross between Zhenshan 97 and IRAT109. The QTLs for aboveground traits and root traits are placed on the left and right sides of the chromosomes, respectively. QTLs detected in both years are shown in bold. QTLs in italics indicate that the alleles for increasing trait values are from Zhenshan 97. See Yue et al (2006) for descriptions of traits.

Table 1. Drought-sensitive-resistant mutants identified from screening of a T-DNA insertion mutant library.

ID	Phenotype	Co-segregation with FST[a]	Putative genes mutated
UZ41	Drought-sensitive	Yes	MAPKKK protein kinase
MP37	Drought-sensitive	Yes	EDR1-homolog, protein kinase
EI38	Drought-sensitive	Yes	Receptor-like protein kinase
EL41	Drought-sensitive	Yes	AMP-binding protein
AC54	Drought-sensitive	Yes	Lipoxygenase
CP86	Drought-sensitive	n.c.	Expressed protein
HG37	Drought-sensitive	n.c.	Unknown protein
FR57	Drought-sensitive	n.c.	Kelch domain-containing
HK73	Drought-sensitive, delayed flowering	Yes	Lipase-like protein

[a]n.c. = not confirmed (phenotype is probably co-segregated with flanking sequence tag but this needs to be confirmed).

Identification of stress-associated key genes by mutant screening

The isolation of stress-related genes by a screen mutant library of rice under drought conditions was considered an effective approach since we have been developing a rice T-DNA insertion/activation mutant library as part of the national program on rice functional genomics of China. The library currently has more than 150,000 independent lines (http://RMD.ncpgr.cn). By using our drought-screening facility, we have screened more than 3,000 activation/insertion mutant lines at the anthesis stage and 8,000 mutant lines at the vegetative stage under severe drought stress conditions (i.e., most plants were killed by drought). Among them, 31 and 4 families showed segregation of drought sensitivity and resistance, respectively. Flanking sequences were successfully isolated for 17 of these mutants. However, only 6 of the 17 mutants have been confirmed with co-segregation of T-DNA insertions with the drought sensitivity phenotype. The putative functions of the mutated genes of these mutants are listed in Table 1. All six of these genes have not been reported for their funtions. Functional complementary testing is ongoing for these mutants. Three mutants showed likely co-segregation with the flanking sequences at the single-plant level but they need to be confirmed at the family level.

Identification of drought-responsive genes by comparative expression profiling analysis

Two varieties, Zhenshan 97 and IRAT109, the parents of the RIL population for QTL mapping described above, were chosen for studying genome-wide expression profiles after drought stress at the vegetative (5-leaf) stage and reproductive (panicle heading) stage. Drought stress was applied by stopping watering at the four-leaf stage and 2 weeks before flowering, respectively. The relative water content (RWC) in the leaves was measured daily and seedling morphology also observed. The RWC remained above 93% and no obvious morphology change occurred. Daily sampling started with RWC in leaves at about 92% until it dropped to about 70%. Leaf samples with equivalent RWC for the two genotypes were chosen for expression profiling analysis. Basically, three types of samples representing slight (RWC 90–91%), moderate (RWC 85–87%), and severe (RWC 75–78%) drought stress underwent RNA extraction and DNA chip hybridization.

In the beginning, a cDNA microarray containing 9,216 unique cDNA sequences identified from a normalized cDNA library (Chu et al 2003) was used for expression profiling of drought-stressed rice at the vegetative stage. A total of 482 and 421 genes were up- and down-regulated, respectively, in IRAT109. However, in the irrigated rice Zhenshan 97, only 241 and 237 genes were up- and down-regulated, respectively. A total of 74 differentially expressed genes were detected in the samples from slight and moderate drought stresses. Among them, 40 genes are up-regulated and 34 genes are down-regulated in at least one genotype. Examples of these genes (based on a sequence homology search) are MAPKK, MADS box protein, protein kinase, EF-hand Ca^{2+}-binding protein, ATP-dependent Clp protease, EREBP-like protein, calmodulin, heat shock protein 70, zinc finger-like protein, etc. Under severe drought-stress conditions (RWC 75–78%), nearly 8% of the genes arrayed on the DNA chips were differentially expressed. Among them, 465 and 406 genes were up- and down-regulated, respectively, in at least one genotype. The differentially expressed genes at this time were involved in a wide spectrum of functions, including osmoprotectant synthases, LEA-like proteins, membrane proteins, detoxification enzymes, proteinase inhibitors, protein kinases, and transcription factors. We noticed that the genes specifically up- or down-regulated at the early stage of drought stress are mainly regulatory genes or unknown function sequences.

Recently, we adopted the Affymetrix rice gene chip, which contains all predicted genes in the rice genome, to profile the expression of drought-stressed leaves at the reproductive stage of rice. A total of 2,614 and 2,618 genes were up- and down-regulated, respectively, in IRAT109. However in the irrigated rice Zhenshan 97, 2,246 and 3,087 genes were up- and down-regulated, respectively. Obviously, more genes are up-regulated in upland rice IRAT109 and more genes are down-regulated in irrigated rice Zhenshan 97. Interestingly, a large proportion (33.7%) of the differentially drought-induced genes can be mapped to the drought resistance–related QTL regions. This information is very useful to delimit the candidate genes in the effort to clone genes of drought resistance–related traits.

Genetic transformation and drought testing of stress-responsive genes

Drought-responsive genes were selected based on the expression profiles of the cDNA microarray and gene chip. To date, we have selected 56 drought-responsive genes for rice transformation and functional analysis (if the transgenic rice showed improved drought resistance). Selected genes for transgenic testing have diverse predicted functions, including signal transduction, transcriptional regulation, and enzymes for synthesis of protective molecules or phytohormone. Some of them are unknown for their putative function based on sequence analysis. For a high-throughput testing of a large number of potentially useful genes, a japonica cultivar, Zhonghua 11, with high efficiency of transformation was used for genetic transformation. Transgenic plants have been generated for all the 56 genes (Table 2) and T_1 families of all these genes have been tested for drought resistance. Results showed that overexpression of at least six genes resulted in a significant improvement in drought resistance (Table 2). The predicted proteins of these promising genes are NAC transcription factor SNAC1

Table 2. Selected drought-responsive genes for rice transformation and drought testing.

No.	Predicted function	Drought testing of transgenic families[a]
1	Transcription factor SNAC1	Improved drought and salt tolerance**
2	PGPD14-like	No improvement
3	Putative bZIP	No improvement
4	Putative bZIP (OsbZIP23)	Improved drought resistance*
5	Putative AP2/EREBP	No improvement
6	Putative AP2/EREBP	Improved dehydration resistance*
7	Putative AP2/EREBP	No improvement
8	Putative AP2/EREBP	No improvement
9	MYB protein	No improvement
10	CBL homologue	No improvement
11	NPK1-like	No improvement
12	Cation transporter	No improvement
13	SKIP homolog	Improved drought resistance*
14	Regulator of amylase	No improvement
15	NPK1-like	No improvement
16	Na/H+ antiporter	No improvement
17	NDPK homolog	No improvement
18	SNAC2	Improved salt and cold tolerance*
19	PP2C-like	No improvement
20	PP2C-like	No improvement
21	Acidic phosphatase	No improvement
22	Cold-induced protein	No improvement
23	LEA protein (OsLEA3-1)	Improved drought resistance**
24	LEA protein	Improved drought resistance*
25	Pathogen-induced protein	Improved drought resistance*
26	Drought-induced protein	No improvement
27	C3HC4 zinc finger	No improvement
28	Stress-induced protein	No improvement
29	Proteinase inhibitor (OCPI1)	Improved drought resistance**
30	CIPK12 (protein kinase)	Improved drought resistance**
31	GYF-containing protein	No improvement
32	Universal stress protein	No improvement
33	Harpin-induced protein	No improvement
34	Purine permease	No improvement
35	ADR1 homolog	No improvement
36	GCR1 homolog	No improvement
37	VP1 homolog	No improvement

Continued on next page

Table 2 continued.

No.	Predicted function	Drought testing of transgenic families[a]
38	VP1-interacted protein	No improvement
39	AIP-like protein	No improvement
40	MAPKK family	No improvement
41	MAPKKK family	Improved drought resistance*
42	Auxin-responsive protein	No improvement
43	Zinc finger protein	No improvement
44	Zinc finger protein	No improvement
45	Myb protein	Improved drought resistance*
46	Homeobox protein	No improvement
47	AN1-like family protein	No improvement
48	Transcription factor BTF3	No improvement
49	Myb-related protein Zm1	No improvement
50	CONSTANT-like 6	No improvement
51	Transcriptional coactivator	Need confirmation for cold tolerance
52	Basic helix-loop-helix	No improvement
53	XH/XS-containing protein	Need confirmation for salt tolerance
54	WZF1	No improvement
55	Oshox22	Need confirmation for stress tolerance
56	Oshox24	Need confirmation for stress tolerance

[a]* and ** = significant at $P<0.05$ and $P<0.01$, respectively, by t-test using relative grain yield for evaluation of drought resistance. Drought stress was applied at the anthesis stage as described by Yue et al (2006). More than 30 T1 families for each construct, 20 plants in each family, were tested.

(Hu et al 2006), CBL-interacting protein kinase (Xiang et al 2007), a bZIP transcription factor (Xiang et al, submitted), a homolog of SKIP protein (Hou et al, submitted), LEA protein (Xiao et al 2007), and a chymotrypsin-like proteinase inhibitor (Huang et al 2007).

Among these genes, *SNAC1* is the most promising gene for improving stress resistance. SNAC1 is a transcription factor belonging to the NAM, ATAF, and CUC (NAC) family. In leaves, this gene is induced specifically in guard cells by dehydration stress (Hu et al 2006). The SNAC1 protein contains a nuclear localization signal located in the N-terminus that can sufficiently mediate nuclear targeting of the protein, a DNA binding domain in the middle that can bind to the NAC-recognized sequence, and a C-terminus transactivation domain with 241–271 amino acids indispensable for activation activity. Transgenic rice overexpressing the *SNAC1* gene showed significantly enhanced drought resistance (22–34% higher seed setting than a control) in the field under severe drought-stress conditions at the reproductive stage, while showing no phenotypic change or yield penalty. The transgenic rice was more sensitive to abscisic acid and lost water more slowly by closing more stomatal pores and maintained turgor pressure at a significantly lower level of relative water content than the wild type (Hu et al 2006). The transgenic rice also showed significantly improved drought and salt tolerance (80% higher survival rate than a control) at the vegetative stage.

DNA chip analysis revealed that more than 150 genes are up-regulated (>2.1-fold) in the *SNAC1*-overexpressing rice plants. Our data suggest that *SNAC1* holds promise for improving drought and salinity tolerance in rice.

Problems and perspectives

There are several obstacles to developing irrigated rice with both drought resistance and high yield. First, of all the cultivars and hybrids that have been bred in the last several decades for high yield in optimal water conditions, they are not adapted to water-deficit conditions. Second, little has been done to explore the germplasm for increasing the drought resistance of irrigated rice, although such germplasm may be readily available. Third, drought resistance is a very complex trait and our knowledge on it in rice is still limited. Last but most important, more accurate and reliable protocols for phenotyping drought resistance of irrigated rice need to be established.

Next, we will focus on a more integrated strategy for identifying genes for drought resistance in irrigated rice. In this strategy, the identification of drought-resistance germplasm and mining of drought-resistance alleles for functionally known genes will be emphasized. Different but related approaches will be used to identify important genes or pathways for drought resistance of rice. These approaches include, but are not limited to, mutant screening of stress-associated candidate genes, comparative genomic expression and proteomics profiling of additional drought-resistant and drought-sensitive genotypes, and knowledge-based mining of candidate genes for improving drought resistance. Besides drought resistance–associated genes, genes associated with water-use efficiency will also be targeted.

Two approaches will be followed for developing drought-resistant cultivars. The first approach will be the introduction of drought-resistance genes (or alleles) identified from various germplasm accessions to the background of elite rice by marker-assisted selection and conventional breeding. The second approach will be the transformation of hybrid parents or drought-tolerance-improved NILs with genes that have been identified as having a significant effect on improving drought resistance. Such genes will be identified essentially from functional genomics studies.

References

Ali ML, Pathan MS, Zhang J, Bai G, Sarkarung S, Nguyen HT. 2000. Mapping QTLs for root traits in a recombinant inbred population from two *indica* ecotypes in rice. Theor. Appl. Genet. 101:756-766.

Apse MP, Aharon GS, Snedden WS, Blumwald E. 1999. Salt tolerance conferred by overexpression of a vacuolar Na^+/H^+ antiport in *Arabidopsis*. Science 285:1256-1258.

Blum A. 1996. Crop responses to drought and the interpretation of adaptation. Plant Growth Reg. 20:135-148.

Blum A. 2002. Drought stress and its impact. http://plantstress.com/Articles/index.asp.

Champoux MC, Wang G, Sarkarung S, Mackill DJ, O'Toole JC, Huang N, McCouch SR. 1995. Locating genes associated with root morphology and drought avoidance in rice via linkage to molecular markers. Theor. Appl. Genet. 90:969-981.

Chandra Babu R, Nguyen BD, Chamarerk V, Shanmugasundaram P, Chezhian P, Jeyaprakash P, Ganesh SK, Palchamy A, Sadasivam S, Sarkarung S, Wade LJ, Nguyen HT. 2003. Genetic analysis of drought resistance in rice by molecular markers: association between secondary traits and field performance. Crop Sci. 43:1457-1469.
Chu ZH, Peng KM, Zhang LD, Zhou B, Wei J, Wang SP. 2003. Construction and characterization of a normalized whole-life-cycle cDNA library of rice. Chinese Sci. Bull. 48:229-235.
Garg AK, Kim JK, Owens TG, Ranwala AP, Choi YD, Kochian LV, Wu RJ. 2002. Trehalose accumulation in rice plants confers high tolerance levels to different abiotic stresses. Proc. Natl. Acad. Sci. USA 99:15898-15903.
Hu H, Dai M, Yao J, Xiao B, Li X, Zhang Q, Xiong L. 2006. Overexpressing a NAM, ATAF, and CUC (NAC) transcription factor enhances drought resistance and salt tolerance in rice. Proc. Natl. Acad. Sci. USA 103:12987-12992.
Huang Y, Xiao B, Xiong L. 2007. Characterization of a stress responsive proteinase inhibitor gene with positive effect in improving drought resistance in rice. Planta 226:73-85.
Iuchi S, Kobayashi M, Taji T, Naramoto M, Seki M, Kato T, Tabata S, Kakubari Y, Yamaguchi-Shinozaki K, Shinozaki K. 2001. Regulation of drought resistance by gene manipulation of 9-*cis*-epoxycarotenoid dioxygenase, a key enzyme in abscisic acid biosynthesis in *Arabidopsis*. Plant J. 27:325-333.
Jang IC, Oh SJ, Seo JS, Choi WB, Song SI, Kim CH, Kim YS, Seo HS, Choi YD, Nahm BH, Kim JK. 2003. Expression of a bifunctional fusion of the *Escherichia coli* genes for trehalose-6-phosphate synthase and trehalose-6-phosphate phosphatase in transgenic rice plants increases trehalose accumulation and abiotic stress tolerance without stunting growth. Plant Physiol. 131:516-524.
Kamoshita A, Wade L, Ali M, Pathan M, Zhang J, Sarkarung S, Nguyen H. 2002. Mapping QTLs for root morphology of a rice population adapted to rainfed lowland conditions. Theor. Appl. Genet. 104:880-893.
Kasuga M, Liu Q, Miura S, Yamaguchi-Shinozaki K, Shinozaki K. 1999. Improving plant drought, salt, and freezing tolerance by gene transfer of a single stress-inducible transcription factor. Nat. Biotechnol. 17:287-291.
Lilley JM, Ludlow MM, McCouch SR, O'Toole JC. 1996. Locating QTLs for osmotic adjustment and dehydration tolerance in rice. J. Exp. Bot. 47:1427-1436.
Lynch JP. 1995. Root architecture and plant productivity. Plant Physiol 109:7-13.
O'Toole J C, Bland WL. 1987. Genetic variation in crop plant root system. Adv. Agron. 41:91-145.
Price AH, Steele KAB, Moore J, Barraclough PP, Clark LJ. 2000. A combined RFLP and AFLP linkage map of upland rice (*Oryza sativa* L.) used to identify QTLs for root-penetration ability. Theor. Appl. Genet. 100:49-56.
Qin X, Zeevaart JA. 2002. Overexpression of a *9-cis-epoxycarotenoid dioxygenase* gene in *Nicotiana plumbaginifolia* increases abscisic acid and phaseic acid levels and enhances drought resistance. Plant Physiol. 128:544-551.
Ray JD, Yu L, McCouch SR, Champoux MC, Wang G, Nguyen HT. 1996. Mapping quantitative trait loci associated with root penetration ability in rice (*Oryza sativa* L.). Theor. Appl. Genet. 92:627-636.
Robin S, Pathan MS, Courtois B, Lafitte R, Carandang S, Lanceras S, Amante M, Nguyen HT, Li Z. 2003. Mapping osmotic adjustment in an advanced back-cross inbred population of rice. Theor. Appl. Genet. 107:1288-1296.

Thompson AJ, Jackson AC, Symonds RC, Mulholland BJ, Dadswell AR, Blake PS, Burbidge A, Taylor IB. 2000. Ectopic expression of a tomato *9-cis-epoxycarotenoid dioxygenase* gene causes over-production of abscisic acid. Plant J. 23:363-374.

Tripathy JN, Zhang J, Robin S, Nguyen HT. 2000. QTLs for cell-membrane stability mapped in rice (*Oryza sativa* L.) under drought stress. Theor. Appl. Genet. 100:1197-1202.

Xiang Y, Huang Y, Xiong L. 2007. Characterization of stress-responsive *CIPK* genes in rice for stress tolerance improvement. Plant Physiol. 144:1416-1428.

Xiao B, Huang Y, Tang N, Xiong L. 2007. Over-expression of a *LEA* gene in rice improves drought resistance under the field conditions. Theor. Appl. Genet. 115:36-45.

Xiong L, Yang Y. 2003. Disease resistance and abiotic stress tolerance in rice are inversely modulated by an abscisic acid-inducible mitogen-activated protein kinase. Plant Cell 15:745-759.

Yadav R, Courtois B, Huang N, McClaren G. 1997. Mapping genes controlling root morphology and root distribution in a doubled-haploid population of rice. Theor. Appl. Genet. 94:619-632.

Yue B, Xiong L, Xue W, Xing Y, Luo L, Xu C. 2005. Genetic analysis for drought resistance of rice at reproductive stage in field with different types of soil. Theor. Appl. Genet. 111:1127-1136.

Yue B, Xue W, Xiong L, Yu X, Luo L, Cui K, Jin D, Xing Y, Zhang Q. 2006. Genetic basis of drought resistance at reproductive stage in rice: separation of drought resistance from drought avoidance. Genetics 172:1213-1228.

Zhang J, Zheng HG, Aarti A, Pantuwan G, Nguyen TT, Tripathy JN, Sarial AK, Robin S, Babu RC, Nguyen BD, Sarkarung S, Blum A, Nguyen HT. 2001. Locating genomic regions associated with components of drought resistance in rice: comparative mapping within and across species. Theor. Appl. Genet. 103:19-29.

Zheng BS, Yang L, Zhang WP, Mao CZ, Wu YR, Yi KK, Liu FY, Wu P. 2003. Mapping QTLs and candidate genes for rice root traits under different water-supply conditions and comparative analysis across three populations. Theor. Appl. Genet. 107:1505-1515.

Zhu JK. 2002. Salt and drought stress signal transduction in plants. Annu. Rev. Plant Physiol. Plant Mol. Biol. 53:247-273.

Notes

Author's address: National Key Laboratory of Crop Genetic Improvement, Huazhong Agricultural University, Wuhan 430070, China, tel: 86-27-87281536; fax: 86-27-87287092; e-mail: lizhongx@mail.hzau.edu.cn.

Acknowledgments: This work was supported by grants from the National Program on the Development of Basic Research, the Animal and Plant Functional Genomics of China, the National Natural Science Foundation of China, and the Rockefeller Foundation.

SNP discovery at candidate genes for drought responsiveness in rice

Kenneth L. McNally, Ma. Elizabeth Naredo, and Jill Cairns

Characterization of single nucleotide polymorphisms (SNP) in candidate genes for drought tolerance is a promising approach for identifying alleles that are associated with drought phenotypes. For SNP discovery, we have used the technique of EcoTILLING contrasting diverse varieties to both japonica variety Nipponbare and indica variety IR64. Our germplasm panel of 1,536 *Oryza sativa* varieties covers the variety groups and eco-cultural types and includes parents of mapping populations used for drought QTL analyses. A set of candidate genes for drought tolerance was identified through convergent evidence taking into account genome annotation, function, expression, and localization with a yield-component QTL under water stress. These genes include DREB2A, ERF3, trehalose-6-phosphate phosphatase, and actin depolymerizing factor among others. EcoTILLING of a set of 900 of the *Oryza sativa* lines for 1,800 bp of coding and regulatory region of ERF3 identified a range of putative SNPs. Sequence confirmation from selected lines for each of the observed patterns identified 31 SNPs and short indels that grouped into nine haplotypes corresponding to variety types. Within-group association tests for drought-related traits were performed with one of the indica subgroups found to have a significant association with yield stability during water stress. Putative mismatches have been identified at 10 other candidate gene loci. Sequencing of these mismatches is under way to verify the SNP setting the stage for additional association tests of the SNPs with drought phenotypes. In addition to SNP analysis of individual candidate genes, we will take a genome-wide approach to relate SNP haplotypic variation with expression polymorphism and phenotypes. Through a multiple-partner SNP discovery project, we expect to obtain genome-wide SNP haplotype data for 20 diverse rice genotypes, some of which are known to have contrasting phenotypic responses to drought stress. The combination of SNP and drought stress-transcriptome data may reveal causal relationships among SNP haplotypic blocks, expression polymorphisms, and adaptive responses to drought.

The identification of allelic variants for genes related to drought tolerance is a promising approach for rice improvement for this most serious abiotic stress. This paper describes our development of the tools and resources to make this feasible as part of our efforts to undertake allele mining. We define the process of allele mining as the search for novel allelic variation at genes of interest in large collections of germplasm varieties (Hamilton and McNally 2005). The main aim of allele mining is to enhance the value of conserved germplasm by identifying the best donor alleles for traits of interest to crop improvement from a wider range of sources than have been used to date in breeding programs. The advent of new molecular tools for high-throughput genotyping, such as the characterization of single nucleotide polymorphisms (SNPs), has allowed such schemes to be possible. Furthermore, access to genome sequences for both japonica (Nipponbare, high-quality BAC-by-BAC, http://rgp.dna.affrc.go.jp/) and indica (93-11, whole-genome shotgun with 6.25X coverage, http://rice.genomics.org.cn/rice/index2.jsp) types of rice facilitates the development of genotyping tools suitable for surveying allelic variation (IRGSP 2005, Yu et al 2005).

Functional genomics is crucial for identifying which genes are the best choices of candidate genes involved in the expression or development of drought tolerance. Hence, allele mining for drought tolerance operates at the interface between functional genomics and crop improvement, aiming to take the products of functional genomics, apply them to genetic resources to identify the best donor alleles, and then deliver these to breeding programs. As such, allele mining can be viewed as a prebreeding activity in which results from physiology and genomics are integrated into genebank research activities.

Allele mining relies on several inputs: the identification of a suitable sample of germplasm such as a core collection (see next section); production of genomic DNAs and sufficient seed for replicated phenotyping for the core collection (third section); fingerprinting with anonymous markers to define the population structure in order to avoid finding spurious associations where the genes underlying a trait do not share a common history (fourth section); definition of the candidate gene targets likely to be involved in drought tolerance (fifth section); and implementation of techniques for high-throughput genotyping of these candidate genes (sixth section). After these components are in place, allele mining involves three main activities—genotyping by molecular markers at candidate gene loci (continued in the sixth section), replicated phenotyping for traits related to the target traits (seventh section), and identification of the best alleles at a particular locus with the trait through the use of association genetics (eighth section). Each of these steps will be described in more detail in the subsequent sections. Figure 1 depicts the flow of these activities in the process of allele mining. Furthermore, allele mining seeks to use genotyping as the entry point wherein once certain genotypes have been associated with particular phenotypes, subsets of germplasm carrying those genotypes from broader collections can be chosen for selective phenotyping, thus eliminating the need to phenotype every accession.

Following the identification of the best donor alleles for a particular trait, the germplasm source and molecular markers for its detection can be delivered to breeding programs, allowing the favorable alleles to be introgressed through marker-assisted

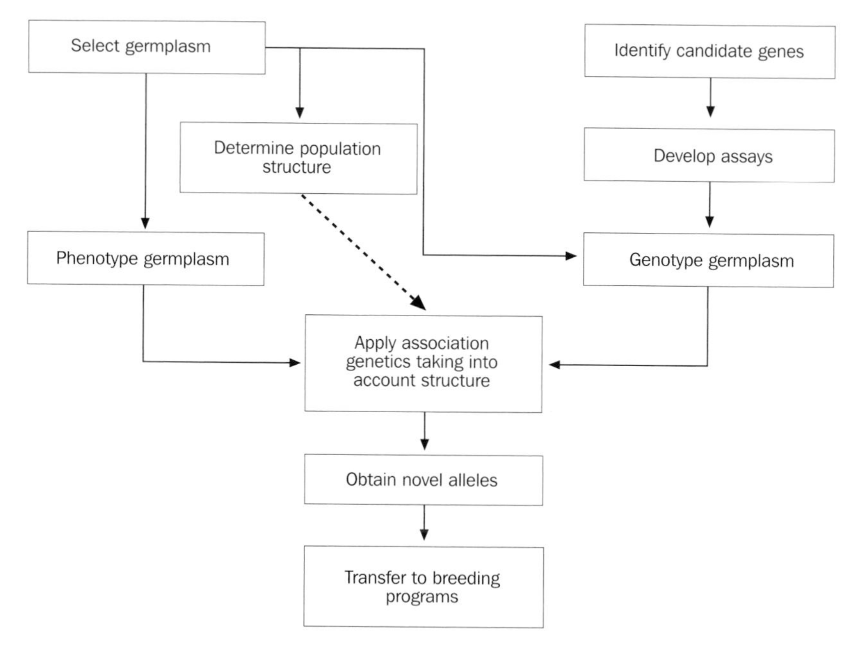

Fig. 1. Diagram of the allele mining process.

selection schemes. With the components of allele mining in place, we are now poised to go beyond the gene, delivering novel alleles, to marker-assisted breeding programs for the improvement of drought tolerance. Through a better understanding of the rice germplasm pool and an enhanced use of this genetic diversity, all those who are dependent on rice as their staple will benefit.

Defining a core collection

The International Rice Genebank Collection (IRGC) is the world's largest, with more than 115,000 registered and incoming accessions of the Asian cultivated rice, *Oryza sativa*; the African cultivated rice, *O. glaberrima*; and 22 wild relatives within the genus *Oryza*. The majority of the accessions contained in the IRGC have not contributed to modern high-yielding varieties; hence, the IRGC is a largely untapped reservoir of potentially useful alleles. In preparation for allele mining, sets of *Oryza* germplasm were chosen from the *O. sativa* and non-*O. sativa* sections of the IRGC to comprise core collections for these gene pools. Sir Otto Frankel (1984) argued that a core collection should represent "with a minimum of repetitiveness, the genetic diversity of a crop species and its relatives," whereas Brown (1988) stated that "a better collection is one that is rationalized, refined and structured, around a small, well-defined and representative core." Brown and Schoen (1994) suggested that core collections be based on "structured or stratified random sampling" where samples are drawn from a target collection that has been divided into groups that are genetically, geographically, or ecologically distinct. The usual intent for defining a core collection is from a management perspective, allowing savings in cost while maintaining the bulk of the genetic diversity within a species in the active collection of a genebank. Our strategy for defining a core collection from the IRGC was not intended as a means to modify management practice, but only to reduce the number of accessions to a workable level for high-throughput genotypic and phenotypic characterization.

Several different strategies have been proposed for the design of core collections and reviewed by Hodgkin et al (1995) and van Hintum et al (2000). The approach we adopted for the design of a rice core collection was a semistratified selection scheme using species and eco-cultural types, characterization data, geographical information, and estimates of usage and deployment based on tracing pedigrees in the International Rice Information System (IRIS, www.iris.cgiar.org; Bruskiewich et al 2003). We began by selecting the accessions in groups of germplasm previously defined by Bonman and Glaszmann (cited by Vaughan 1991), Virk et al (1996), and Gregorio (IRRI, unpublished) as well as those IRGC accessions referenced by McCouch in the Gramene database (www.gramene.org). These widely distributed and extensively characterized accessions served as a base set. Other accessions were added to this base set, balancing the aforementioned criteria from the IRGC hierarchically organized into groups by species, country of origin, and variety name. Random selection from these groups was performed so that from 100% to 8% of the accessions from each country were chosen across the name-space of varieties for *Oryza sativa*. We have also included additional accessions from the IRGC or non-IRGC varieties nominated by plant breeders, physiologists, and plant pathologists to eliminate potential gaps in the core collection. The resulting *O. sativa* core collection consists of 11,000 diverse accessions and spans the range of variety groups and eco-cultural types for improved and traditional varieties, landraces, and breeding lines. For non-*O. sativa* accessions, from 100% to 20% of the accessions for a species were chosen. For related species, the non-*O. sativa* core collection consists of 2,000 *O. glaberrima* and wild relatives from the IRGC, which includes all of the species and genome types from the closely related AA to the distant HHKK genomes (e.g., from *O. rufipogon* to *O. schlechteri*).

These core collections are for high-throughput genotyping that can be achieved using pooled samples. For medium- to low-throughput genotyping and phenotyping from the *O. sativa* core, we have further sampled a minicore collection of 1,536 accessions. This minicore collection includes all of the accessions in the special-purpose sets from Bonman/Glaszmann, Virk, Gregorio, and McCouch and a random sampling from the other accessions in the *O. sativa* core. The *O. sativa* minicore collection is embedded within the rice composite collection used in the Generation Challenge Program (GCP, www.generationcp.org) as accessions 1 to 1,536 with passport information available from the GCP central repository.

Processing the core collection

We have established a leaf bank for the *O. sativa* and non-*O. sativa* core collections by harvesting 10 g fresh-weight tissue from 10–15 plants per accession, followed by lyophilization. Plants were grown in the screenhouse in seedling boxes, in culture solution, or in field plots. Morpho-agronomic characters, such as sheath color, were used to monitor intra-accession variability. Accessions appearing to be mixtures were separated into types. The leaf bank of lyophilized tissue is stored at 4 °C in the IRGC medium-term storage facility. The leaf bank serves as the source material for genomic DNA preparation using the modified CTAB procedure of Fulton et al (1995). Concurrent with production of the leaf bank, we have produced seed from one-third of the *O. sativa* core collection accessions, including all of the minicore accessions for replicated phenotyping studies.

Determining population structure

The population structure of a core set will influence the ability to effectively use association genetics. Accessions that are distantly related and in which many thousands of meioses separate them may have diverged such that genes no longer share the same function or separate loci have converged so that they influence a particular trait in a similar fashion. For these cases, spurious associations may be detected unless the effect of population structure is taken into account. Association genetics is discussed in greater detail in the eighth section. Fingerprinting by anonymous markers (SSR) sampled from across the genome allows determination of the population structure (Pritchard et al 2000, Thornsberry et al 2001). As part of the activities in the Generation Challenge Program for rice genotyping, we have fingerprinted the minicore collection with an SSR marker panel of 50 SSRs chosen from those optimized by Coburn et al (2002) for semiautomated genotyping with fluorescent labeling. The resulting groups consisted of indica, aus, aromatic, tropical japonica, and temperate japonica with good correspondence to isozyme variety groups as well as those groupings in published studies using SSRs (Garris et al 2005) and SNPs (Caicedo et al 2007). These data will be used to stratify the minicore collection into historically related subsets prior to attempting association analyses.

Candidate genes for drought response/tolerance

We have used convergent evidence from functional annotation; expression analyses including transcriptomics, proteomics, or metabalomics; co-localization with QTLs; and shifts in allele frequencies under selection to identify candidate genes likely to be involved in the expression or regulation of drought tolerance. Those genes (or loci) with multiple lines of support were given a higher weight for consideration as targets for involvement in drought tolerance.

Early expression data obtained for a cDNA microarray of the IR64 panicle library that involved hybridization of mRNA from IR64 and Azucena under field drought and well-watered conditions indicated 96 clones with significant changes in their expression (Kathiresan et al 2006). These clones were aligned to rice sequence data taking into account a range of the best hits so that cross-hybridization of mRNAs to clones belonging to multigene families was not discounted. These positions were then cross-referenced to the intervals for a set of QTLs for yield under drought stress and other drought-related traits consolidated by Lafitte et al (2004) across mapping populations and trials allowing the expressed sequence tag (EST) clones to be ranked.

Candidate genes for drought response/tolerance that have been used for Eco-TILLING are shown in Table 1. In the case of actin depolymerizing factor, proteomics data played a role in its choice (Salekdeh et al 2002). We have designed primers for locus-specific detection of the coding sequence and upstream regulatory regions using Primer3. Electronic PCR was used to confirm whether or not primer designs were locus-specific by alignment of primers to the Nipponbare reference genome.

High-throughput genotyping

The high-throughput genotyping methods we are currently using for allele mining or intend to apply in the near future are methods for the detection of single nucleotide polymorphisms (SNPs) or small insertions or deletions. The use of SNPs as the preferred marker will allow the eventual assignment of function to different alleles.

Targeted SNP detection

We have implemented the technology of EcoTILLING for the targeted detection of SNPs by mismatch detection at specific gene loci through collaboration with the developers of the technique, Drs. Steven Henikoff (Fred Hutchinson Cancer Research Center) and Luca Comai (University of Washington) (McCallum et al 2000). EcoTILLING is an application of the reverse genetics technique of tilling (Targeting Induced Local Lesions IN Genomes) to detect natural variation at known genes (Comai et al 2004). EcoTILLING allows a relatively rapid survey using primers designed from reference sequence data without the need for *de novo* sequencing for assay design. The methodology as developed by Henikoff and Comai relies on the use of automated platforms and fluorescent labeling for the detection of fragments resulting from the cleavage of heteroduplex mismatch molecules. At IRRI, we have developed an adaptation for tilling and EcoTILLING in which the detection of mismatches is accomplished on agarose gels (Raghavan et al 2007). Agarose-based detection is now our method of choice.

EcoTILLING is being used to screen the minicore collection at the drought candidate genes given in Table 1. Primers for these genes are used in reactions contrasted to either the japonica type Nipponbare or the indica type IR64 as reference lines. Mismatches (putative SNPs) have been detected at all loci that have been screened. From 4 to 9 mismatches per locus have been detected depending on whether the contrast was to either IR64 or Nipponbare. Figure 2 shows the mismatches detected in a subset of the IRRI minicore collection for one of the actin depolymerizing factors on chromosome 2 and compares fluorescent detection on the LiCor automated genotyper with agarose-based detection. Similar patterns were observed on both platforms. Since the agarose-based method allows us to achieve higher throughput at reduced cost, we have used it for our routine screening. In the case of ERF3, EcoTILLING for a set of 900 of the *O. sativa* lines for 1,800 bp of coding and regulatory region of ERF3

Table 1. Drought candidate gene targets for allele mining.

Gene	Bioprocess	Functional evidence	Primer_name	Primer_sequence	Amplicon (bp)
DREB2	Drought-responsive element binding protein 2	Transformation of homolog results in enhanced survival under stress	DREB2upL	ACgTCAAAACCCAACCCCAACCAT	988
			DREB2upR	gATTgAATCAgggCCATCgCTTTTC	
			DREB2-16cdL	CCgTTgATTgCTgATAgCCTCCTTgA	969
			DREB2-16cdR	TgAAATATTCCTATTgACCCgCAgCA	
ERF3	AP2 domain TF	Repressor of ABA-induced response	EREBP-cdL	ACACCCAAACCCAACCTCCCAAAAC	984
			EREBP-cdR	CATCgCCggCATgATCTgTTCTAAT	
TPP	Trehalose 6-phosphate phosphatase	Up-regulated under drought on IR64 panicle arrays	TPPL	ggCACACTgTCgCCTATTgTggATg	947
			TPPR	gTTTACgAgCCgTgCgACCAgTTTC	
14-3-3	Signal cascade		CG18-1L	ggCATgCTTgATTTgggACATgAgA	1,005
			CG18-1R	TCTCCTCATACCTCTCggCCTgCTC	
MAPK	Transcriptional control		MAPk-1L	CACCATCTCCTTCAgCCTCCgTTTC	1,006
			MAPk-1R	CACACCTCCACCCCAATCAAATTCC	
SucSase	Sucrose synthase		Suc-L	CgCTCAgCgAgTgCgTgAgACTATT	1,004
			Suc-R	ggCgATTgTACACTgCACATgATgg	
BZIP	Transcriptional control		Bzip-L	ATggCATCggAgATgAgCAAgAACg	991
			Bzip-R	TTCggAgCAACATgTgTACCTgCTTT	
ADF	Actin depolymerizing factor	Up-regulated in proteomics of drought	ADFCH02aL	CCTCCTTgCAgggCCTgTCTgT	954
			ADFCH02aR	TggCAgTgAggCATTCTCATAAAA	
			ADFCH02bL	AAggAACAATgCTgCTTTTggTCA	878
			ADFCH02bR	ACCATATgACgCCACggCCTTT	

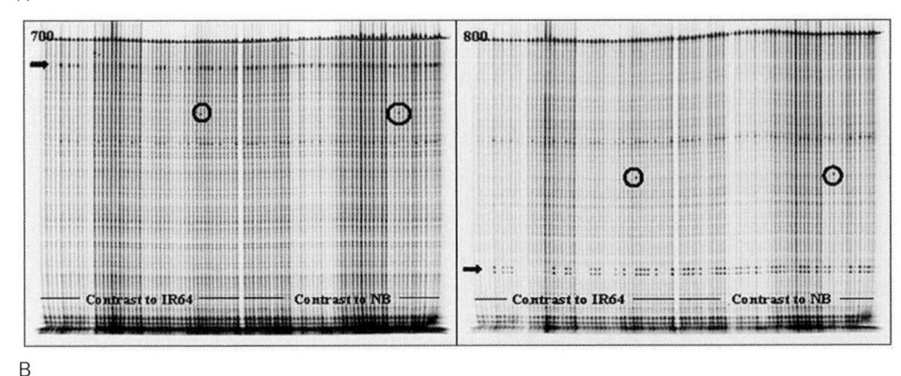

Fig. 2. EcoTILLING of actin depolymerizing factor 2a (ADF) on *Oryza sativa*. (A) LiCor images for ADF2a on panel of 48 accessions contrasted to IR64 and Nipponbare (NB). Arrows indicate a mismatch common to most japonica and indica contrasts. Circles indicate a rare mismatch. (B) Detection of mismatches on agarose for 24 accessions contrasted to IR64 and NB. Black arrows indicate common mismatch and white arrows an infrequent mismatch.

identified a range of putative SNPs. Sequence confirmation from selected lines for each of the observed patterns identified 31 SNPs and short indels that grouped into nine haplotypes corresponding to variety types (Wang 2005).

For the other loci, distinct mismatch banding patterns are assigned different haplotype scores for each of the contrasts to IR64 and Nipponbare. Representative accessions for each of the mismatch patterns have been sequenced to establish the identities of the nucleotide differences and convert the mismatches to SNPs relative to the Nipponbare genome. For association genetics, we are comparing association tests to the mismatch patterns with those to haplotypes defined by SNP identities.

Genome-wide SNP detection

IRRI, along with partners in the International Rice Functional Genomics Consortium, has embarked on a project with Perlegen Sciences for genome-wide SNP discovery in a diverse set of rice varieties with funding from IRRI, the Generation Challenge Program, and USDA-CSREES (McNally et al 2006). This project will use high-density oligomer arrays to re-sequence the nonrepetitive regions of the genome through hybridization (Olivier et al 2001, Patil et al 2001). Perlegen Sciences (www.perlegen.com) has successfully used this technology to identify SNPs and haplotype regions in the human (Hinds et al 2005), mouse (Frazer et al 2007), and *Arabidopsis* (Clark et al 2007) genomes.

The 20 varieties for re-sequencing were chosen for their diversity, use as donors in breeding programs, or previous use in genetic studies. The varieties include representatives from temperate and tropical japonica, aromatic, aus, deepwater, and indica types. We have included the temperate japonica variety Nipponbare as a control since SNP discovery is relative to its genomic sequence. Although the Nipponbare genome is about 390 Mb, funding was available to cover only 100 Mb for SNP discovery. Therefore, we filtered the Nipponbare genome for repetitive sequence and chose 100 Mb of the nonrepetitive regions that maximally covered annotated gene models. These regions are dispersed throughout the genome, with multiple segments occurring in almost every 100-kb window across the genome.

Hybridization of the 20 varieties to arrays led to the identification of more than 260,000 nonredundant SNPs by Perlegen's model-based algorithms. We are in the process of applying novel informatic methods using machine-learning approaches that were developed for the analysis of the *Arabidopsis* project (Clark et al 2007) to extend the SNP collection. Applying the condition that a minimum of one SNP occurs in a region of 100 kb indicates that the Perlegen model-based predictions cover about 93% of the genome. Estimates of the extent of linkage disequilibrium (LD) in rice indicate that LD extends to 100 kb or longer (Garris et al 2003, Olsen et al 2006). More recently, Mather et al (2007) have estimated LD to range from about 75 kb for indica to 150 kb for tropical japonica based on the decay across five genomic regions. The extent of LD determines the effective size of recombination blocks carrying similar haplotypes or patterns of alleles. Hence, the state of one or a few SNPs within the haplotype block can serve as a proxy for the states of the other SNPs within the block. Such a collection of tag SNPs will allow whole-genome scanning across haplotype blocks as described by Hirschorn and Daly (2005) and would substantially improve the power of association tests. Genotyping a wider collection of rice varieties with a tag SNP collection will enable whole-genome scans to be applied in association studies with detailed phenotypes for traits of interest such as those for drought tolerance.

Phenotyping traits related to drought for association genetics

The counterpart to high-throughput genotyping is phenotyping of traits expected to be associated with the expression of and/or regulation by candidate genes. A broad-based phenotyping approach for drought response was taken because detailed molecular or physiological characterization closely linked to the expression of candidate genes on a large scale for diverse accessions was not feasible. More importantly, this type of approach will also help dissect the practical value of putative drought-responsive genes in the development of varieties with improved tolerance. The effect of different haplotypes of candidate genes will be reflected in overall plant performance for water deficit. Yield is the ultimate trait affected by drought; however, a wide range of secondary traits is often used to evaluate plant response to drought stress (Fukai and Cooper 1995). The following groups of secondary traits were chosen as indicators of plant response and tolerance of water deficit: biomass accumulation and plant growth (plant height and tillering), plant water status (relative water content and leaf water potential), tolerance (leaf drying and cell membrane stability), appearance (leaf rolling and chlorophyll content), and phenology (flowering delay). Within the minicore collection, large genotypic variation was established for all traits.

The entire minicore collection was phenotyped over a period of 3 years in both upland and lowland managed field screens. To minimize genotype × year (environment) interaction, a large number of checks were included each season to allow the effect of year to be assessed, and incorporated into the data analysis. The response of yield and secondary traits associated with performance under stress is dependant on the stage of the plant during the drought period. To manage large differences in duration within diverse germplasm, vegetative-stage drought stress was imposed, thereby preventing differences in performance under drought stress related only to phenology. All 1,536 accessions of the minicore collection have now been characterized for response to drought stress. These data will be combined with genotype data, and significant associations between haplotypes and drought response will be established. An initial test of association with ERF3 haplotypes was performed using phenotypic data from the first two seasons of screening; one of the indica subgroups was found to have a significant association with yield stability during water stress (Wang 2005).

Association genetics

Association genetics tests the correlation between the presence of SNPs or other types of allelic/sequence variation and their contribution to various traits. Central to these tests are measures of the LD between loci. The structure of LD and use of association genetics in plants has been reviewed recently (Flint-Garcia et al 2003, Gupta et al 2005). The two most common measures of LD are r^2 and D', which measure the difference between observed and expected haplotype frequencies at two loci, but differ in their scaling of this difference. If we consider alleles at two loci, over time, recombination will result in individuals in which the linkage between the loci is broken (low LD) or, if the two loci are closely linked, recombination will be infrequent (high LD). Regions of high LD define haplotype blocks. In the case of self-pollinating species such as rice, these haplotype blocks may be relatively large. Current estimates in rice indicate that they are of the order of 75 to 150 kb (Garris et al 2003, Olsen et al 2006, Mather et al 2007). Since LD measurement depends on allele frequencies and recombination between loci, all stochastic processes observed in populations will affect their measurement. The population structure will exert a particularly strong effect. To avoid spurious associations between loci that are not historically related, the effect of population structure must be taken into account. Fingerprinting by anonymous markers (SSR) sampled from across the genome allows population structure to be determined (Pritchard et al 2000, Thornsberry et al 2001). The population can then be stratified into groups prior to performing the association tests.

The simplest method for an association test is to perform a one-way ANOVA between phenotypes partitioned according to their shared haplotype. It should be noted that computational techniques for association studies are undergoing rapid development (Beaumont and Rannala 2004, Hirschhoren and Daly 2005, Posada et al 2005, Wang et al 2005), and more advanced methods are becoming increasingly available. Although detection of associations between two loci is straightforward, the tests to determine multilocus interactions, where each locus has a small contribution to a trait (QTL), are computationally more difficult (Gupta et al 2005, Wang et al 2005). For determining multipoint associations, approaches based on haplotypes, allele frequencies at many individual loci, multifactor-dimensionality reduction, or novel methods will be necessary (Hahn et al 2003, Johnson 2004, Lou et al 2003).

Perspectives for understanding drought tolerance

We now have established resources for allele mining—a core collection with associated leaf and DNA banks, candidate gene targets for drought tolerance, a genome-wide collection of SNPs—and have implemented high-throughput genotyping for SNPs targeting specific candidate genes and phenotyping for traits involved in drought tolerance. The use of association genetics on the sets of genotypic and phenotypic data promises to identify the best alleles for different components of drought tolerance that can then be delivered to breeding programs using marker-assisted selection strategies. One looming issue is the complexity of phenotyping diverse germplasm with very different phenologies. Hence, phenotyping sets of germplasm with narrower ranges of phenology may be necessary as well as the phenotyping of drought component traits that may more closely reflect the action of a specific candidate gene or genes.

Specific SNPs associated with drought tolerance in the minicore collection could be applied for genotyping a wider collection. Choosing those varieties showing novel alleles followed by targeted phenotyping of only those with novel alleles would demonstrate whether or not the observed allelic variation is predictive of a drought phenotype. This final step will verify the use of allele mining as an efficient means to identify useful variation without the need for expensive and time-consuming phenotyping at the onset.

One drawback of the candidate gene approach is the requirement for prior knowledge that a particular gene is or might be important in a trait of interest. To overcome this limitation, whole-genome scans using tag SNPs can be accomplished on sets of germplasm that have been phenotyped in detail. Association tests on this scale could identify regions in the genome corresponding to haplotype blocks that are involved in drought tolerance. Although the specific genes underlying a trait may not be clearly identified, the involvement of a region in the trait is sufficient to begin introgressing the region in marker-assisted programs. The effect of these genes or regions can then be tested in different genetic backgrounds, confirming their importance in the expression of drought tolerance. Another promising approach is the combination of expression analyses with whole-genome SNP genotypes. Integrating the allelic variation with transcriptional variation may reveal causal relationships among SNP haplotypic blocks, expression polymorphisms, and adaptive responses to drought.

References

Beaumont MA, Rannala R. 2004. The Bayesian revolution in genetics. Nature Rev. Genet. 5:251-261.

Brown AHD. 1988. Core collections: a practical approach to genetic resources management. CSIRO, Division of Plant Industry, Canberra, Australia.

Brown AHD, Schoen DJ. 1994. Optimal sampling strategies for core collections of plant genetic resources. In: Loeschcke V et al, editors. Conservation genetics. Basal (Switzerland): Birkhuser Verlag. p 357-370.

Bruskiewich RM, Cosico AB, Eusebio W, Portugal AM, Ramos LM, Reyes MT, Sallan MAB, Ulat VJM, Wang X, McNally KL, Sackville Hamilton R, McLaren CG. 2003. Linking genotype to phenotype: the International Rice Information System (IRIS). Bioinformatics 19:i63-i65.

Caicedo AL, Williamson SH, Hernandez RD, Boyko A, Fledel-Alon A, York TL, Polato NR, Olsen KM, Nielsen R, McCouch SR, Bustamante CD, Purugganan MD. 2007. Genome-wide patterns of nucleotide polymorphism in domesticated rice. PLoS Genet. 3:e163.

Clark RM, Schweikert G, Toomajian C, Ossowski S, Zeller G, Shinn P, Warthmann N, Hu TT, Fu G, Hinds DA, Chen H, Frazer KA, Huson DH, Schölkopf B, Nordborg M, Rätsch G, Ecker JR, Weigel D. 2007. Common sequence polymorphisms shaping genetic diversity in *Arabidopsis thaliana*. Science 317:338-342.

Coburn JR, Temnykh SV, Paul EM, McCouch SR. 2002. Design and application of microsatellite marker panels for semiautomated genotyping of rice (*Oryza sativa* L.). Crop Sci. 42:2092-2099.

Comai L, Young K, Till BJ, Reynolds SH, Greene EA, Codomo CA, Enns LC, Johnson JE, Burtner C, Odden AR, Henikoff S. 2004. Efficient discovery of DNA polymorphisms in natural populations by EcoTILLING. Plant J. 37:778-786.

Flint-Garcia SA, Thorsnberry JM, Buckler ES 4th. 2003. Structure of linkage disequilibrium in plants. Annu. Rev. Plant Biol. 54:357-374.

Frankel OH. 1984. Genetic perspectives of germplasm conservation. In: Arber W, Llimensee K, Peacock WJ, Starlinger P, editors. Genetic manipulation: impact on man and society. Cambridge (UK): Cambridge University Press, Part III, Paper No. 15.

Frazer KA, Eskin E, Kang HM, Bogue MA, Hinds DA, Beilharz EJ, Gupta RV, Montgomery J, Morenzoni MM, Nilsen GB, Pethiyagoda CL, Stuve LL, Johnson FM, Daly MJ, Wade CM, Cox DR. 2007. A sequence-based variation map of 8.27 million SNPs in inbred mouse strains. Nature 448:1050-1053.

Fukai S, Cooper M. 1995. Development of drought-resistant cultivars using physio-morphological traits in rice. Field Crops Res. 40:67-86.

Fulton TM, Chunwongse J, Tanksley SD. 1995. Microprep protocol for extraction of DNA from tomato and other herbaceous plants. Plant Mol. Biol. Rep. 13:207-209.

Garris AJ, McCouch SR, Kresovich S. 2003. Population structure and its effect on haplotype diversity and linkage disequilibrium surrounding the xa5 locus of rice (*Oryza sativa* L.). Genetics 165:759-769.

Garris AJ, Tai TH, Coburn J, Kresovich S, McCouch SR. 2005. Genetic structure and diversity in *Oryza sativa* L. Genetics 169:1631-1638.

Gupta PK, Rustgi S, Kulwal PL. 2005. Linkage disequilibrium and association studies in higher plants: present status and future prospects. Plant Mol. Biol. 57:461-485.

Hahn LW, Ritchie MD, Moore JH. 2003. Multifactor dimensionality reduction software for detecting gene-gene and gene-environment interactions. Bioinformatics 19:376-382.

Hamilton NRS, McNally KL. 2005. Unlocking the genetic vault. Rice Today, September 2005. p 32-33.

Hinds DA, Stuve LL, Nilsen GB, Halperin E, Eskin E, Ballinger DG, Frazer KA, Cox DR. 2005. Whole-genome patterns of common DNA variation in three human populations. Science 307:1072-1079.

Hirschhorn JN, Daly MJ. 2005. Genome-wide association studies for common diseases and complex traits. Nature Rev. Genet. 6:95-108.

Hodgkin T, Brown AHD, van Hintum TJL, Morales EAV. 1995. Core collections of plant genetic resources. Chichester (UK): John Wiley & Sons.

IRGSP (International Rice Genome Sequencing Project). 2005. The map-based sequence of the rice genome. Nature 436:793-800.

Johnson T. 2004. Multipoint linkage disequilibrium mapping using multilocus allele frequency data. http://homepages.ed.ac.uk/tobyj/ .

Kathiresan A, Lafitte HR, Chen J, Mansueto L, Bruskiewich R, Bennett J. 2006. Gene expression microarrays and their application in drought stress research. Field Crops Res. 97:101-110.

Lafitte HR, Price AH, Courtois B. 2004. Yield response to water deficit in an upland rice mapping population: associations among traits and genetic markers. Theor. Appl. Genet. 109:1237-1246.

Lou XY, Casella G, Littell RC, Yang MCK, Johnson JA, Wu R. 2003. A haplotype-based algorithm for multilocus linkage disequilibrium mapping of quantitative trait loci with epistasis. Genetics 163:1533-1548.

Mather KA, Caicedo AL, Polato NR, Olsen KM, Nielsen R, McCouch SR, Purugganan MD. 2007. The extent of linkage disequilibrium in rice (*Oryza sativa* L.). Genetics 177:2223–2232.

McCallum CM, Comai L, Greene E, Henikoff S. 2000. Targeting induced local lesions IN genomes (TILLING) for plant functional genomics. Plant Physiol. 123:439-442.

McNally KL, Bruskiewich R, Mackill D, Leach JE, Buell CR, Leung H. 2006. Sequencing multiple and diverse rice varieties: connecting whole-genome variation with phenotypes. Plant Physiol. 141:26-31.

Olivier M, Bustos VI, Levy MR, Smick GA, Moreno I, Bushard JM, Almendras AA, Sheppard K, Zierten DL, Aggarwal A, Carlson CS, Foster BD, Vo N, Kelly L, Liu X, Cox DR. 2001. Complex high-resolution linkage disequilibrium and haplotype patterns of single-nucleotide polymorphisms in 2.5 Mb of sequence on human chromosome 21. Genomics 78:64-72.

Olsen KM, Caicedo AL, Polato N, McClung A, McCouch S, Purugganan MD. 2006. Selection under domestication: evidence for a sweep in the rice Waxy genomic region. Genetics 173:975-983.

Patil N, Berno AJ, Hinds DA, Barrett WA, Doshi JM, Hacker CR, Kautzer CR, Lee DH, Marjoribanks C, McDonough DP, Nguyen BT, Norris MC, Sheehan JB, Shen N, Stern D, Stokowski RP, Thomas DJ, Trulson MO, Vyas KR, Frazer KA, Fodor SP, Cox DR. 2001. Blocks of limited haplotype diversity revealed by high-resolution scanning of human chromosome 21. Science 294:1719-1723.

Posada D, Maxwell TJ, Templeton AR. 2005. TreeScan: a bioinformatics application to search for genotype/phenotype associations using haplotype trees. Bioinformatics 21:2130-2132.

Pritchard JK, Stephens M, Rosenberg NA, Donnelly P. 2000. Association mapping in structured populations. Am. J. Human Genet. 37:170-181.

Raghavan C, Naredo MEB, Wang HH, Atienza G, Liu B, Qiu FL, McNally KL, Leung H. 2007. Rapid method for detecting SNPs on agarose gels and its application in candidate gene mapping. Mol. Breed. 19:87-101.
Salekdeh GH, Siopongco J, Wade LJ, Ghareyazie B, Bennett J. 2002. Proteomic analysis of rice leaves during drought stress and recovery. Proteomics 2:1131-1145.
Thornsberry JM, Goodman MM, Doebley J, Kresovich S, Nielsen D, Buckler ES 4th. 2001. Dwarf8 polymorphisms associate with variation in flowering time. Nature Genet. 28:286-289.
van Hintum TJL, Brown AHD, Spillane C, Hodgkin T. 2000. Core collections of plant genetic resources. IPGRI Technical Bulletin No. 3. Rome (Italy): International Plant Genetic Resources Institute.
Vaughan DA. 1991. Choosing rice germplasm for evaluation. Euphytica 54:147-154.
Virk PS, Ford-Lloyd BV, Jackson MT, Pooni H, Clemeno TP, Newbury HJ. 1996. Predicting quantitative traits in rice using molecular markers and diverse germplasm. Heredity 76:296-304.
Wang H. 2005. Application of EcoTILLING to relate molecular variation in a rice ethylene responsive element factor (ERF3) gene to drought stress responses. M.S. thesis, University of the Philippines Los Baños. 86 p.
Wang WYS, Barratt BJ, Clayton DG, Todd JA. 2005. Genome-wide association studies: theoretical and practical concerns. Nature Rev. Genet. 6:109-118.
Yu J, Wang J, Lin W, Li S, Li H, et al. 2005. The genomes of *Oryza sativa*: a history of duplications. PLoS Biol. 3:e38.

Notes

Authors' addresses: K.L. McNally and M.E. Naredo, T.T. Chang Genetic Resources Center; J. Cairns, Crop and Environmental Sciences Division, International Rice Research Institute, DAPO Box 7777, Metro Manila, Philippines.
Acknowledgments: The authors gratefully acknowledge funding from IRRI, the Rockefeller Foundation, the Generation Challenge Program, BMZ, and USDA-CREES for contributing to these activities.

Research activities on drought tolerance of rice at JIRCAS

Takashi Kumashiro and Kazuko Yamaguchi-Shinozaki

This article summarizes recent research activities aiming at developing drought-tolerant rice at the Japan International Research Center for Agricultural Sciences (JIRCAS). We are using two approaches, molecular and conventional, to reach the targeted goal. In the molecular approach, characterizations of transcriptional factors such as DREB, AREB, and stress tolerance of the resultant transformants will be introduced. In the conventional approach, the current status of screening for drought-tolerant germplasm will be presented.

In line with its mission to contribute to global agriculture, especially in developing regions through research in agricultural sciences, the Japan International Research Center for Agricultural Sciences (JIRCAS) has been conducting many research activities in various areas of agriculture. Among traits needed for such regions, JIRCAS gives top priority to abiotic stress tolerance in a number of crops, for the following three reasons: (1) the magnitude of damage from abiotic stresses has been reported to be much larger than that of biotic stresses in terms of yield loss; (2) because of global warming, abiotic stresses, such as drought, would be more severe in developing regions; and (3) through long-term in-house research on the elucidation of molecular mechanisms of abiotic stress tolerance, JIRCAS has accumulated research assets that can be used to create abiotic stress-tolerant crops. In April 2006, JIRCAS started a new research project according to a newly-defined 5-year mid-term plan. In this article, we would like to explain the research activities dealing with drought tolerance of rice.

Elucidation of molecular mechanisms of abiotic stress tolerance

JIRCAS's molecular research group has been conducting research to elucidate the molecular mechanisms of abiotic stress tolerance built into plants, and has revealed the complex regulatory network of gene expression in response to drought, high salinity, and cold stress in a model plant, *Arabidopsis*.

The promoter of an *Arabidopsis* drought-inducible, high salinity-inducible, and cold-inducible *RD29A* (responsive to dehydration, *29A*) gene encoding an LEA (late embryogenesis abundant)-like protein has been found to contain two major *cis*-acting elements, the ABA-responsive element (ABRE) and the dehydration-responsive element (DRE)/C repeat (CRT), that are involved in stress-inducible gene expression. Transcription factors belonging to the AP2/ERF (APEETALA2/ethylene-responsive element binding factor) family that bind to DRE/CRT have been isolated and termed DREB1/CBF and DREB2. The conserved DNA-binding motif of DREB1/CBF and DREB2 is A/GCCGAC. The *DREB1/CBF* genes are quickly induced by cold stress and their products activate the expression of target stress-inducible genes. The DREB2 gene is induced by dehydration, leading to the expression of various genes that are involved in drought stress tolerance.

DREB1

Gene constructs consisting of the constitutive cauliflower mosaic virus 35S promoter linked to the coding region of the *DREB1A* gene were introduced into *Arabidopsis*. The resultant transformants (3-week-old plants) were subjected to freezing (–6 °C for 2 days), drought (withholding water for 2 weeks), and salt (600 mM NaCl for 2 h) and showed an increased level of tolerance of freezing, drought, and high salt stress. In the case of the 35S promoter, transformants showed retarded growth. The use of a stress-inducible promoter such as *Arabidopsis rd29A*, however, circumvents the negative effect on the growth retardation of the transformed *Arabidopsis*. RNA gel blot and microarray analyses of the transformants revealed that *DREB1A* can activate more than 40 target stress-inducible genes under stress conditions. Overexpression of *DREB1A* activated strong expression of the target genes, which, in turn, increased tolerance of freezing and drought stress (for a review, see Nakashima and Yamaguchi-Shinozaki 2005, Umezawa et al 2006).

The orthologs of *DREB1A* have been found in many crop plants such as canola, broccoli, tomato, alfalfa, wheat, barley, and maize. These orthologs found in the rice genome were termed *OsDREB1*. It has been shown that overexpression of *OsDREB1A* in *Arabidopsis* also resulted in higher tolerance of drought, high salinity, and freezing stress (Dubouzet et al 2003). *OsDREB1A* and *OsDREB1B* were introduced into japonica rice cultivars Nipponbare and Kitaake under the control of constitutive promoters, either maize ubiquitin or the 35S promoter. These transgenic rice plants showed growth retardation under normal growth conditions similar to what was observed in transgenic *Arabidopsis* overexpressing *DREB1* under control of the 35S promoter. The 17-day-old seedlings of these transgenic rice plants were exposed to various stresses, including drought (dehydration for 9 days), high salinity (250 mM NaCl for 3 days), and low temperature (2 °C for 93 h). Whereas the wild-type plants died under these stress treatments, the transgenic lines showed improved tolerance of these stresses in terms of plant survival (Ito et al 2006).

DREB2

Arabidopsis DREB2 also carries a conserved ERF/AP2 DNA binding domain and recognizes the DRE sequence, just like *DREB1*. Whereas expression of *DREB1* genes is induced by cold stress but not by drought and high salinity, expression of *DREB2* genes is induced by drought and high salt stress (Liu et al 1998). When a native form of *DREB2A* under the control of a constitutive promoter was introduced into *Arabidopsis*, the transgenic plants did not show improved tolerance of various stresses nor did they show any growth retardation (Liu et al 1998). Recently, a constitutive active form of *DREB2A* has been successfully produced by deleting the negative regulatory domain. When this constitutive active form of *DREB2A* under the control of the 35S promoter was introduced into *Arabidopsis*, the transgenic plants showed growth retardation under normal growth conditions and improved drought (dehydrated for 14 days) tolerance, but only a slight change in freezing (–6 °C for 30 h) tolerance (Sakuma et al 2006).

Recently, Qin et al (2007) isolated a *DREB2A* homolog from maize termed *ZmDREB2A*. Transcripts of *ZmDREB2A* were accumulated following cold, dehydration, salt, and heat stresses in maize seedlings. Unlike *DREB2A* of *Arabidopsis*, *ZmDREB2A* produced two forms of transcript, of which only the functional form was significantly induced by stress. Furthermore, the ZmDREB2A protein exhibited high transactivation activity compared with DREB2A in *Arabidopsis* protoplasts, suggesting that protein modification is not necessary for ZmDREB2A to be active. The *ZmDREB2A* gene under control of the 35S promoter was introduced into *Arabidopsis*. Drought treatment by withholding water for 10 days to these transgenic plants revealed an elevated level of drought tolerance. Although the constitutive expression of *ZmDREB2A* resulted in a dwarf phenotype of the transformants, the use of stress-inducible promoter *RD29A* circumvented the dwarfism.

AREB

Molecular elucidation work has been extended to other transcription factors such as AREBs (abscisic acid-responsive element binding proteins). *AREB1* is a basic domain/leucine zipper transcription factor that binds to the abscisic acid (ABA)-responsive element (ABRE) motif in the promoter region of ABA-inducible genes. Fujita et al (2005) found that expression of the intact *AREB1* gene alone is insufficient to lead to the expression of downstream genes under normal growth conditions. To overcome the masked transactivation activity of AREB1, we have created an activated form of AREB1 by deleting the regulatory sequence linking the transcriptional activation and DNA-binding domains of the protein. A constitutive active form of AREB1 under control of the constitutive promoter 35S was introduced into *Arabidopsis*. Transformants of *Arabidopsis* with the activated form of AREB1 exhibited ABA hypersensitivity and eight genes with two or more ABRE motifs in their promoter regions were greatly up-regulated. Three-week-old plants of these transgenic plants carrying the activated form of *AREB1* underwent drought treatment by withholding water for 12 days. Whereas almost all of the wild-type plants withered completely, nearly all the transgenic plants survived this drought treatment.

Potential of *DREB* genes for developing abiotic stress–tolerant crops

Through extensive work on the molecular elucidation of stress tolerance of plants, we have identified useful genes for transcription factors, *DREB*s and *AREB*s.

Results from the evaluation of transformed *Arabidopsis* and rice at the greenhouse level indicate that genes for transcription factors, including *DREB1* and *OsDREB1*, an active form of *DREB2A*, *ZmDREB2A* from maize, and an active form of *AREB1*, have great potential to generate crops that show tolerance of abiotic stresses, such as drought, high salinity, and freezing at the field level.

To investigate whether various kinds of transgenic crops carrying these potential genes exhibit tolerance of abiotic stresses, JIRCAS has started collaborative research with several international research institutes (International Rice Research Institute—IRRI, International Center for Tropical Agriculture—CIAT, International Maize and Wheat Improvement Center—CIMMYT, International Crops Research Institute for the Semi-Arid Tropics—ICRISAT, and International Center for Agricultural Research in the Dry Areas—ICARDA) covering many different economically important crops such as rice, wheat, soybean, lentil, groundnut, and chickpea.

Since an ultimate goal of this collaborative research is to generate elite transformants carrying, ideally, a single copy of a transgene, a transformation protocol should enable us to generate a large number of independent primary transformants. The reason why a large number of independent transformants is necessary derives from the fact that any transformation system currently available is not a perfect system in the sense that the following parameters cannot be controlled: (1) intactness of the transgene (without modification or fragmentation of the transgene), (2) the copy number of the transgene (related to the number of loci for the transgene), (3) the location of insertion of the transgene in the host genome (resulting in a positional effect), and (4) suppression of somaclonal variation during the dedifferentiation phase (if a tissue culture phase is involved).

Although examples of large-scale transformation are limited, one example can be found in a paper by Hu et al (2003). In an attempt to generate elite transformants of herbicide-tolerant wheat, they found the frequency of such elite lines as 0.8% and 0.2% of primary transformants, by *Agrobacterium* and biolistic transformation, respectively. Therefore, a high-throughput transformation system is certainly one of the most important prerequisites for obtaining elite transformants.

JIRCAS recently started its internal transformation work aiming at the improvement of drought tolerance of rice in Africa, including NERICA (New Rice for Africa), which is an interspecific variety between *Oryza glaberrima* and *O. sativa*. Since there has been no prior work on transformation systems of NERICA, we are now working on establishing a transformantion system that would fulfill the requirements listed above.

Screening of rice germplasm for drought tolerance

Drought tolerance is a complex trait. According to Mitra (2001), the expression of drought tolerance depends on the interaction of different morphological (earliness, reduced leaf area, leaf rolling, wax content, rooting system, and reduced tillering), physiological (reduced transpiration, high water-use efficiency, stomatal closure, and osmotic adjustment), and biochemical (accumulation of compatible solutes and polyols) characteristics (Bohnert et al 1995). Tobita et al (2001) found that xylem exudation rate, which was expressed by water volume trapped overnight from cut internode of rice plants, can be used as a criterion to discriminate drought-tolerant and drought-susceptible rice varieties. With regard to the roots, the importance of a deep root system has been repeatedly emphasized as a critical trait in rainfed rice (Nguyen et al 1997, Ito et al 1999).

Among the traits related to drought tolerance, we have selected rooting depth as an indicator of drought tolerance. A wide range of rice germplasm consisting of the core collection of IRRI, the core collection of the National Institute of Agrobiological Sciences (NIAS) in Japan, upland rice varieties provided by the Ibaraki Agricultural Center, and 86 lines of *O. glaberrima* from WARDA were first evaluated in Bouaké, Côte d'Ivoire, for drought tolerance at the seedling stage using the standard evaluation system (IRRI 1996). The varieties classified as drought tolerant were subjected to measurement of root length in the field. Six drought-tolerant varieties, including Azucena and Black Gora, showed deeper roots than drought-sensitive varieties, indicating correlation between drought tolerance and rooting depth. Under this evaluation, NERICA 1-4 and its parental lines were not considered to have deeper roots in the experimental field at WARDA (Sakagami and Tsunematsu 2003).

Furthermore, root depth at the reproductive stage was evaluated for approximately 600 rice germplasm accessions, including *O. glaberrima*. Based on the deepest root length at 2 weeks after heading, the top 100 of the best germplasm accessions have been selected. The second and third rounds of the evaluation are now ongoing at the WARDA-Nigeria Station. After identifying germplasm with deeper roots, correlation with drought tolerance at the reproductive stage will be further examined. Also, hybridization between germplasm with deep roots and with a shallow root system will be carried out for the development of a segregating population, which will be used for QTL analysis for rooting depth.

Evaluation methods for drought tolerance

The most practical criteria for drought tolerance would be yield under drought conditions. However, yield alone is not easy to measure precisely. Many studies have been conducted to find reliable criteria for drought tolerance. The physiological traits to be considered as potential selection criteria for improving yield should be genetically correlated with yield and should have a greater heritability than yield itself. Ideal traits are those whose measurement is fast, accurate, and inexpensive. So far, however, it is clear that none of the physiological traits meets all of these ideal requirements (Tuberosa 2004).

With respect to an evaluation platform, uniform drought conditions are difficult to prepare in field conditions. Although the use of a rainout shelter might offer more uniform conditions in terms of controlling water supply, the heterogeneity of soil composition as well as heterogeneity of water content in soil remain problems to be overcome.

Hydroponic systems would render a more uniform evaluation platform for drought conditions by adding a substance that creates water-deficit conditions. By using polyethylene glycol, Cabuslay et al (2002) detected genotypic differences in response to water deficit. We are now investigating whether and to what extent such a hydroponic system could be used as a simple and reliable method to evaluate drought tolerance using rice germplasm reported to be tolerant and susceptible to drought.

Future directions

Variation seems to be quite wide in terms of drought tolerance among existing rice germplasm. Furthermore, several genes showing potential to enhance drought tolerance have been identified and isolated. Thus, there is potential to further increase genetic variability in drought tolerance of rice.

A reliable evaluation method for drought tolerance under greenhouse conditions is initially required for the evaluation of transformants in the early generation, as well as for screening a large number of rice germplasm accessions. The history of plant breeding in a number of crops shows that breeding practices have been very successful when reliable selection criteria and evaluation protocols for a given trait are used. Therefore, what we definitely need for developing drought-tolerant varieties is a reliable evaluation protocol that enables us to identify plants carrying genetic constituents for drought tolerance.

References

Bohnert HJ, Nelson DE, Jensen RG. 1995. Adaptations to environmental stresses. Plant Cell 7:1099-1111.

Cabuslay GS, Ito O, Alejar AA. 2002. Physiological evaluation of responses of rice (*Oryza sativa* L.) to water deficit. Plant Sci. 163:815-827.

Dubouzet JG, Sakuma Y, Ito Y, Kasuga M, Dubouzet EG, Miura S, Seki M, Shinozaki K, Yamaguchi-Shinozaki K. 2003. OsDREB genes in rice, *Oryza sativa* L., encode transcription activators that function in drought-, high-salt- and cold-responsive gene expression. Plant J. 33:751-763.

Fujita Y, Fujita M, Satoh R, Maruyama K, Parvez MM, Seki M, Hiratsu K, Ohme-Takagi M, Shinozaki K, Yamaguchi-Shinozaki K. 2005. AREB1 is a transcription activator of novel AREB-dependent ABA signaling that enhances drought stress tolerance in *Arabidopsis*. Plant Cell 17:3470-3488.

Hu T, Metz S, Chay C, Zhou HP, Biest N, Chen G, Cheng M, Feng X, Radionenko M, Lu F, Fry J. 2003. *Agrobacterium*-mediated large scale transformation of wheat (*Triticum aestivum* L.) using glyphosate selection. Plant Cell Rep. 21:1010-1019.

IRRI (International Rice Research Institute). 1996. Standard evaluation system for rice. Los Baños (Philippines): IRRI.

Ito O, O'Toole J, Hardy B, editors. 1999. Genetic improvement of rice for water-limited environments. Workshop on Genetic Improvement of Rice for Water-Limited Environments, International Rice Research Institute, Los Baños, Philippines, 1-3 Dec. 1998. Manila (Philippines): International Rice Research Institute. 353 p.

Ito Y, Katsura K, Maruyama K, Taji T, Kobayashi M, Seki M, Shinozaki K, Yamaguchi-Shinozaki K. 2006. Functional analysis of rice DREB/CBF-type transcription factors involved in cold-responsive gene expression in transgenic rice. Plant Cell Physiol. 47:141-153.

Liu Q, Kasuga M, Sakuma Y, Abe H, Miura S, Yamaguchi-Shinozaki K, Shinozaki K. 1998. Two transcription factors, DREB1 and DREB2, with an EREBP/AP2 DNA binding domain separate two cellular signal transduction pathways in drought- and low-temperature-responsive gene expression, respectively, in *Arabidopsis*. Plant Cell 10:1391-1406.

Mitra J. 2001. Genetics and genetic improvement of drought resistance in crop plants. Curr. Sci. 80:758-763.

Nakashima K, Yamaguchi-Shinozaki K. 2005. Molecular studies on stress-responsive gene expression in *Arabidopsis* and improvement of stress tolerance in crop plants by regulon biotechnology. JARQ 39:221-229.

Nguyen HT, Babu RC, Blum A. 1997. Breeding for drought resistance in rice: physiology and molecular genetics considerations. Crop Sci. 37:1426-1434.

Qin F, Kakimoto M, Sakuma Y, Maruyama K, Osakabe Y, Tran LSP, Shinozaki K, Yamaguchi-Shinozaki K. 2007. Regulation and functional analysis of ZmDREB2A in response to drought and heat stresses in *Zea mays* L. Plant J. 50:54-69.

Sakagami J, Tsunematsu H. 2003. Cultivar difference in drought tolerance on deep-rooted ability. Jpn. J. Crop Sci. 72 (suppl. 1):106-107.

Sakuma Y, Maruyama K, Osakabe Y, Qin F, Seki M, Shinozaki K, Yamaguchi-Shinozaki K. 2006. Functional analysis of an *Arabidopsis* transcription factor, DREB2A, involved in drought-responsive gene expression. Plant Cell 18:1292-1309.

Tobita S, Ookawa T, Audebert AY, Jones MP. 2001. Xylem exudation rate: a proposed screening criterion for drought resistance in rice. 6th Symp. Intl. Soc. Root Research. p 310-312.

Tuberosa R. 2004. Molecular approaches to unravel the genetic basis of water use efficiency. In: Bacon MA, editor. Water use efficiency in plant biology. Oxford (UK): Blackwell Pub. p 228-301.

Umezawa T, Fujita M, Fujita Y, Yamaguchi-Shinozaki K, Shinozaki K. 2006. Engineering drought tolerance in plants: discovering and tailoring genes to unlock the future. Curr. Opinion Biotechnol. 17:113-122.

Notes

Authors' address: Biological Resources Division, Japan International Research Center for Agricultural Sciences (JIRCAS), Owashi, Tsukuba, Ibaraki, Japan.

GM technology for drought resistance

Philippe Hervé and Rachid Serraj

Stress responses and adaptation in crops involve complex mechanisms since the plants must respond to variable stress occurrence parameters and interacting environmental factors. However, experiments describing the enhancement of stress resistance by a transgenic approach are usually restricted to the modulation of the plant response to one stress factor. In this note, we review examples from the recent literature and focus, in more detail, on the methodology and the results of seven published studies that describe the enhancement of drought resistance in rice using various strategies (transcription factors, cell metabolism, water fluxes, and reactive oxygen species scavenging). Although the transformation methodology is somehow similar, the protocols for plant evaluation and the parameters used to assess plant resistance are diverse and difficult to compare. The low number of independent transgenic lines and the poor assessment of a gene effect versus an overall effect due to gene insertion and somaclonal variation could be major drawbacks. The relevance of some screening methods for drought is questionable from a breeding perspective and only a few studies provide actual data about enhanced drought resistance under field conditions. In addition, it is noteworthy to observe that most of the experiments are based on the overexpression of a single gene. If the use of transgenic crops could potentially enhance drought responses of plants by promoting changes in the genome, there is clearly an important need to define or redefine the major steps and criteria to obtain better crop performance in the field. We summarize in this paper some of the major steps and key criteria to identify better rice cultivars with enhanced drought resistance using GM technology.

Gene modification (GM) and genetic engineering are now often proposed as a solution for increasing crop yields worldwide, particularly in less-developed areas that are threatened by food insecurity and low crop productivity (Nelson et al 2007, Zhang 2007). However the scientific debate over the potential of GM crops in the improvement of crop stress resistance is still highly litigious, as the opinions vary from highly optimistic to extremely skeptical (Marris 2008). While several public and private crop-science companies started to invest heavily in complex genetic traits such as drought resistance as part of their GM research portfolio, the overall experience of

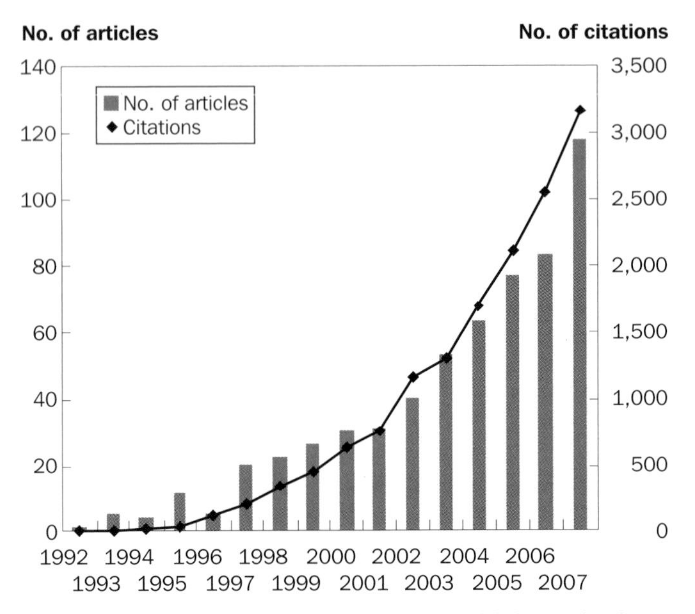

Fig. 1. Papers and citations found in the Web of Science database, including "drought" and "transgenic" as topic keywords, between 1992 and 2007.

four decades of crop physiological research on stress adaptation highlights the great challenge in translating such research into actual crop yield improvement (Sinclair et al 2004).

In a survey of the recent literature using the ISI Web of Science database, we found that the annual number of papers and citations on GM for "drought tolerance" has increased almost exponentially during the last 15 years, to more than 120 papers and 3,100 citations in 2007 (Fig. 1). These studies, all reported in well-reputed international journals, mainly focus on model plant species such as *Arabidopsis* (365 articles), tobacco (248), and rice (123), with fewer studies on maize (66), wheat (38), and barley (29). Most of the rice GM work is carried out in five major countries: China (30% of the publications), Japan (18.5%), the United States (18.5%), India (17%), and South Korea (11%). Although it is becoming more and more difficult to keep track of all that is being published in this field and review it, it is noteworthy that a large majority of these studies have surprisingly similar scientific hypotheses and common genealogies of the underlying concepts and methodologies. As reported by Passioura (2006), hundreds of patents claim gene inventions and sequences that may improve "drought tolerance." Most of these studies report the positive effect of genes involved in stress signaling and metabolic pathways using plant evaluation protocols that are generally far away from an agronomic context, with no immediate prospects for producing suitable GM crops that could greatly improve drought adaptation and/or water productivity in the field.

Because of their high level of integration and multitude of interactions, crop stress responses and adaptation mechanisms are highly complex. At a given time, a single plant must respond to several abiotic and biotic environmental factors while ensuring development and housekeeping functions. At the genetic and molecular levels, this complexity has been illustrated by both the identification of multiple stress-related QTLs (Bernier et al 2007, Yue et al 2006, Zeng et al 2006) and genomics/transcriptome analysis (Bohnert et al 2006, Kathiresan et al 2006). Changes in the abundance of the transcripts at the whole-genome level have confirmed the role of downstream structural genes involved in drought response mechanisms but also unraveled genes encoding regulatory proteins such as transcription factors or protein kinase/phosphatases. This last group of genes is assumed to play a key role because they regulate other downstream stress-inducible genes or proteins (Yamaguchi-Shinozaki and Shinozaki 2006). The regulatory genes are thus an entry point of a gene network.

Most of the experiments using genetic engineering or GM technology are based on the hypothesis that higher resistance can be achieved by adding or modulating the activity of one or a limited number of key components. The overexpression of downstream genes and more recently regulatory genes has been reported to enhance drought resistance in plants or crops for the past few years. Although some of these experiments aimed at demonstrating a positive effect of a gene, they barely provided validation of the results at a larger scale or in field conditions. In this note, we reviewed the recent literature on GM rice under drought, and focused on seven different case studies of papers describing enhanced drought resistance in rice. We decided not to comment on the scientific strategy used by the authors but focused instead on the transformation methodology, the protocols for plant evaluation, and the parameters used to assess plant resistance under drought. Finally, we attempted to summarize some of the major steps and key criteria for identifying better rice cultivars with enhanced drought resistance using GM technology.

Some lessons from recently published work

With more than 123 published papers reporting studies on GM rice under drought, we decided on an in-depth analysis focusing on contrasting case studies based on seven papers published between 2002 and 2006. We compared the methodology and results of experiments that described the overexpression in rice of downstream genes encoding the aquaporin RWC3 (Lian et al 2004), arginine decarboxylase (Capell et al 2004), superoxide dismutase (Wang et al 2005), trehalose-6-phosphate synthase/phosphatase (Garg et al 2002), and the late abundant embryogenesis protein HVA1 (Babu et al 2004) or transcription factors CBF3/DREB1A (Oh et al 2005) and NAC1 (Hu et al 2006). The genes used in these experiments are structural proteins, enzymes, and transcription factors and their roles or effects had been previously demonstrated in other species or rice cultivars. We have chosen papers attempting different strategies and it is clear that, in all cases, the authors aimed at conferring drought resistance to rice by recombinant DNA technology with prior knowledge of putative gene effects. The overall transformation methodology is summarized in Table 1. The most

Table 1. Transformation methodology.

Article	Lian et al (2004)	Capell et al (2004)	Oh et al (2005)	Wang et al (2005)	Garg et al (2002)	Hu et al (2006)	Babu et al (2004)
Journal	*Plant cell physiology*	*PNAS*	*Plant cell physiology*	*Journal of Plant Physiology*	*PNAS*	*PNAS*	*Plant Science*
Gene	*RWC3* aquaporin	*adc* (arginine decarboxylase)	*CBF3/ABF3*	MnSOD (superoxide dismutase)	*OtsA and OtsB*	*SNAC1*	*HVA1*
Known junction	Aquaporin	Polyamine biosynthesis	Transcription factor	Reactive oxygen species scavenging	Trehalose biosynthesis	Transcription factor	LEA protein, membrane
Gene source	Rice(?)	Datura	*Arabidopsis*, rice	Pea	*E. coli*	Upland rice IRAT109	Barley
DNA	Genomic	cDNA	Genomic/DNA	cDNA	Genomic	cDNA	cDNA
Promoter	SWAP2	Maize Ubi-1	Maize Ubi-1	SWAP-2	ABRC1-Actin1-HVA22	CaMV35S	Actin1
Genotype	Zhonghua 11	n.d.	Nakdong	Zhonghua-11	Pusa Basmati PB-1	Nipponbare	Nipponbare
Type of variety	Japonica	n.d.	Japonica	Japonica	Indica	Japonica	Japonica
Plasmid	pCambia1301	n.d.	n.d.	pCambia1301	pSB11	pCambia1301	n.d.
Gene transfer	*Agrobacterium*	*Agrobacterium*	*Agrobacterium*	*Agrobacterium*	*Agrobacterium*	*Agrobacterium*	*Agrobacterium*
Selectable marker	Hygromcyin	–	bar	Hygromcyin	bar	Hygromcyin	bar
Analysis of transgenic	PCR	–	Southern; RNA blots	PCR	–	PCR, northern, Southern	Herbicide, Southern
Number of transgenic plants	14	50	15/20	15	29	29	63

Table 2. Design of the phenotypic screening.

Article	Lian et al (2004)	Capell et al (2004)	Oh et al (2005)	Wang et al (2005)	Gang et al (2002)	Hu et al (2006)	Babu et al (2004)
Number of transgenic lines	14	50	15/20	15	22 (29)	29	63
Number of transgenic lines analyzed (T1)	n.d.	–	–	2	–	5	3
Population size	n.d.	–	–	n.d.	–	20 (fertility); 6 (mRWC)	–
Number of transgenic lines analyzed (T2)	–	10 (1)	–	–	–	5 (?)	2
Population size	–	n.d.	–	–	–		
Number of transgenic lines analyzed (T4)	–	–	3 (homozygous)	–	6 (2 with published data)	70 –	n.d. –
Population size	–	–	18 (Fv/Fm): 36 (survival rate)	–	15	–	–
Control	Wild type	Wild type	Wild type	Wild type	Wild type	Wild type	Non-transgenic
	–	–	–	–	–	No expression of transgenic line	–
	–	–	–	–	–	Isogenic line	–

remarkable features are the diverse gene sources, the use of cDNA driven by an inducible or constitutive promoter, the use of *Agrobacterium tumefaciens* for gene transfer, and the use of a japonica cultivar because of its ease for transformation. The vectors and the gene cassettes are commonly and broadly used for crop GM technology. The rice transformation methodology is based on the transformation of callus tissues and selection with either antibiotic or herbicide agents. Surprisingly, the number of independent primary events reported in these studies is relatively low although the transformation efficiency is high with the cultivars used.

As shown in Table 2, the design of the phenotypic screening is often insufficiently described but the population size is small or unknown, which does not allow a good assessment of the results. As highlighted above, the number of independent events that are reported in these publications for drought resistance evaluation is very low. It is, however, necessary to assess a significant number of events in order to take into account possible position effects of the transgene. It is well known that the expression level of a gene (or transgene in that case) can be regulated by its position in the genome (structural regulation). Somaclonal drag caused by tissue culture is another important unknown change in the genome that must be taken into consideration since it may influence the overall effect of a transgene. Another common feature is the use of wild-type plants as controls except by Hu et al (2006), who reported the use of isogenic lines. The comparison of transgenic lines versus wild-type plants is very common but it does not take into account possible somaclonal drag induced by the transformation protocol. In the case of rice, the re-activation of transposons that may occur during in vitro tissue culture may affect overall plant performance and cannot be neglected. For T1 or T2 populations, it is thus more appropriate to compare transgenic lines versus null segregants or isogenic lines as described in this chapter.

Tables 3 and 4 summarize the phenotypic screening protocols. All studies described plant evaluation in greenhouse conditions using pots or hydroponic cultures. Only Hu et al (2006) reported an extensive screening including paddy field/rainout shelter and field evaluation. In these studies, drought stress is imposed by water-withholding or replacing water by a PEG solution. As recently discussed by Bhatnagar-Mathur et al (2008), the use of PEG in hydroponics can be useful to test certain responses of plants under a given osmotic potential (Pilon-Smits et al 1999), but it offers relatively different conditions than those in the soil where the water reservoir is finite, and the dynamics of soil drying are an inherent part of the stress-response mechanisms.

It is also noteworthy to indicate that the size of the pots differs greatly from one study to another and that the intensity and the timing of drought treatment are thus very different (data not shown). These studies clearly used different drought treatments to assess the resistance of the transgenic lines and ad hoc measurement to assess drought stress level is not provided. Sixteen different parameters are monitored but only six of them are used in at least two studies: relative water content (RWC), leaf water potential, photosynthesis rate, chlorophyll fluorescence, transpiration rate, and survival rate or plant recovery rate after re-watering. Important physiological parameters such as yield components, spikelet fertility, or root biomass, however, are used in only one study. Furthermore, even the simplest visual observation of leaf rolling to monitor drought stress level is not systematically reported.

As shown in Table 4, there is a similar trend of the different parameters measured under different managed-stress treatments, whatever the strategy is. A higher RWC, reduced leaf water potential, higher photosynthesis rate and chlorophyll fluorescence, reduced transpiration rate, and higher recovery rate were observed. In each study, the correlation between the overexpression of a gene and a physiological or biochemical response of the plant or plant cell under stress has been established and the authors did note improved drought resistance in the transgenic lines based on observed parameters. It remains questionable, however, what significance such reported drought resistance can have on crop performance and yield under stressed and nonstressed conditions. It would also be relevant to assess the same parameters and the performance of all these transgenic lines under similar drought treatment. The next and most important step is an evaluation of the lines under paddy field/field conditions since it is difficult to extrapolate the reported data from pot studies and hydroponics to field performance,

Table 3. Physiological and morphological parameters used in the phenotypic screening.

Article	Lian et al (2004)	Capell et al (2004)	Oh et al (2005)	Wang et al (2005)	Garg et al (2002)	Hu et al (2006)	Babu et al (2004)
Stage	4 weeks old	8 weeks old	4 weeks old	5 weeks old	5 weeks old	Reproductive anthesis	6 weeks old
Drought-prone field						✓	
Paddy field–rainout shelter						✓	
Soil-greenhouse			✓		✓	✓	✓
Hydroponic-greenhouse	✓	✓		✓			
Relative water content (RWC)						✓	✓
Leaf water potential	✓						✓
Leaf osmotic potential							✓
Minimum RWC						✓	
Photosynthesis rate				✓		✓	
Chlorophyll fluorescence (Fv/Fm)			✓		✓		
Soluble carbohydrate level					✓		
Stomata closure						✓	
Transpiration rate	✓					✓	
Relative cumulative transpiration	✓						
Leaf rolling		✓					
Root osmotic potential	✓						
Spikelet fertility						✓	
Root biomass							✓
Shoot biomass							✓
Survival rate (recovery rate)			✓			✓	

Table 4. Comparison of the measurements of parameters used by at least two studies.

Article		Lian et al (2004)	Capell et al (2004)	Oh et al (2005)	Wang et al (2005)	Garg et al (2002)	Hu et al (2002)	Babu et al (2004)
Stage		4 weeks old	8 weeks old	4 weeks old	5 weeks old	5 weeks old	Reproductive anthesis	6 weeks old
Relative water content (RWC)	Transgenic						62% (5 days)	92% (28 days)
	Control						50% (5 days)	51% (28 days)
Leaf water potential (Mpa)	Transgenic	–1 (88%)						–1.6 (62%)
	Control	–1.14						–2.6
Photosynthesis rate (% decrease vs initial rate)	Transgenic				86%		75%	
	Control				67%		75%	
Chlorophyll fluorescence (Fv/Fm) (photooxidative damage)	Transgenic			0.55		0.55		
	Control			0.35		0.3		
Transpiration rate	Transgenic						0.9	
	Control							
Relative cumulative transpiration (% initial value)	Transgenic	55%						
	Control	42%						
Survival rate (recovery rate) after re-watering	Transgenic			83/58%			50%	
	Control			0/8%			10%	

in particular for yield and biomass accumulation. Unfortunately, the selected studies did not provide any yield or agronomic data.

Finally, phenotypic evaluation of the transgenic lines under normal irrigation was not reported in these studies except by Hu et al (2006). For example, the higher stomatal closure observed under drought in transgenic lines overexpressing SNAC1 is also observed under normal irrigation (Hu et al 2006) and the authors reported that the photosynthesis rate was not affected. It is very important to evaluate the transgenic lines under normal irrigation since better drought resistance may, in some cases, impair the overall performance of a crop under optimal conditions.

Key issues and possible roadmap for drought GM breeding

Several key issues and questions arise from the above review, including the pertinence of the choice of the target candidate genes, the transformation protocols, experimental and statistical designs, and, most importantly, the relevance of the screening protocols and criteria for breeding programs.

Drought survival vs. crop performance

Although substantial research efforts on GM crops for drought have so far been devoted to "drought tolerance genes" focusing on survival under severe stress, several authors have repeatedly demonstrated the little scope this strategy has for crop improvement (Serraj and Sinclair 2002, Sinclair et al 2004, Bhatnagar-Mathur et al 2008). On the other hand, a dehydration avoidance strategy is likely to be relevant as a more general approach to relieve agricultural drought and maintain crop performance, before drought survival develops. In rice, long-term multilocation drought studies demonstrated that rainfed lowland rice is mostly a drought avoider. The genotypes that produce higher grain yield under drought are those able to maintain better plant water status around flowering and grain setting (Fukai et al, this volume). With a few exceptions, GM studies in rice have been focusing on plant survival and tolerance traits, rather than harnessing the dehydration avoidance mechanisms, which may have a better scope for improving rice productivity in drought-prone rainfed environments.

Finally, there are two general targets for increasing crop yield in the drought-prone rainfed environments: (1) increase the overall capacity of plants to produce harvestable yield and (2) ameliorate resistance to abiotic stresses. The main challenge for deploying GM technologies successfully for stress environments is not different from other breeding approaches and is to what extent any improvement for a target environment compromises the yield potential of the crop. Farmers are more interested in crop performance and yield stability than in drought tolerance per se; therefore, it is crucial to measure systematically the variations in biomass production and yield components that result from gene modification.

Experimental design and preselection of events

The positive effect of a transgene or a combination of transgenes in a given cultivar requires the evaluation of several primary events. For a complex trait such as drought, we recommend the phenotypic evaluation of at least 15 independent single-copy events in the T1 generation. Special attention must be given to replication and statistical design, in order to increase trait heritability, increase the statistical power of the experimental comparison, and reduce the probabilities of Type I and Type II errors. Since the amount of T1 seed from each primary event is often very limited, it is important to strengthen the first phenotypic screening by more events. The more events showing a positive trend under stress conditions, the more likely the gene may confer drought resistance to the cultivar.

Current GM technologies for both indica and japonica rice cultivars are very efficient and there is no major technical bottleneck in producing a large number of primary events provided that the facilities do not have any space limitation (Hervé and Kayano 2006, Hiei and Komari 2006). Single-copy or insert events with at least 100 T1 seeds should be the preferred material for phenotypic evaluation and we argue that expression analysis is not necessarily required at an early stage (T0 plants) because a transgene-expression study of primary events does not provide very informative data. We thus favor systematic phenotypic screening of all single-copy events without preselection of the events based on transgene expression. Such expression analysis could rather be done during the second screening in order to establish a possible correlation between the phenotype (stress resistance level) and the expression level of the transgene. Since space for screening is often a major bottleneck, a powerful and relevant molecular screening at an early stage could include expression analysis of a selected set of genes that are involved in the mechanisms of action being targeted by the transgene. By doing so, the molecular screening allows selecting the events based on the functional drought resistance mechanism rather than the transgene expression only. One must, however, keep in mind that expression screening at an early stage may be inappropriate if one aims at using a drought-inducible promoter to drive the expression of the transgene. Finally, it is important to analyze isogenic lines for each event in order to exclude any unknown effect that is not due to the insertion of the transgene. Once it has been demonstrated that the null segregant population does not differ from the wild type during the first screening, only the wild-type population can be used as a control in the next generations.

Drought screening criteria

It is essential to link the drought phenotypic screening of GM rice to breeding at a very early step. A comparative phenotypic evaluation of transgenic lines with several breeding lines, both susceptible and resistant, is highly recommended for the first screening. Such a screening would not only allow a precise monitoring of the applied drought stress level but also identify a competitive advantage, if any, of the transgenic lines versus the promising breeding lines. This obviously requires large-scale infrastructure but it may speed up decision making about the beneficial effect of a gene. Finally, the performance of the events should be assessed under both stress and normal irrigation in order to identify off-types, and reveal whether any transgene that provides a biomass and seed yield production advantage over a wild type or a null event in a water-deficient condition impairs yield potential in a normal nonstressed season.

Both the genetic background into which a gene is inserted and the biophysical environments in which the transformed plants are grown and evaluated will have large impacts on gene expression and potential value. As it is becoming more and more evident that there are no universal "drought tolerance" traits (Hammer and Jordan, this volume), it is important to take G × E into account while screening GM plants under drought. Any putative drought resistance trait is unlikely to be important under all water-deficit scenarios due to the high levels of G × E interactions generally observed in the phenotypic expression of component traits, and their impacts on crop productivity. A drought trait that might offer substantial benefit in one weather scenario of developing drought, for example, early closure of stomata, might well result in a negative response in another scenario (Sinclair and Muchow 2001). One way to overcome the large G × E limitations is to understand the basic processes accounting for the drought trait and how the mechanism reacts under a range of weather scenarios. Simulation models can also help overcome G × E limitations by combining a mechanistic understanding of a drought trait with a range of weather scenarios. Breeding for specific drought resistance characters can thus be targeted to those geographical regions that would have the highest probability of frequent yield increases. One possibility for overcoming the complexities of G × E of putative drought resistance traits is to adopt a reverse physiology approach, which starts from the measurement of plant performance under drought (Fig. 2).

The parameters to be measured during the first phenotypic screenings are obviously important. Because of the inherent technical bottleneck of GM technology (limited amount of seeds, evaluation in confined environment, etc.), it is crucial to monitor the most relevant morphological and physiological parameters. However, since it is difficult to predict crop yield under drought field conditions from artificial growth conditions (pots and/or hydroponics), one must start with the end in mind, and first evaluate GM plants for performance under realistic soil drying similar to that occurring in the field. Also, because of the large numbers of transgenic events generated in a high-throughput transformation program, it is more efficient to discard a maximum of plants from the early steps, and keep only those showing promising responses. The most robust and integrative selection criteria are biomass accumulation and yield performance. At an early step of the evaluation, one may want to assess the impact on plant growth of water deficit and nondestructive measurements could be the preferred methodology to relate crop performance. Parameters such as plant phenology, canopy growth and temperature measurements with imagery, leaf rolling, tillering ability, root biomass, and spikelet fertility are relatively simple parameters to be measured for a large number of events and plants (Serraj, this volume). A correct assessment would require two cycles of screening. We argue that one successful approach would require at least two cycles of large-scale screening of events with two cycles of phenotypic evaluation of a limited number of events. It would require a minimum of 2 years to perform such preliminary evaluation. Although the large-scale screening would demonstrate a gene effect, a more precise phenotyping would allow identifying/characterizing the most suitable events to be further evaluated in

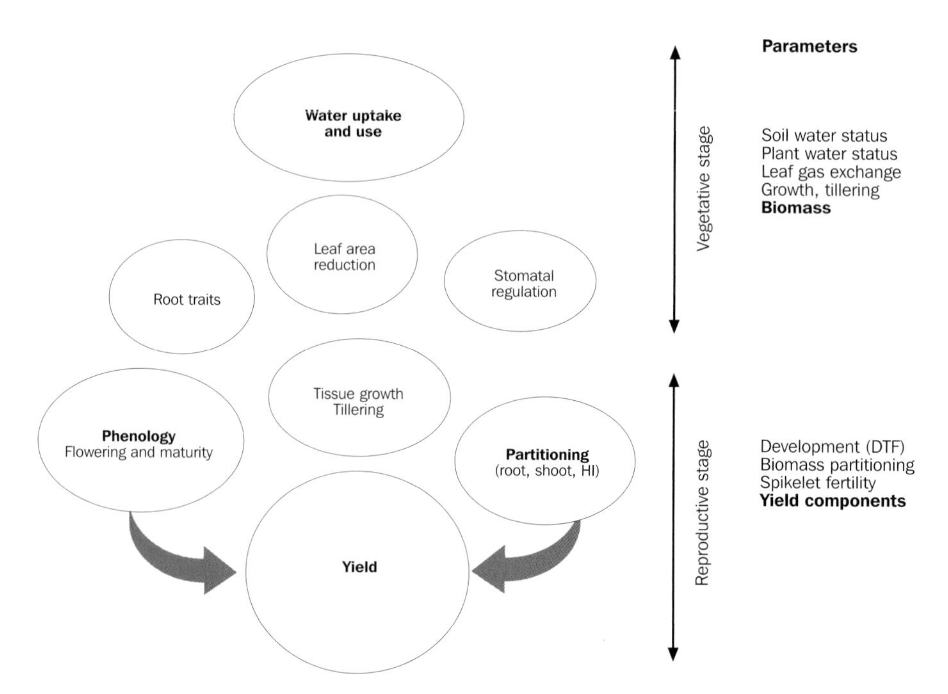

Fig. 2. Physiological framework for drought screening: analysis of crop growth and yield components. DTF = days to flowering, HI = harvest index.

field conditions. As already mentioned, it is essential to include both susceptible and resistant cultivars during the early screening since the application of the drought treatment and re-watering decision would require a visual inspection of well-known cultivars. These reference cultivars are the key controls for a linkage of improved lines by GM technology with breeding.

GM approach and reverse drought physiology

We have recently established a new drought screening facility and procedure for transgenic research on drought at IRRI. Because of the biosafety requirements, it is logistically challenging to perform early drought screening of large populations of transgenic events under field conditions. It is thus necessary to establish a robust and reliable procedure that allows the identification of likely successful transgenic events that can be further evaluated in field conditions. In order to design the screening procedure, the two major considerations were thus (1) a screening facility to mimic field conditions and (2) a screening protocol to realistically demonstrate and validate a gene effect.

A containment screenhouse facility with two independent 1-m-deep soil beds of equal surface was chosen, which allows a simultaneous screening under both

Fig. 3. Design of a screenhouse for drought screening of GM rice. (A) General view, with a drought-stressed drained plot covered as rainout shelter (left) and flooded well-watered control (right). (B) Each plot is equipped with three drainage pipes (broken lines) that are connected at each end of the bed to a sump pit.

irrigated and drought conditions (Fig. 3A). The drought-screening facility is equipped with four main systems:

(1) A screenhouse is equipped with a rainout shelter installed only above the bed used for the drought treatment to prevent any rainfall event during the screening. The roof and surrounding screenhouse divider are made of a double layer of mesh to satisfy biosafety requirements. Incandescent lamps are installed to provide supplemental lighting if necessary.

(2) Environmental parameters such as air temperature, relative humidity, and vapor pressure deficit (VPD) are monitored continuously through data-loggers to take into account the genotype × environment interactions in each screening experiment.

(3) A plastic sheet is placed up to 1-m depth, surrounding the drought bed, to prevent water seepage and percolation from adjacent flooded plots.

(4) Each soil bed is equipped with a drainage system consisting of 90-cm-deep pipes that are connected at each end of the bed to a sump pit (Fig. 3B). The drainage pipes are wrapped in geotextile fabric and surrounded with small gravel to avoid soil particles from clogging the geotextile. The gravel is placed as a padding to create the needed slope for the pipes. Finally, the pipes are further surrounded with gravel to make at least a 10-cm distance between the pipes and the soil layer. This overall design facilitates gravitational flow of the drained water toward the pits at both ends, which allows one to gradually reduce the soil moisture of the top 50-cm soil layer.

We have established a screening protocol that may facilitate early drought evaluation of a large number of events, based on the following experimental procedure: an alpha-lattice design with four replications of 17 blocks and five plots each. The material tested in each experiment consists of at least 20 independent single-copy events, five control varieties, including the recipient of the transgene, one drought-sensitive variety, two drought-resistant check varieties, the upland-adapted cultivar Apo that is also used as border plants, and two treatments (irrigated and drought). The control plot is maintained under flooded conditions and the drought-stressed treatment is imposed by draining the water and gradually drying the soil. Periods of managed water deficits are imposed with precise parameters of stress timing, duration, and severity. Soil water deficits are imposed a few weeks after transplanting throughout the period bracketing the flowering and grain-setting stages, with soil moisture levels decreasing from fully saturated to minimal levels of the fraction of transpirable soil water (FTSW) (Serraj, this volume). After the water-deficit treatment period, plants are generally re-watered and kept under well-watered conditions until physiological maturity.

One of the major difficulties in drought field screening, which results generally in low heritability and high coefficients of variation, is related to the spatial variation of soil moisture due to patchiness of soil draining and drying. Thus, our drought screening protocol dedicates much importance to improving the precision of soil drying. Piezometer access tubes are placed in each drought replication block to monitor water table levels across the drought bed. Soil moisture profiles are monitored using capacitance probes (Diviner-2000) at different depths in the range of 10 to 70 cm, and by placing 16 soil tensiometers at 15–30-cm depth to monitor the soil water tension in the drought plot daily throughout the dry-down period.

Plants are continuously monitored for phenology and plant water status and scored for drought stress symptoms. Plant water status parameters (i.e., leaf water potential and relative water content) are measured twice a week during the stress period. For selected sets of transgenic events and their corresponding nulls and checks, leaf gas exchange measurements (i.e., photosynthesis, stomatal conductance, C_i, and transpiration) are measured using the LiCor6400 photosynthesis system twice a week during the stress period. At the end of the dry-down experiment, plants are harvested and evaluated for biomass accumulation and yield components. Additional parameters measured are plant height, tillers, and panicle number at flowering. Panicles are threshed and total number of spikelets per plant is determined in addition to spikelet sterility percentage. Plant tissue samples are collected during critical periods of the dry-down for various biochemical and hormonal analyses. Finally, to better integrate the physiological phenotyping parameters, we adopted a reverse physiology approach, which consists of focusing the drought evaluation on biomass accumulation and performance under drought (Fig. 3), and investigated the underlying drought response mechanisms only in those events and constructs that were showing promising trends in terms of plant growth and performance under drought.

The first generation of transgenic events (T1) is compared to the control varieties under drought conditions, which allows us to identify the best-performing events for each gene construct. Isogenic null segregant T1 plants for each event are grown under irrigated conditions to identify any particular off-type lines (leading to yield penalty under optimal conditions). Although it is possible to evaluate null segregants under drought conditions, it may be logistically preferred to optimize the use of the drought plot by increasing the number of events or plants per event to be screened under stress conditions. This first screening allows us to identify some transgenic lines exhibiting both an optimal performance under optimal or irrigated conditions (no visible negative somaclonal drag in both the transgenic plants and the isogenic null segregant progeny) and the best performance of transgenic plants under stress conditions (possible gene-construct effect). For a subset of selected events from the first screening, the second generation (T2) of transgenic and isogenic null segregants is evaluated in both irrigated and drought conditions. Under drought conditions, the comparison of transgenic plants and their isogenic null segregants for each event allows us to validate a gene effect.

The first drought-screening experiment using the GM rice screenhouse facility and approach described above was carried out during the dry season of 2007 to evaluate the effects of *osDREB1* gene constructs provided by JIRCAS. We successfully achieved a gradual reduction in soil moisture and we observed a significant reduction in the soil moisture of the top 50-cm soil layer (data not shown). The drought intensity is sufficient to trigger drought symptoms matching those observed in rainfed lowland field conditions. With this screenhouse facility, our phenotypic data also showed that the calculated yields under irrigated and drought conditions are similar to the ones obtained in field conditions. For example, the 2007 trials showed that the average yield of rice transformants in the background elite indica variety IR64 varied in the range of 9.0–30.5 g plant^{-1} and 1.5–12.5 g plant^{-1} under irrigated and drought treatments, respectively, corresponding to an equivalent of 1.8–6.1 and 0.3–2.5 t ha^{-1}. These yields were similar to those observed in nontransgenic lines grown under open field conditions at IRRI during the same year. These data also suggest that our design of the screening in a containment facility can precisely mimic open field conditions and can sustain a robust screening procedure.

Concluding remarks

Skepticism about the potential of improving complex traits by GM technology may come from the way in which such research has been conducted or reported so far. In this short note, we reviewed the recent literature and analyzed seven different case studies describing enhanced drought resistance in rice by overexpressing a single gene. Although each paper provided supportive evidence that the transgenic lines showed better growth under drought treatment, the overall criticism of these studies is the low number of events and a lack of convincing data reporting the evaluation of transgenic material. This is likely due to a lack of linkage with breeding and to major technical bottlenecks. It is thus becoming urgent to fully integrate GM technology within breeding programs and link it with proper physiological dissection to assess transgenic lines like any other cultivars. It is also necessary to connect the introgression or stacking of any trait to the most promising advanced breeding lines or stress-sensitive rice mega-varieties. A substantial amount of research has been devoted to improving drought resistance using transgenic plants, especially in the private sector, and it is more than likely that it will be possible to contribute to sustaining yield of rice breeding lines under adverse environments thanks to GM technology. However, the bottleneck has remained the screening procedure. We suggest an early screening in a deep soil bed in a containment facility that can offer the advantage of mimicking field conditions, leading to a robust preselection of events, based on biomass accumulation and performance under both well-watered and drought conditions. This can then pass through multiple-location field trials, which would provide the ultimate validation of any improved variety. A key take-home message is that one must deploy GM technologies like any other breeding approach and it is critical to assess to what extent any improvement for a target environment compromises the yield potential of the crop across a range of environments. By doing so, GM technologies may successfully become part of a simple solution to a complex trait.

References

Chandra Babu R, Zhang J, Blum A, David Hod T-H, Wue R, Nguyen HT. 2004. *HVA1*, a LEA gene from barley, confers dehydration tolerance in transgenic rice (*Oryza sativa* L.) via cell membrane protection. Plant Sci. 166:855-864.

Bernier J, Kumar A, Venuprasad R, Spaner D, Atlin GN. 2007. A large-effect QTL for grain yield under reproductive-stage drought stress in upland rice. Crop Sci. 47:507-518.

Bhatnagar-Mathur P, Vadez V, Sharma KK. 2008. Transgenic approaches for abiotic stress tolerance in plants: retrospect and prospects. Plant Cell Rep. 27:411-424.

Bohnert HJ, Gong Q, Li P, Ma S. 2006. Unraveling abiotic stress tolerance mechanisms: getting genomics going. Curr. Opin. Plant Biol. 9:180-188.

Capell T, Bassie L, Christou P. 2004. Modulation of the polyamine biosynthetic pathway in transgenic rice confers tolerance to drought stress. Proc. Natl. Acad. Sci. USA 101(26):9909-9914.

Garg AK, Kim J-K, Owens TG, Ranwala AP, Choi YD, Kochian LV, Wu RJ. 2002. Trehalose accumulation in rice plants confers high tolerance levels to different abiotic stresses. Proc. Natl. Acad. Sci. USA 99(25):15898-15903.

Hervé P, Kayano T. 2006. Japonica rice varieties (*Oryza sativa* Nipponbare and others). In: Wang K, editor. Agrobacterium Protocols. Second edition. Vol. I. Humana Press. p 213-222.

Hiei Y, Komari T. 2006. Improved protocols for transformation of Indica rice mediated by *Agrobacterium tumefaciens*. Plant Cell Tissue Organ Cult. 85:271-283.

Hu H, Dai M, Yao J, Xiao B, Li X, Zhang Q, Xiong L. 2006. Overexpressing a NAM, ATAF, and CUC (NAC) transcription factor enhances drought resistance and salt tolerance in rice. Proc. Natl. Acad. Sci. USA 103(35):12987-12992.

Kathiresan A, Lafitte HR, Chen J, Mansueto L, Bruskiewich R, Bennett J. 2006. Gene expression microarrays and their application in drought stress research. Field Crops Res. 97:101-110.

Lian H-L, Yu X, Ye Q, Ding X-S, Kitagawa Y, Kwak S-S, Su W-A, Tang Z-C. 2004. The role of aquaporin RWC3 in drought avoidance in rice. Plant Cell Physiol. 45(4):481-489.

Marris E. 2008. More crop per drop. Nature 452:273-277.

Nelson DE, Repetti PP, Adams TR, Creelman RA, Wu J, Warner DC, Anstrom DC, Bensen RJ, Castiglioni PP, Donnarummo MG, Hinchey BS, Kumimoto RW, Maszle DR, Canales RD, Krolikowski KA, Dotson SB, Gutterson N, Ratcliffe OJ, Heard JE. 2007. Plant nuclear factor Y (NF-Y) B subunits confer drought tolerance and lead to improved corn yields on water-limited acres. Proc. Natl. Acad. Sci. USA 104(42):16450-16455.

Oh S-J, Song SI, Kim YS, Jang H-J, Kim SY, Kim M, Kim Y-K, Nahm BH, Kim J-K. 2005. *Arabidopsis CBF3/DREB1A* and *ABF3* in transgenic rice increased tolerance to abiotic stress without stunting growth. Plant Physiol. 138:341-351.

Passioura J. 2006. Increasing crop productivity when water is scarce: from breeding to field management. Agric. Water Manage. 80:176-196.

Pilon-Smits EAH, Terry N, Sears T, van Dun K. 1999. Enhanced drought resistance in fructan-producing sugar beet. Plant Physiol Biochem. 37:313-317.

Serraj R, Sinclair TR. 2002. Osmolyte accumulation: can it really help increase crop yield under drought conditions? Plant Cell Environ. 25:333-341.

Sinclair TR, Muchow RC. 2001. System analysis of plant traits to increase grain yield on limited water supplies. Agron. J. 93:263-270.

Sinclair TR, Purcell LC, Sneller CH. 2004. Crop transformation and the challenge to increase yield potential. Trends Plant Sci. 9(2):70-75.

Wang F-Z, Wanga Q-B, Kwon S-Y, Kwak S-S, Su W-A. 2005. Enhanced drought tolerance of transgenic rice plants expressing a pea manganese superoxide dismutase. J. Plant Physiol. 162:465-472.

Yamaguchi-Shinozaki K, Shinozaki K. 2006. Transcriptional regulatory networks in cellular responses and tolerance to dehydration and cold stresses. Annu. Rev. Plant Biol. 57:781-803.

Yue B, Xue W, Xiong L, Yu X, Luo L, Cui K, Jin D, Xing Y, Zhang Q. 2006. Genetic basis of drought resistance at reproductive stage in rice: separation of drought tolerance from drought avoidance. Genetics 172:1213-1228.

Zeng H, Zhong Y, Luo L. 2006. Drought tolerance genes in rice. Funct. Integr. Genomics 6:338-341.

Zhang Q. 2007. Strategies for developing Green Super Rice. Proc. Natl. Acad. Sci. USA 104(42):16402-16409.

Notes

Authors' address: International Rice Research Institute, DAPO Box 7777, Metro Manila, Philippines.

Acknowledgments: The authors would like to thank Dr. Bas Bouman and Ms. Olivyn Angeles for their advice in designing the drainage system in the screenhouse for the drought screening.

Biotechnology and transposon-tagging for improving drought resistance in rice for Indonesia

I.H. Slamet-Loedin, S. Purwantomo, P.B.F. Ouwerkerk, S. Nugroho, and R. Serraj

Transgenic methods offer a complementary approach for classical breeding to improve the tolerance of plants toward biotic and abiotic stresses. The objective of our study at RC Biotechnology LIPI in Indonesia, in collaboration with the Institute of Biology of Leiden University (The Netherlands), is to explore the use of HD-Zip transcription factors in improving the performance of rice under dry conditions. Enhancing drought resistance entails transgenic expression of HD-Zip transcription factors involved in drought response of rice. Earlier, seven HD-Zip I and II genes were identified in rice that were analyzed for drought-responsiveness. One of the two dehydration-repressed HD-Zip genes was induced by 4 hours of flooding, as a treatment opposite to drought. This Oshox4 gene from the HD-Zip gene family was selected for study in transgenic plants. Transgenic Nipponbare lines overexpressing this particular gene developed smaller leaves and exhibited a reduction in senescence compared with the controls. Transformation of Indonesian rice with Oshox4 and characterization of transgenic lines, including drought phenotyping, are now being carried out. Experiments carried out in *Arabidopsis* showed that overexpression of this particular HD-Zip gene can confer resistance to drought. Our second approach aims at identifying novel drought-resistance genes and is based on insertional mutagenesis in rice using gene trap and activation tag constructs.

Rainfall patterns in different areas in the Indonesian archipelago have a wide variation, ranging from below 1,000 mm year^{-1} to 5,000 mm year^{-1}. Drought-prone areas where the total rainfall is below 1,000 mm year^{-1} have three different patterns. The simple-wave pattern includes the islands of Sumbawa and Flores and South Sulawesi, the fluctuated pattern includes Luwuk and Palu, while the so-called double-wave pattern includes part of the islands of Flores and Sumba. In these areas, mostly upland rice is cultivated. Large areas in Indonesia have rainfall of 1,000–2,000 mm per year with the simple pattern of a wet season with monthly rainfall of 200–300 mm from November to March and low rainfall (below 100 mm) occurring from April to October. This pattern stretches from the western part of Indonesia up to the border of eastern Indonesia and includes Aceh Sigli, Palangkaraya, Banten, Karawang and Indramayu,

Pati, Lasem, Gunung kidul, West Nusa Tenggara, Bali, Lombok, and South Papua (RCAH 2003). The majority of lowland irrigated rice grows in these locations. Crops in these areas are very prone to long dry seasons.

During the El Niño of 1997 and 1998, significant areas were seriously affected by drought. A study undertaken by Levine and Yang (2006) on rainfall data and rice production in Indonesian districts from 1993 to 1999 confirmed that rainfall is positively related to rice production in local areas, which resulted in better economic conditions and eventually led to improved health (Maccini and Yang 2006). In contrast, drought has a strong impact on agricultural output and poverty. The long dry season in 2002 affected 348,512 ha of rice area and, of this, production on 41,690 ha failed completely (166,760 tons). In 2003, the total area planted for rice was 11.5 million hectares. Of this area, 450,339 ha were affected by drought (USDA FAS 2003). Naylor et al (2007) analyzed climate variability and its impact on rice in Java and Bali areas, which account for 55% of the country's total rice production. Data collected over a 20-year period showed that a 30-day delay in monsoon onset caused rice production to fall by 6.5–11%. This resulted in a drop in rice production of 540,000–580,000 tons in the January-April period. Production declines of this magnitude, when scaled up to the country as a whole, such as in 1997-98, resulted in a situation in which Indonesia had to import 5.8 million t of rice, which was 20% of the total world rice trade that particular year. Recently, in 2006, a severe drought took place at the end of the growing season, resulting in harvest failure in several major rice-growing areas in Java and delayed planting, causing a significant increase in the price of rice. In line with the decrease in harvested area, milled rice production in 2007 was also estimated to decrease to 33.7 million t from 34.96 million t in 2006 (USDA FAS 2007).

Increasing scarcity of resources such as land and water due to the population increase and consequent urbanization will inevitably cause larger deficits in future rice production in Indonesia. Conversion of irrigated rice fields to nonagriculture between 1981 and 1999 reached 1.6 million hectares, whereas the replaced lands had a lower productivity. A study by Irawan (2004) showed that the production lost due to land conversion even after land replacement on the island of Java from 1978 to 1998 reached 4.7 million t per annum. Increasing rice production in Indonesia in the near future would be difficult to reach if Indonesia needed to rely on production only in irrigated lowland and rainfed lowlands, not only because of the land conversion rate but also because productivity tends to be stagnant. Uplands and tidal swamp lands are the major prospective areas for the extension of rice culture in Indonesia. Data of 2001 showed that 59.6% of Indonesian land is irrigated land, 27.3% is rainfed lowland, whereas only 8.3% is tidal swamp and 4.8% is upland rice-producing area (Pasandaran et al 2004). The average production of upland rice in Indonesia was 2.2 t ha^{-1} in 1998, which increased to 2.4 t ha^{-1} in 2002, and it is grown mainly in Sumatera, Java, Bali, and Kalimantan, and on a smaller scale in Sulawesi and Nusa Tenggara (Faqi et al 2004). The area outside Java has more potential to grow upland rice. In principle, a sustainable upland rice production system will provide an important alternative to reach the goal of increasing rice production by 1.8% per year, which is required to meet the yield gap. However, upland rice production has drawbacks that need to be overome. In fact, the major challenge for upland rice production is dehydration stress, weed control, rice blast disease, and the realization that current upland rice cultivars yield less.

A range of strategies has been explored to tackle drought resistance in plants. Slow progress has taken place, however, in developing rice varieties adapted to water stress. The lack of desirable progress is attributable to the fact that tolerance of abiotic stress is a very complex multigenic trait influenced by coordinated and differential expression of a network of genes (Garg et al 2002), though progress has been made in analyzing the complex cascades of genes, particularly the specificity and cross-talk in stress-signaling pathways (Yamaguchi-Shinozaki and Shinozaki 2006). Our hypothesis is that a combination of marker-assisted breeding, functional gene analysis using genomics approaches, including analysis of mutated populations, and transgenic technology are necessary to support classical breeding programs. This will ultimately lead to the required increase in annual rice production on less agricultural land, with less water consumption per ton of rice produced.

RC Biotechnology, in collaboration with Dr. P.B.F. Ouwerkerk, the Institute of Biology, Leiden University (The Netherlands); Dr. N. Upadhyaya, CSIRO–Plant Industry, Canberra, Australia; and Dr. A. Pereira, Virginia Bioinformatics Institute, Virginia (formerly at Plant Research International, Wageningen, The Netherlands), explores the use of a multidisciplinary approach to tackle the problem of drought resistance in rice. One strategy was to use homeobox transcription factors of the homeo domain–leucine-zipper (HD-Zip) class in molecular breeding for drought resistance in rice. The second one is a genomics approach aimed at the generation of *Ac/Ds* enhancer trap and activation tag populations and screening of these for new drought-resistance genes.

Perspectives for using transcription factor genes for drought-resistance breeding

Water stress causes adverse effects on the growth of rice and its productivity by initiating a cascade of plant responses at the molecular and cellular levels as well as the physiological level (Chaves et al 2003). Molecular mechanisms regulating responses of plant genes to water stress include the sensing mechanisms of osmotic stress, modulation of stress signals to cellular signals, transduction of cellular signals to the nucleus, transcriptional control of stress-inducible genes, and the function and cooperation of stress-inducible genes allowing dehydration-stress tolerance. Genes expressed during dehydration stress are roughly classified into two groups. The first group includes functional proteins that are involved in dehydration-stress tolerance and cellular adaptation such as chaperones, LEA (late embryogenesis abundant) proteins, osmotin, antifreeze proteins, mRNA-binding proteins, and key enzymes for osmolyte biosynthesis such as proline, water channel proteins, sugar and proline transporters, detoxification enzymes, enzymes for fatty acid metabolism, proteinase inhibitors, ferritin, and lipid-transfer proteins. The second group includes regulatory proteins with functions in gene expression and signal transduction in stress response that include transcription factors from various classes (Yamaguchi-Shinozaki and Shinozaki 2001, 2006). Several approaches were attempted to improve the stress tolerance of plants by gene transfer. One promising approach is the use of genes encoding protein factors that are involved in the regulation of gene expression and signal transduction and function in stress response that can regulate many stress-inducible genes involved in stress tolerance. However, experiments describing the enhancement of stress resistance by a transgenic approach are usually restricted to the modulation of the plant response to one stress factor, and most of the experiments are generally based on the overexpression of a single gene (Hervé and Serraj, this volume). Although the transformation methodology is somehow becoming standard, the protocols for plant evaluation and the parameters used to assess plant resistance are diverse and difficult to compare. The low number of independent transgenic lines and the poor assessment of a gene effect versus an overall effect due to gene insertion and somaclonal variation could be major drawbacks. The relevance of some screening methods for drought is questionable from a breeding perspective and only a few studies provide actual data about enhanced drought resistance under field conditions. If the use of transgenic crops could potentially enhance drought responses of plants by promoting changes in the genome, there is clearly an important need to define or redefine the major steps and criteria to obtain better crop performance in the field (Hervé and Serraj, this volume).

It is now possible to use transgenic approaches to improve abiotic stress tolerance in agriculturally important crops with far fewer target traits than had been anticipated. It was shown that overexpression of the AP2/EREBP factors CBF1, DREB1a, and CBF4 resulted in drought/salt/cold/heat tolerance in *Arabidopsis* (Jaglo-Ottosen et al 1998, Kasuga et al 1999, Yamaguchi-Shinozaki and Shinozaki 2001, Haake et al 2002) and in tobacco (Kasuga et al 2004). Overexpression of the stress-responsive gene *SNAC1* significantly enhances drought resistance in transgenic rice (22–34% higher seed setting than in a control) in the field under severe drought-stress conditions at the reproductive stage (Hu et al 2006). Overexpression of a zinc-finger-type transcription factor (*SCOF-1* from soybean) increased cold tolerance in tobacco without affecting normal growth and increased freezing tolerance in *Arabidopsis* (Kim et al 2001). These examples show that modification of transcription factor levels can successfully produce stress tolerance.

Members of several different classes of transcription factors have been implicated in stress responses, including MYC, MYB, bZIP, AP2, homeodomain, and zinc-finger proteins. In this project, we have chosen to concentrate on one particular homeobox class that encodes for the so-called homeodomain–leucine-zipper (HD-Zip) proteins. Homeobox genes were originally identified as cross-hybridizing DNA sequences shared among several genes that control the body plan of *Drosophila*, but were then found in many genes controlling cell fate during the development of animals, plants, and fungi (Gehring 1987). HD-Zip proteins have been found exclusively in plants, and are characterized by the presence of a DNA-binding homeodomain with a closely linked leucine-zipper motif that functions in protein-protein dimer formation (Ruberti et al 1991, Schena and Davis 1992). Homeodomain-containing (HD) proteins occur only in eukaryotes and contain a characteristic HD of 61 amino acids, function as transcription factors, and are generally associated with the regulation of developmental processes.

The *Arabidopsis* genome contains 43 HD-Zip genes and, based on amino acid criteria, they are grouped into four families (HD-Zip I to IV) (Sessa et al 1994). Experiments with GFP-tagged proteins show nuclear localization of HD-Zip proteins, which is consistent with their function as transcription factors (Scarpella et al 2000, Deng et al 2006). Different HD-Zip proteins are known as repressors or activators (Meijer et al 1997, 2000, Steindler et al 1999, Ohgishi et al 2001, Himmelbach et al 2002). Apart from dicots, HD-Zip genes were also found in monocots, including rice (Meijer et al 1997, 2000). For one of these, Oshox1, an HD-Zip family II protein, a function was found in vascular cell fate specification (Scarpella et al 2000, 2002). However, evidence is accumulating that many other HD-Zip proteins are related to the regulation of developmental adaptation to environmental stress conditions. For example, the HD-Zip gene *Athb-2* from *Arabidopsis* is proposed to function in shade-avoidance response based on observations that this gene is inducible by far red–rich light and that its overexpression supports accelerated elongation (Carabelli et al 1993, 1996, Schena et al 1993, Steindler et al 1999) and *Hahb*-4, a member of *Helianthus annuus* subfamily I of HD-Zip proteins, in transgenic *Arabidopsis thaliana* plants exhibits a strong tolerance of water stress (Cabello et al 2007).

Moreover, expression of many HD-Zip genes is affected by drought stress. These include *Athb-6*, *Athb-7*, and *Athb-12* from *Arabidopsis*, whose expression is also inducible by abscisic acid (ABA) (Lee and Chun 1998, Lee et al 2001, Söderman et al 1996, 1999, Hjellstrom et al 2003). Furthermore, in the resurrection plant (*Craterostigma plantagineum*), an extremely drought-resistant species normally growing in South Africa, seven HD-Zip genes (*CpHB-1* to *-7*) have been identified that all respond to drought stress, either by up- or down-regulation (Frank et al 1998, Deng et al 2002). ABA plays a central role in regulating plant responses to drought, including stomata closure (Leung and Giraudat 1998). It was found that *CpHB-2*, *-6*, and *-7* are induced by both drought and ABA and that *CpHB-1* is inducible only by drought. This suggests that some HD-Zip proteins function in ABA-dependent and -independent drought-responsive signaling pathways. More evidence for a link between HD-Zips and drought and ABA comes from sunflower, in which the family I HD-Zip gene *Hahb-4* was found to be strongly inducible by both stimuli (Gago et al 2002, Manavella et al 2006). Furthermore, *Athb-7* and *-12* from *Arabidopsis*, which are both inducible by drought, are clearly operating in an ABA-dependent signaling pathway, because, in the ABA-insensitive mutants *abi1* and *abi2*, induction of *Athb-7* and *-12* is absent after treatment by ABA. Mutations in ABI1 and ABI2 affect sensitivity toward ABA in both seeds and vegetative parts and encode related serine/threonine protein phosphatases (Leung et al 1997). Other evidence for interactions between ABA signaling and an HD-Zip protein is *Athb-6,* which was found to be able to interact on a protein-protein level with ABI1. In addition, ABA-mediated induction of *Athb-6* is impaired in *abi1* (Himmelbach et al 2002).

So far, little was known about the downstream target genes that are regulated by HD-Zip proteins and the physiological and developmental processes in which they operate. But, recently, Deng et al (2006) identified the ABA- and drought-responsive group 2 LEA/dehydrin gene *CdeT6-19* as a potential target for *CpHB-7* in *C. plantagineum*. CpHB-7 protein is able to bind to a conserved HD-Zip recognition motif in the promoter of *CdeT6-19*. Furthermore, *CpHB-7* overexpressing plants show reduced sensitivity toward ABA during seed germination and stomata closure. These lines were used as tools in a strategy involving a combination of genome mining for HD-Zip binding-site-containing genes and macroarrays to identify target genes in *Arabidopsis*. The results show that CpHB-7 modifies the expression of a number of known ABA-responsive target genes as a negative regulator (Deng et al 2006). Another HD-Zip target gene identification study was done with microarray analysis of transgenic *Arabidopsis* overexpressing *Hahb-4* from sunflower (Manavella et al 2006). This study indicated that this particular HD-Zip is part of ethylene signaling pathways because overexpression leads to repression of genes involved in ethylene signaling and synthesis and also identified genes known to be involved in protection toward osmotic stress, including an arginine decarboxylase and betaine-aldehyde dehydrogenase. In agreement with this, analysis of the *Hahb-4* overexpressors showed enhanced drought resistance and *Hahb-4* expression levels showed an inverse correlation with ethylene sensitivity. In summary, consistent with expression patterns in response to drought and ABA treatment, several HD-Zips seem to function in signaling pathways. This is most likely only the case for a subset of HD-Zips. Other HD-Zips are probably involved in other processes. For instance, for *Oshox1*, a function was found in vascular cell fate specification (Scarpella et al 2000, 2002) but there is no indication that there might be a relation with drought response.

Characterization of drought-responsive HD-Zip genes in rice

Based on the relation between HD-Zip gene expression and drought stress, we decided to investigate whether drought-responsive HD-Zip genes also exist in rice and whether these genes could be used to improve drought resistance since it has been shown that overexpression of HD-Zips in *Arabidopsis* can result in enhanced drought-resistance (Dezar et al 2005, Manavella et al 2006). Initially, seven HD-Zip genes were identified in rice, named *Oshox1–7* (Meijer et al 1997, 2000). Database-mining efforts have shown that the rice genome in total contains 26 HD-Zip family I and II genes (unpublished results). Northern blot analysis was performed to determine whether the expression of these genes is affected by dehydration or by the plant hormone ABA, which often mediates dehydration responses. When seedlings were transferred from moist to dry filter paper, the expression of several HD-Zip genes was down- or up-regulated at the RNA level after 24–48 h. Expression of a known dehydration-responsive control gene, *salT* (Claes et al 1990, Garcia et al 1998), increased gradually, indicating that drought treatment was effective. The relation of rice HD-Zip genes to other stress treatments and to the stress hormone ABA was examined too. This showed that, like in other species, expression of several HD-Zip genes correlates with several different stress conditions in rice. This knowledge is valuable to determine whether there are more HD-Zip genes suitable for future testing in transgenic plants or validation with mutants in order to investigate the effects of altered HD-Zip expression on drought resistance.

Complete annotation (Oshox1–Oshox33) of the HD-Zip families I, II, and III from rice was conducted and compared with *Arabidopsis* in a phylogeny reconstruction. We found that a number of HD-Zip genes appear to be unique in rice. Most HD-Zip genes were broadly expressed in mature plants and seedlings, but others showed more organ-specific patterns. Like in other plants, a subset of eight HD-Zip I and II genes was found to be regulated by drought stress and we demonstrated that these genes are differentially regulated in drought-sensitive versus drought-tolerant rice cultivars. It also seems that the level of expression is associated with the level of drought resistance (probably drought tolerance) of the respective rice variety. We chose the eight selected subsets of HD-Zip genes based on earlier detailed studies to look at drought response under field conditions using Q-PCR. The results showed that all eight tested HD-Zip genes were drought-responsive under evaluation at the rainout shelter facilities when drought treatment was applied for 2 weeks at the reproductive stage (Agalou et al 2008).

One of the dehydration-repressed HD-Zip genes identified in our study, *Oshox4*, was induced by 4 hours of flooding as an opposite treatment to the drought condition. This gene was selected for study in transgenic plants (Agalou et al 2008). Growth of Nipponbare plants overexpressing this gene under control of the CaMV 35S promoter was much reduced. Two months after transfer to the greenhouse, control transgenics (expressing a GUS reporter gene) already had formed multiple-flowering tillers, whereas overexpressors (confirmed by northern analysis) had produced only a few small leaves and in most cases had not yet started to form tillers. The most severely affected overexpressors never formed tillers and the shoot displayed a prolonged vegetative phase in that it continued to produce leaves (more than 15 compared with 5–6 in the control) instead of entering the reproductive stage. Other expressors had formed the first flowering tillers 4–6 months after transfer to the greenhouse, when seed maturation in the controls was already completed. It is clear that the CaMV 35S-HD-Zip Nipponbare plants cannot be used directly for transgenic breeding purposes since the phenotypes are too strong and the seed set is too low, which hampers validation for further drought resistance.

Therefore, we have adopted a different strategy in which the HD-Zip gene is expressed under the control of a stress-inducible promoter in order to minimize side-effects on the phenotype and maximize the stress-tolerance effect (Kasuga et al 1999). The new construct will be driven by the promoter of the *salT* gene, which was used as a control gene for northern blot analyses of drought-treated plants. The *salT* gene and its promoter have been isolated and were shown to be rapidly induced by drought and other abiotic-stress conditions (Claes et al 1990, Garcia et al 1998). Using this construct, transgenic lines of Rojolele (Indonesian javanica variety) were obtained and are currently being analyzed. Progenies of three transgenic events along with tissue culture–derived plants and a wild-type control were selected for a dry-down experiment using the fraction of soil transpiration water (FSTW) as a stress covariable in a pot experiment in the greenhouse (Bhatnagar-Mathur et al 2007). Fourteen plants of each line were grown in the controlled medium and half of them were exposed to well-watered (WW) conditions and the other half to drought stress (DS) at the reproductive stage for 14 days. Initial results showed that the percentage of filled grains under DS conditions in comparison with WW conditions on average was 57% in the wild type, whereas the individual average of three transgenic events was 45.1%, 74.1%, and 81.1%. Experiments will be repeated on a larger scale to give insight into the question of the potential validity of this gene in molecular breeding programs for drought resistance, or strategies with different (inducible or tissue-specific) promoters or other candidate genes need to be employed.

To investigate the expression pattern of the drought-responsive rice HD-Zip in more detail, transgenic plants were made using a promoter-reporter gene construct. The promoter region of the rice HD-Zip gene (approximately 2 kb of sequence upstream of the start codon) was obtained via PCR, fused to the *gusA* reporter gene, and introduced into Nipponbare rice. The reporter gene expression pattern was analyzed under normal growth conditions and under drought-stress conditions. These experiments revealed in which tissues and cell types the HD-Zip gene is expressed under different environmental conditions. In summary, in leaves, the promoter conferred expression exclusively in the phloem and is down-regulated by drought, but, in roots, expression was in all cell types. Finally, the HD-Zip promoter itself can be useful as a biotechnological tool in modifying the expression of transgenes under drought conditions. It is conceivable that certain transgenes (e.g., genes that improve yield or grain quality) could have beneficial effects under normal growth conditions, but it is better to shut such genes temporarily off during drought stress when the plant has to use all its energy to overcome the stress situation. Therefore, the HD-Zip promoter may provide a tool to repress transgenes under drought conditions. No promoters with similar characteristics are currently available.

Perspectives for the discovery of new drought-resistance genes in rice via transposon-tagging

Insertional mutagenesis is still among the most widely used methods for gene discovery and gene functional analyses. The insertion mutagen can generate knockout mutations by blocking the expression of a gene, resulting in mutant phenotypes. The mutant gene tagged by the insertion can be isolated, which will subsequently lead to the identification of the wild-type gene. The ability of the well-characterized maize transposon *Ac/Ds* to transpose in heterologous systems offers possibilities for transposon tagging and provides an efficient tool for identifying genes (Baker et al 1986, Pereira and Saedler 1989). In the maize *Ac/Ds* system, the autonomous *Activator* (*Ac*) transposon has the ability to induce its own transposition and that of the nonautonomous dissociation (*Ds*) element (McClintock 1947). The *Ac/Ds* system has been used widely as an efficient tool for gene tagging in a variety of dicots such as tobacco, *Arabidopsis*, carrot, potato, tomato, soybean, lettuce, and flax (Haring et al 1991, Sundaresan 1996). The activity of transposon *Ac/Ds* has also been shown in other monocots, including wheat (Takumi et al 1999) and barley (Koprek et al 2000, Scholz et al 2001). In rice, *Ac/Ds* transposition has been described (Izawa et al 1991, Shimamoto et al 1993) and has been shown to be an efficient tool for functional genomics (Izawa et al 1997, Chin et al 1999, Nakagawa et al 2000, Greco et al 2001a,b, 2003, van Enckevort et al 2004, Kim et al 2004, Upadhyaya et al 2006).

Insertion tags can be designed for knockout mutagenesis, gene expression profiling such as gene/enhancer traps, and activation tagging. In knockout mutagenesis, the function of the knockout gene is expected to be disrupted, resulting in a mutant phenotype. However, knockout mutagenesis has limited use as the majority of genes display no obvious phenotype (Burns et al 1994), among other reasons due to functional redundancy, in which one or more homologous genes can substitute for the same functions. In *Arabidopsis,* it is estimated that less than 5% of the genes will reveal a visible phenotype upon disruption (Long et al 1993) and similar behavior is expected from rice. The addition of a reporter gene (such as *gusA*, GFP, or luciferase) in the transposon T-DNA construct will complement knockout mutagenesis and may help in unraveling the function of a tagged gene, which cannot be predicted by knockout mutagenesis only (Springer 2000). The reporter gene will trap and display the pattern of expression of the neighboring tagged genes. The pattern of the expression could reveal the gene function. Another approach by adding strong enhancer sequences in the insert could also complement knockout mutagenesis (Weigel et al 2000, Marsch-Martinez et al 2002). Activation tagging will result in overexpression of tagged genes, providing an efficient way to obtain and study overexpression phenotypes. These two complementary transposon-tagging approaches provide powerful tools in studying gene function, especially those with overlapping functions.

Rice mutant populations harboring a gene trap and activation tag are being generated. For this, a two-component *Ac/Ds* system was incorporated to enable fast development of a large mutant population. Two different sets of plasmids are used, one for the gene-trap system and the other for the activation-tag system. The rice transformation was performed by the *Agrobacterium*-mediated method (Hiei et al 1994). The rice mutant populations obtained will be used for phenotypic screening. So far, more than 1,500 independent mutagenic lines have been obtained and grown in a greenhouse at RCBt Cibinong, Bogor. Stable insertion lines with transposed *Ds* (and *Ac* segregated out) are being generated and will be used to screen for phenotypes, including drought phenotype. Five hundred stable insertion lines are targeted to be screened for drought at the vegetative and reproductive stage by intermittent soil drying.

References

Agalou A, Purwantomo S, Övernas E, Johannesson H, Zhu X, Estiati A, De Kam RJ, Engström P, Slamet-Loedin IH, Zhu Z, Wang M, Xiong L, Meijer AH, Ouwerkerk PBF. 2008. A genome-wide survey of HD-Zip genes in rice and analysis of drought-responsive family members. Plant Mol. Biol. 66:87-103.

Baker B, Schell J, Lorz H, Fedoroff N. 1986. Transposition of maize controlling element 'Activator' in tobacco. Proc. Natl. Acad. Sci. USA 83:4844-4848.

Bhatnagar-Mathur P, Devi MJ, Reddy DS, Lavanya M., Vadez V, Yamaguchi Shinozaki K, Sharma KK. 2007. Stress inducible expression of At DREB1A in transgenic peanut (*Arachis hypogaea* L.) increases transpiration efficiency under water-limiting conditions. Plant Cell Rep. 26:2071-2082.

Burns N, Grimwade R, Macdonald PB, Choi EY, Finberg K, Roeder GS, Snyder M. 1994. Large scale analysis of gene expression, protein localization and gene disruption in *Saccharomyces cerevisiae*. Genes Dev. 8:1087-1105.

Cabello JV, Dezar CA, Manavella PA,·Chan RL. 2007. The intron of the *Arabidopsis thaliana* COX5c gene is able to improve the drought tolerance conferred by the sunflower Hahb-4 transcription factor. Planta 226:1143-1154.

Carabelli M, Morelli G, Whitelam G, Ruberti I. 1996. Twilight-zone and canopy shade induction of the ATHB-2 homeobox gene in green plants. Proc. Natl. Acad. Sci. USA 93:3530-3535.

Carabelli M, Sessa G, Baima S, Ruberti I. 1993. The *Arabidopsis* Athb-2 and -4 genes are strongly regulated by far-red-rich light. Plant J. 4:469-479.

Chaves MM, Maroco JP, Pereira JS. 2003. Understanding plant response to drought: from genes to the whole plant. Funct. Plant Biol. 30:239-264.

Chin HG, Choe MS, Lee SH, Park SH, Koo JC, Kim NY, Lee JJ, Oh BG, Yi GH, Kim SC, Choi HC, Cho MJ, Han CD. 1999. Molecular analysis of rice plants harboring an *Ac/Ds* transposable element-mediated gene trapping system. Plant J. 19:615-623.

Claes B, Dekeyser R, Villarroel R, Van Den Bulcke M, Bauw G, Van Montagu M, Caplan A. 1990. Characterization of a rice gene showing organ-specific expression in response to salt stress and drought. Plant Cell 2:19-27.

Deng X, Phillips J, Bräutigam A, Engström P, Johannesson H, Meijer AH, Ouwerkerk PBF, Ruberti I, Salinas J, Vera P, Iannacone R, Bartels D. 2006. A homeodomain leucine zipper gene from *Craterostigma plantagineum* regulates abscisic acid responsive gene expression and physiological responses. Plant Mol. Biol. 61:469-489.

Deng X, Phillips J, Meijer AH, Salamini F, Bartels D. 2002. Characterization of five novel dehydration-responsive homeodomain leucine zipper genes from the resurrection plant *Craterostigma plantagineum*. Plant Mol. Biol. 49:601-610.

Dezar CA, Gago GM, Gonzalez DH, Chan RL. 2005. *Hahb-4*, a sunflower homeobox-leucine zipper gene, is a developmental regulator and confers drought tolerance to *Arabidopsis thaliana* plants. Transgenic Res. 14:429-440.

Faqi AM, Toha HM, Baharsyah JS. 2004. Potency of upland rice and food security. In: Kasryno F, Pasandaran E, Fagi AM, editors. Indonesian paddy and rice economy. Indonesian Agency of Agricultural Research and Development. p 347-372.

Frank W, Phillips J, Salamini F, Bartels D. 1998. Two dehydration inducible transcripts from the resurrection plant *Craterostigma plantagineum* encode interacting homeodomain-leucine zipper proteins. Plant J. 15:413-421.

Gago GM, Almoguera C, Jordano J, Gonzalez DH, Chan RL. 2002. *Hahb-4*, a homeobox-leucine zipper gene potentially involved in abscisic acid-dependent responses to water stress in sunflower. Plant Cell Environ. 25:633-640.

Garcia AB, de Almeida Englar J, Claes B, Villarroel R, Van Montagu M, Gerats T, Caplan A. 1998. The expression of the salt-responsive gene *salT* from rice is regulated by hormonal and development cues. Planta 207:172-180.

Garg AK, Kim JK, Owens TG, Ranwala AP, Do Choi Y, Kochian LV, Wu RJ. 2002. Trehalose accumulation in rice plants confers high tolerance levels to different abiotic stresses. Proc. Natl. Acad. Sci. USA 99(25):15898-15903.

Gehring WJ. 1987. Homeotic genes, the homeobox, and the spatial-organization of the embryo. Harvey Lectures 81:153-172.
Greco R, Ouwerkerk PBF, de Kam RJ, Sallaud C, Favalli C, Colombo L, Guiderdoni E, Meijer AH, Hoge JHC, Pereira A. 2003. Transpositional behaviour of an *Ac/Ds* system for reverse genetics in rice. Theor. Appl. Genet. 108:10-24.
Greco R, Ouwerkerk PBF, Sallaud C, Kohli A, Colombo L, Puigdomenech P, Guiderdoni E, Christou P, Hoge JHC, Pereira A. 2001a. Transposon insertional mutagenesis in rice. Plant Physiol. 125:1175-1177.
Greco R, Ouwerkerk PBF, Taal AJC, Favalli C, Beguiristain P, Puigdomenech P, Colombo L, Hoge JHC, Pereira A. 2001b. Early and multiple *Ac* transposition in rice suitable for efficient insertional mutagenesis. Plant Mol. Biol. 46:215-227.
Haake V, Cook D, Riechmann JL, Pineda O, Thomashow MF, Zhang JZ. 2002. Transcription factor CBF4 is a regulator of drought adaptation in *Arabidopsis*. Plant Physiol. 130:639-648.
Haring MA, Romens CMT, Nijkamp HJJ, Hille J. 1991. The use of transgenic plants to understand transposition mechanisms and to develop transposon tagging strategies. Plant Mol. Biol. 16:449-461.
Hiei Y, Ohta S, Komari T, Kumashiro T. 1994. Efficient transformation of rice (*Oryza sativa* L.) mediated by *Agrobacterium* and sequence analysis of the boundaries of the T-DNA. Plant J. 6:2271-2282.
Himmelbach A, Hoffmann T, Leube M, Hohener B, Grill E. 2002. Homeodomain protein ATHB6 is a target of the protein phosphatase ABI1 and regulates hormone responses in *Arabidopsis*. EMBO J. 21:3029-3038.
Hjellström M, Olsson ASB, Engström P, Söderman EM. 2003. Constitutive expression of the water deficit-inducible homeobox gene *ATHB7* in transgenic *Arabidopsis* causes a suppression of stem elongation growth. Plant Cell Environ. 26:1127-1136.
Hu H, Dai M, Yao J, Xia, B, Li X, Zhang Q, Xiong L. 2006. Overexpressing a NAM, ATAF, and CUC (NAC) transcription factor enhances drought resistance and salt tolerance in rice. Proc. Natl. Acad. Sci. USA 103:12987-12992.
Irawan B. 2004. Land conversion in Java and impact towards rice production. In: Kasryno F, Pasandaran E, Fagi AM, editors. Indonesian paddy and rice economy. Indonesian Agency of Agricultural Research and Development. p 277-293.
Izawa T, Miyazaki C, Yamamoto M, Terada R, Iida S, Shimamoto K. 1991. Introduction and transposition of the maize transposable element *Ac* in rice (*Oryza sativa*). Mol. Gen. Genet. 227:391-396.
Izawa T, Ohnishi T, Nakano T, Ishida N, Enoki H, Hashimoto H, Itoh K, Terada R, Wu C, Miyazaki C, Endo T, Iida S, Shimamoto K. 1997. Transposon tagging in rice. Plant Mol. Biol. 35:219-229.
Jaglo-Ottosen KR, Gilmor SJ, Zarka DG, Schabenberger O, Thomashow MF. 1998. *Arabidopsis* CBF1 over-expression induces COR genes and enhances freezing tolerance. Science 280:104-106.
Kasuga M, Liu Q, Miura S, Yamaguchi-Shinozaki K, Shinozaki K. 1999. Improving plant drought, salt, and freezing tolerance by gene transfer of a single stress-inducible transcription factor. Nature Biotechnol. 17:229-230.
Kasuga M, Miura S, Shinozaki K, Yamaguchi-Shinozaki K. 2004. A combination of the *Arabidopsis DREB1A* gene and stress-inducible rd29A promoter improved drought- and low-temperature stress tolerance in tobacco by gene transfer. Plant Cell Physiol. 45:346-352.

Kim JC, Lee SH, Cheong YH, Yoo CM, Lee SI, Chun HJ, Yun DJ, Hong JC, Lee SY, Lim CO, Cho MJ. 2001. A novel cold-inducible zinc finger protein from soybean, SCOF-1, enhances cold tolerance in transgenic plants. Plant J. 25:247-259.

Kim CM, Piao HL, Park SJ, Chon NS, Je BI, Sun B, Park SH, Park JY, Lee EJ, Kim MJ, Chung WS, Lee KH, Lee YS, Lee JJ, Won YJ, Yi G, Nam MH, Cha YS, Yun DW, Eun MY, Han C. 2004. Rapid, large-scale generation of *Ds* transposant lines and analysis of the *Ds* insertion sites in rice. Plant J. 39:252-263.

Koprek T, McElroy D, Louwerse J, Williams-Carrier R, Lemaux PG. 2000. An efficient method for dispersing *Ds* element in the barley genome as a tool for determining gene function. Plant J. 24:253-264.

Lee Y-H, Chun J-Y. 1998. A new homeodomain-leucine zipper gene from *Arabidopsis thaliana* induced by water stress and abscisic acid treatment. Plant Mol. Biol. 37:377-384.

Lee YH, Oh HS, Cheon CI, Hwang IT, Kim YJ, Chun JY. 2001. Structure and expression of the *Arabidopsis thaliana* homeobox gene *Athb-12*. Biochem. Biophys. Res. Commun. 284:133-141.

Leung J, Giraudat J. 1998. Abscisic acid signal transduction. Annu. Rev. Plant Physiol. Plant Mol. Biol. 49:199-222.

Leung J, Merlot S, Giraudat J. 1997. The *Arabidopsis ABSCISIC ACID INSENSITIVE 2 (ABI2)* and *ABI1* genes encode homologous protein phosphatases 2C involved in abscisic acid signal transduction. Plant Cell 9:759-771.

Levine DIL, Yang D. 2006. A note on the impact of local rainfall on rice output in Indonesian districts. Mimeo. University of California, Berkeley, and University of Michigan.

Long D, Martin M, Sundberg E, Swomburne J, Puangsomlee P, Coupland G. 1993. The maize transposable element system *Ac*/*Ds* as a mutagen in *Arabidopsis*: identification of an albino mutation induced by *Ds* insertion. Proc. Natl. Acad. Sci. USA 90:10370-10374.

Maccini S, Yang D. 2006. Under the weather: health, schooling, and socioeconomic consequences of early life rainfall. Mimeo. University of Michigan.

Manavella PA, Arce AL, Dezar CA, Bitton F, Renou FP, Crespi M, Chan RL. 2006. Cross-talk between ethylene and drought signalling pathways is mediated by the sunflower Hahb-4 transcription factor. Plant J. 48:125-137.

Marsch-Martinez N, Greco R, Van Arkel G, Herrera-Estrella L, Pereira A. 2002. Activation tagging using the *En-I* maize transposon system in *Arabidopsis*. Plant Physiol. 129:1544-1556.

McClintock B. 1947. Cytogenetic studies of maize and *Neurospora*. Carnegie Inst. Wash. Yearbook 46:146-152.

Meijer AH, de Kam RJ, d'Erfurth I, Shen W, Hoge JHC. 2000. HD-Zip proteins of families I and II from rice: interactions and functional properties. Mol. Gen. Genet. 263:12-21.

Meijer AH, Scarpella E, van Dijk EL, Qin L, Taal AJ, Rueb S, Harrington SE, McCouch SR, Schilperoort RA, Hoge JHC. 1997. Transcriptional repression by Oshox1, a novel homeodomain leucine zipper protein from rice. Plant J. 11:263-276.

Nakagawa Y, Machida C, Machida Y, Toriyama K. 2000. Frequency and pattern of transposition of the maize transposable element *Ds* in transgenic rice plants. Plant Cell Physiol. 41:733-742.

Naylor RL, Battisti D, Vimont DJ, Falcon WP, Burke MP. 2007. Assessing risks of climate variability and climate change for Indonesian rice agriculture. Proc. Natl. Acad. Sci. USA 104(19):7752-7757.

Ohgishi OM, Oka A, Giorgio Morelli G, Ida Ruberti I, Aoyama T. 2001. Negative autoregulation of the *Arabidopsis* homeobox gene *ATHB-2*. Plant J. 25:389-398.

Pasandaran E, Irianto G, Zuliasri N. 2004. Effectiveness and potency of irrigation development for rice improvement. In: Kasryno F, Pasandaran E, Fagi AM, editors. Indonesian paddy and rice economy. Indonesian Agency of Agricultural Research and Development. p 277-293.

Pereira A, Saedler H. 1989. Transpositional behaviour of the maize *En/Spm* element in transgenic tobacco. EMBO J. 8:1315-1321.

RCAH (Research Centre for Agroclimate and Hydrology). 2003. Atlas Sumberdaya Iklim Pertanian Indonesia.

Ruberti I, Sessa G, Lucchetti S, Morelli G. 1991. A novel class of plant proteins containing a homeodomain with a closely linked leucine zipper motif. EMBO J. 10:1787-1791.

Scarpella E, Boot KJM, Rueb S, Meijer AH. 2002. The procambium specification gene *Oshox1* promotes polar auxin transport capacity and reduces its sensitivity towards inhibition. Plant Physiol. 130:1349-1360.

Scarpella E, Rueb S, Boot KJM, Hoge JHC, Meijer AH. 2000. A role for the rice homeobox gene *Oshox1* in provascular cell fate commitment. Development 127:3655-3669.

Schena M, Davis RW. 1992. HD-Zip protein members of *Arabidopsis* homeodomain protein superfamily. Proc. Natl. Acad. Sci. USA 89:3894-3898.

Schena M, Lloyd AM, Davis RW. 1993. The *HAT4* gene of *Arabidopsis* encodes a developmental regulator. Genes Dev. 7:367-379.

Scholz S, Lorz H, Lutticke S. 2001. Transposition of the maize transposable element *Ac* in barley (*Hordeum vulgare* L.). Mol. Gen. Genet. 264:653-661.

Sessa G, Carabelli M, Ruberti I. 1994. Identification of distinct families of HD-Zip proteins in *Arabidopsis thaliana*. In: Coruzzi G, editor. Berlin (Germany): Springer-Verlag. p 412-426.

Shimamoto K, Miyazaki C, Hashimoto H, Izawa T, Itoh K, Terada R, Inagaki Y, Iida S. 1993. Transactivation and stable integration of the maize transposable element *Ds*. Mol. Gen. Genet. 239:354-360.

Söderman E, Hjellström M, Fahleson J, Engström P. 1999. The HD-Zip gene *ATHB6* in *Arabidopsis* is expressed in developing leaves, roots and carpels and up-regulated by water deficit conditions. Plant Mol. Biol. 40:1073-1083.

Söderman E, Mattsson J, Engström P. 1996. The *Arabidopsis* homeobox gene *Athb-7* is induced by water deficit and by abscisic acid. Plant J. 10:375-381.

Springer P. 2000. Gene traps: tools for plant development and genomics. Plant Cell 12:1007-1020.

Steindler C, Matteucci A, Sessa G, Weimar T, Ohgishi M, Aoyama T, Morelli G, Ruberti I. 1999. Shade avoidance responses are mediated by the ATHB-2 HD-Zip protein, a negative regulator of gene expression. Development 126:4235-4245.

Sundaresan V. 1996. Horizontal spread of transposon mutagenesis: new uses for old elements. Trends Plant Sci. 1:184-190.

Takumi S, Murai K, Mori N, Nakamura C. 1999.Trans-activation of a maize *Ds* transposable element in transgenic wheat plants expressing the *Ac* transposase gene. Theor. Appl. Genet. 98:947-953.

Upadhyaya NM. 2006. Dissociation (*Ds*) constructs, mapped *Ds* launch pads and a transiently-expressed transposase system suitable for localized insertional mutagenesis in rice. Theor. Appl. Genet. 112:1326-1341.

USDA Foreign Agriculture Service. 2003. Indonesia grain and feed drought impact on rice production. www.fas.usda.gov.

USDA Foreign Agriculture Service. 2007. Indonesia grain and feed rice update. www.fas.usda.gov.

van Enckevort LE, Droc G, Piffanelli P, Greco R, Gagneur C, Weber C, Gonzalez VM, Cabot P, Fornara F, Berri S, Miro B, Lan P, Rafel M, Capell T, Puigdomenech P, Ouwerkerk PBF, Meijer AH, Pe' E, Colombo L, Christou P, Guiderdoni E, Pereira A. 2005. EU-OSTID: a collection of transposon insertional mutants for functional genomics in rice. Plant Mol. Biol. 59:99-110.

Weigel D, Ahn JH, Blázquez MA, Borevitz JO, Christensen SK, Fankhauser C, Ferrándiz C, Kardailsky I, Malancharuvil EJ, Neff MM, Nguyen JT, Sato S, Wang ZY, Xia Y, Dixon RA, Harrison RJ, Lamb CJ, Yanofsky MF, Chory J. 2000. Activation tagging in *Arabidopsis*. Plant Physiol. 122:1003-1013.

Yamaguchi-Shinozaki K, Kasuga M, Nakashima K, Sakuma Y, Shinwari ZK, Shinozaki K. 2002. Biological mechanisms of drought stress reponse. In: Iwanaga M, editor. JIRCAS Working Report 23: Genetic engineering of crop plants for abiotic stress. Wallingford (UK): CAB International. p 1-8.

Yamaguchi-Shinozaki K, Shinozaki K. 2001. Improving plant drought, salt and freezing tolerance by gene transfer of a single stress-inducible transcription factor. In: Rice biotechnology: improving yield, stress tolerance and grain quality. No. 236. Novartis Foundation Symposium. Chichester (UK): Wiley. p 176-189.

Yamaguchi-Shinozaki K, Shinozaki K. 2006. Transcriptional regulatory networks in cellular responses and tolerance to dehydration and cold stress. Annu. Rev. Plant Biol. 57:781-803.

Notes

Authors' addresses: I.H. Slamet-Loedin, S. Purwantomo, and S. Nugroho, Research Centre for Biotechnology, Indonesian Institute of Sciences (LIPI), Jl. Raya Bogor Km 46, Cibinong 16911, Indonesia; P.B.F. Ouwerkerk, Institute of Biology, Leiden University, Clusius Laboratory, P.O. Box 9505, 2300 RA Leiden, The Netherlands; R. Serraj, International Rice Research Institute (IRRI), DAPO Box 7777, Metro Manila, Philippines.

Acknowledgment: We thank Dr. A.H. Meijer, IBL Leiden University; Dr. A. Pereira; and Dr. N. Upadhyaya for scientific exchange and Dr. A. Estiati and Glenn Dimayuga (IRRI) for technical assistance. The gene-trap construct was provided by Dr. N. Upadhyaya (CSIRO, Australia). The activation-tag construct was obtained from Dr. A. Pereira (Virginia Bioinformatics Institute, Virginia; formerly at PRI, The Netherlands). This work was supported by the RUTI Programme-Indonesian Government for IHS (2002-04) and SN (2005-07), the SPIN-BIORIN Programme of the Royal Netherlands Academy of Arts and Sciences (KNAW) for SP and PBFO (99-BT-01), and by the EU FP6 INCO-MPC2 project CEDROME (INCO-CT-2005-015468) for PBFO.

Bioinformatics for drought resistance

Victor Jun Ulat, Samart Wanchana, Ramil Mauleon, and Richard Bruskiewich

This paper starts with a general discussion about genomics and bioinformatics research,[1] then proceeds to highlight the status of IRRI-hosted and externally available rice bioinformatics resources for crop genomics research, with a special emphasis on some of the special kinds of data we have pertinent to drought research. The paper will be a bit of a brainstorming exercise on where we could go in the future with bioinformatics to help answer questions on drought resistance. We conclude with a brief discussion about what community-level coordination to drought research can be achieved using available informatics-based collaboration.

Rice is the primary staple cereal for a majority of the world's population, in particular, in Asia. Rice is, by its nature, a hydrophilic crop species, to the extent that rice production is traditionally undertaken in flooded paddy fields and generally consumes a great deal of water in its production cycle. Unfortunately, intensification of rice production over recent decades, coupled with competition for available freshwater supplies by expanding urban populations and industry, has resulted in a relative scarcity of irrigation water for rice production. Amplifying this scarcity of water is the negative impact of climate change on rainfall patterns in many rainfed rice production areas, generally resulting in an increased frequency of drought. Thus, breeding for drought resistance and water-use efficiency in rice is a significant breeding objective for rice research teams worldwide.

[1]The first part of this paper is reproduced, with some modifications, from the paper "C_4 rice: brainstorming from bioinformaticians," published by the principal author, Richard Bruskiewich, in *Charting New Pathways to C_4 Rice*, edited by J.E. Sheehy, P.L. Mitchell, and B. Hardy, 2007, published by the International Rice Research Institute and World Scientific, Singapore. This duplication is merited in the sense that this overview provides important information for drought researchers not fully familiar with the field of genomics and bioinformatics, that the target audience for this second drought paper does not heavily overlap the first, and that two of the authors (Bruskiewich and Wanchana) of the current paper hold the copyright to the earlier paper.

Drought resistance is defined to include drought escape (DE) via a short life cycle or developmental plasticity, drought avoidance (DA) via enhanced water uptake and reduced water loss, and drought tolerance (DT) via osmotic adjustment (OA), antioxidant capacity, and desiccation tolerance (Yue et al 2006).

The genomics paradigm

Seeking a deeper understanding of possible genetic mechanisms for drought resistance in plants is a promising path toward systematic crop improvement to adapt rice to water-scarce conditions. Contemporary efforts in crop genomics are providing significant data and information about the number and response of genes to water deficiency, providing a critical tool for such understanding.

The genome (of an organism) is the entire DNA content of a cell, including all of the genes and all of the intergenic regions.[2] The field of genomics strives to characterize genome structure (by sequencing) and function (by experimental and bioinformatics analysis).

A genomic paradigm in modern germplasm-based crop research is that high-throughput characterization of the structure and function of genomes of diverse germplasm (including specialized collections such as mapping populations and mutants) can reveal by functional annotation (obtained from forward and reverse genetics, gene expression experiments, and similar techniques) many specific targets ("genes") important in many plant traits (Jung et al 2008). Such genes become the targets for high-throughput molecular characterization of germplasm to identify genetic polymorphism ("alleles") causally associated with a desirable phenotype. Genes and alleles with positive trait values can be subsequently applied toward crop improvement using gene transfer technologies, such as marker-assisted selection (MAS) and transgenic transformation.

What is the role of bioinformatics?

Bioinformatics is the discipline integrating the tools and techniques of mathematics and statistics, computing science, information technology, and the natural sciences to capture, analyze, store, represent, and disseminate biological information from a variety of sources to generate results that guide further experimentation or provide valuable insights into biology. As such, bioinformatics is a critical tool for genomics data management and analysis given an explosive growth in the volume and complexity of genomics data.[3]

---

[2]Definition borrowed, with slight modification, from Brown (1999), *Genomes*. Bios Scientific Publishers.

[3]Database repositories of one common biological data type, sequence data, alone have increased at least 10-fold over the past five years, to more than 100 billion (i.e., thousand million) base pairs. See www.ncbi.nlm.nih.gov/Genbank/index.html.

The application of informatics to crop research is not a new activity. The International Rice Research Institute (IRRI) already had an IBM computer storing and analyzing crop data back in the early days of the Institute. However, the onset in the 1990s of high-throughput genome sequencing of plant genomes, the application of new laboratory technologies for probing gene function, and the rapid expansion of the Internet and online databases established bioinformatics as an essential tool for modern crop research (Bruskiewich et al 2006b).

Bioinformatics research and development now offers a rich combination of protocols, tools, databases, and computing infrastructure that can be applied to help answer biological research questions, often at a significant savings of time and laboratory resources. Bioinformatics can integrate information across a diverse collection of crop data about germplasm, genotype, phenotype, cellular expression (of transcripts, proteins, and metabolites), growth characteristics, applied treatments, and environmental conditions. These data are cross-linked and interpreted to develop a complete picture of genome function from DNA sequence, through RNA and protein structures, into biochemical and cell structural characteristics interacting with the plant's environment, giving rise to the observable phenotype of the germplasm.

Of special interest here is the task of linking the impact of specific variations in germplasm DNA sequence ("genotype") upon all these components, resulting in the variation of morphology or behavior ("phenotype") that one observes among distinct germplasm varieties in the field. Significant progress is anticipated in the near future to characterize such molecular variation in a wide range of representative rice germplasm (McNally et al 2006; see www.oryzasnp.org).

Crop (plant) genome sequencing

The first genome sequence of a model plant, *Arabidopsis thaliana* (Meinke et al 1998), was completed in December of 2000 (Pennisi 2000).

Draft sequences of rice were completed in 2002 (Yu et al, Goff et al 2002), with completed chromosome 1 (Sasaki et al 2002) and chromosome 4 (Feng et al 2002). The finished genome was published in 2005 (IRGSP 2005), with the genome annotation available online (Ohyanagi et al 2006, Yuan et al 2003).

Genome sequencing projects for the gene-rich regions of larger crop genomes, such as maize and sorghum, are advancing rapidly and becoming publicly available (Chan et al 2006, Bedell et al 2005). Although the large size of some crop genomes may currently discourage complete genomic sequence characterization, the available crop genome sequence resources and associated functional genomics data already provide a rich foundation for comparative studies (Paterson et al 2005), including comparisons between wild species of rice (OMAP; www.omap.org; Wing et al 2005).

The crop research community is working hard to annotate the function of the resulting crop genomic sequences. For rice, the Rice Annotation Project database provides annotation curated by a team of experts (Itoh et al 2007, Rice Annotation Project 2008). The Institute for Genome Research (TIGR; now renamed the J. Craig Venter Institute, JCVI) provides complementary plant annotation resources across many genomes, including rice (Ouyang et al 2007; www.tigr.org/tdb/e2k1/osa1/). The Gramene comparative grasses resource (www.gramene.org) has worked steadily since 2000 to provide the plant community with curated genetic, genomics, and comparative genomics data for major crop species, including rice, maize, wheat, and many other plant (mainly grass) species (Liang et al 2008).

General observations about sequenced plant genomes

Gene identification is proving to be an especially challenging task in crop genomes. Early estimates of gene count suggested a larger complement of rice (monocot)-specific genes; however, this has subsequently been subject to considerable debate (Wyrwicz et al 2004). In addition, in recent years, increased evidence for noncoding microRNA fragments has complicated the picture (Mallory and Vaucheret 2004). The nature of whole-genome expression is a story remaining to be fully told.

Some things are clear, however. There is significant evidence for ancient whole and segmental genomic duplications, followed by gene loss or divergence (Yu et al 2005). Specific gene families often exhibit differential expansion in different clades (Rabinowicz et al 1999). Most eukaryote (especially plant) genomes are found to be replete with repetitive transposable elements. These latter sequence elements stir up the pot quite a bit (Morgante 2006):

- **Recombination hot spots:** Due to their high copy number and sequence similarity to one another, tandem copies of transposons may be susceptible targets for homologous recombination, creating unequal crossing over, thus creating tandem genome sequence duplications. Genes within such duplicated regions can become paralogs to the original gene, by divergent evolution. Homologous recombination also generates significant structural rearrangements. These may be the norm rather than the exception, even within a species (at the micro-synteny level) and most certainly between species.
- **Knocking out genes:** Transposons can insert to create new gene structures and may even introduce novel genome regulatory signals such as *cis*-element binding sites and chromatin matrix attachment sites.

Novel genes can arise from transposable element insertion in animals and sometimes such novel genes can have a DNA binding character with pleiotropic regulatory impact within the genome (Cordaux et al 2006, Jordan 2006). Although equivalent examples in plant genomes remain to be discovered, some observations suggest the possibility that this process also occurs in plants (Elrouby and Bureau 2000, 2001).

Framework for gene discovery

Accurate and complete functional annotation of genomes remains a daunting task (Jung et al 2008). Efforts to characterize gene function are largely driven by a paradigm of "intersecting evidence" using experimental results about genes from

- **Position:** Quantitative trait locus results link segregating chromosome markers with plant traits.

- **Function:** Rice genome annotation with literature-documented biochemical or sequence homology analysis, combined with genetic dissection of rice mutants, can reveal the role of specified genes in specific biological processes.
- **Expression:** Gene expression experiments at the transcript, protein, and metabolic level can associate the expression of specific genes with specific processes, both constitutively and under conditional treatments.
- **Selection:** Analysis of genetic resource collections (through association genetics) and bulk population selection experiments can identify genomic regions in linkage disequilibrium with traits or alleles conserved under selection pressure in molecular breeding. This again provides additional support for the role of specified loci and alleles in pertinent biological processes and phenotypes.
- **Crop models:** This is a relatively new source of evidence, but it is expected that linking gene systems information with whole-crop models will help support or refute gene candidacy in specific field-level processes, such as drought stress response.

A key objective of contemporary crop bioinformatics research and development is to provide effective information systems for capturing and integrating such intersecting experimental evidence, allowing researchers to take intersection sets of gene candidates, to narrow their focus down to a few key candidate gene loci or alleles for further cost-effective laboratory or field validation.

Perspectives in modern biology

In contemplating the task of engineering drought resistance into rice using genomics, one can view the challenge from various perspectives:

- Evolutionary perspective
- Molecular biology perspective
- Systems biology perspective

The evolutionary perspective

> *"...I submit that all these remarkable findings make sense in the light of evolution: they are nonsense otherwise. ... Seen in the light of evolution, biology is, perhaps, intellectually the most satisfying and inspiring science. Without that light it becomes a pile of sundry facts, some of them interesting or curious but making no meaningful picture as a whole...."*[4]

[4]Quote attributed to the American evolutionary biologist Theodosius Dobzhansky (1900-75) in his article in the *American Biology Teacher*, March 1973.

Genomes can use only what DNA is in front of them at any one time, or have DNA transferred to them from outside by reproductive or infection processes. Genome changes are generally incremental (single or a small number of mutations or rearrangements) except under rare circumstances (e.g., viable whole-genome duplications or hybridizations, hyperactive infection of transposons). Some genome mutations can be irreversible (e.g., chromosomal rearrangements, insertions and deletions of sequence), driving a unidirectional genetic drift process.

Incremental steps (by genome mutation) that increase reproductive fitness may or may not get fixed in a species (population). Intermediates need only be fit enough to survive (reproductively) until they make the transition to a state of significant competitive advantage for a given environment. Nature is patient: it has millions of years to experiment. Evolutionary selection acts on large genetically diverse populations over many generations.

Of particular interest to us is the rapid regulatory evolution possible in some large gene families such as *Myb* (Rabinowicz et al 1999, Dias et al 2003), which may play a key role in modulating key plant metabolic pathways. In general, existing gene products and processes could be exapted[5] at any time to play a novel role in plant function, including drought resistance. This suggests that plants can easily evolve a multitude of diverse strategies to cope with drought (see below).

The molecular biology perspective

Figure 1 is a basic road map to the molecular scope of biology. It is beyond the scope of this paper to dwell upon it—it is self-explanatory. The take-home message is that a full understanding of any biological system requires a complete cataloguing of the relationship between sequence variation at the DNA level and the observed functional variation at the molecular level: in terms of the stability and function of molecular structure, biochemical activity, and gene (product) regulation.

The systems biology perspective

Plants, like all organisms, which are often quite aptly called "living systems," cannot simply be reduced to their component parts (genes, molecules). Rather, all such parts synergistically combine together into complex systems.

It is not surprising, therefore, that, after taking organisms apart by genome sequencing and biochemical analysis, scientists are driven to try to put those systems back together, using bioinformatics, in order to gain a more realistic perspective on the interactions and emergent properties inherent in living organisms. This "systems biology" perspective may embrace a range of systems:

[5]An *exaptation* is a biological *adaptation* in which the biological function currently performed by the adaptation was not the function performed while the adaptation evolved under earlier pressures of *natural selection*.

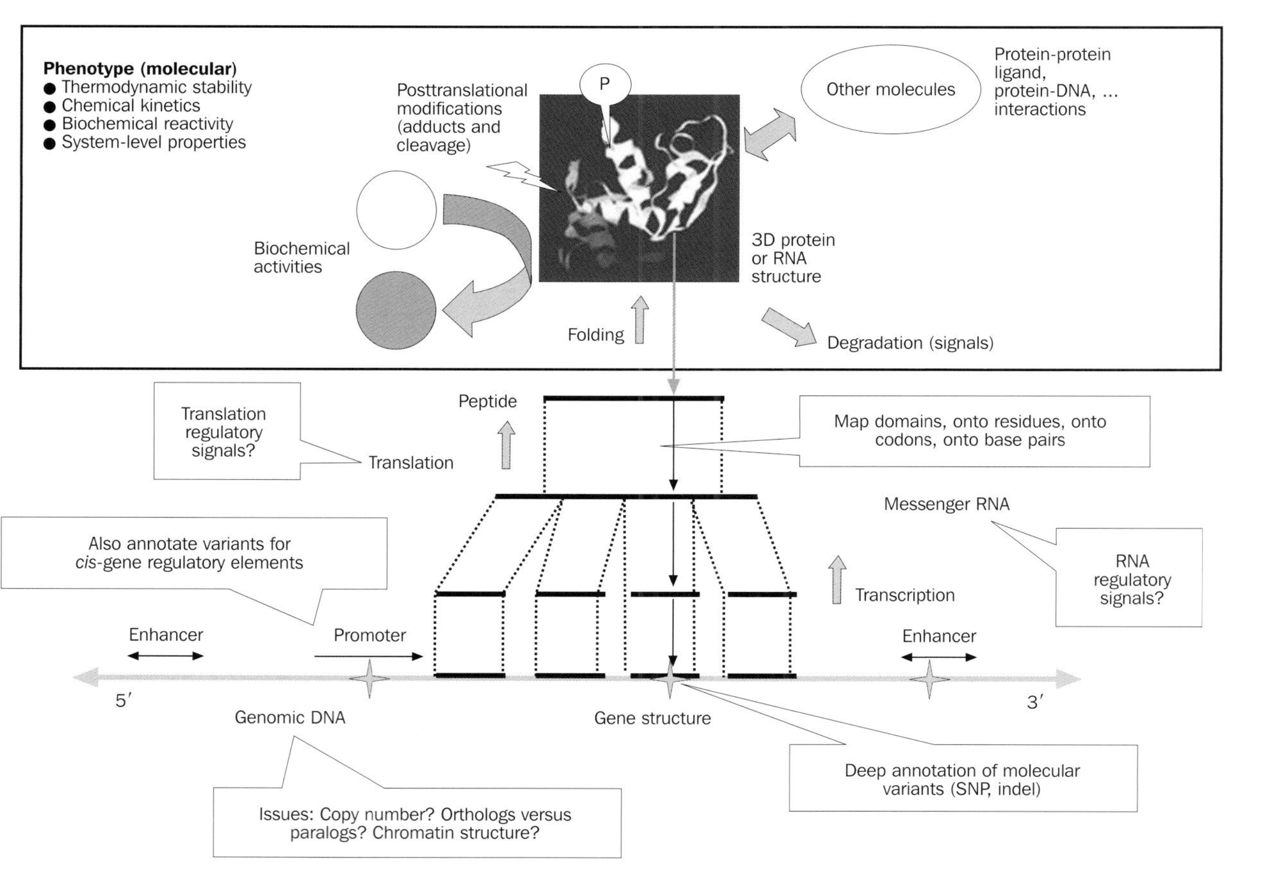

Fig. 1. From genome to molecular phenotype through the central dogma of biology (DNA to RNA to protein).

- **Regulatory networks:** signal transduction, genetic regulation
- **Metabolic networks:** biochemical fluxes
- **Structural networks:** morphology (e.g., chloroplast positioning)
- **Physiological systems:** whole-plant development and function

It is pertinent to note that evolution acts on whole living systems, not really just their component parts. For systems too, optimization is a global, not local, process.

Bioinformatics for drought resistance at IRRI

In this section, we will mention a few of the bioinformatics activities and resources at IRRI that have evolved over the past several years for crop research in general and for drought resistance research in particular.

Drought-stressed panicle library

Some years ago, IRRI commissioned the development and sequencing of an approximately 10,000-clone drought-stressed panicle expressed sequence tag (EST) library. The sequence annotation for this library is publicly available online at www.iris.irri.org/clone. Among the information available are sequence, annotation, and trace files for each clone. This library was subsequently arrayed on microarray slides using an IRRI-hosted printing facility and various hybridization experiments were undertaken. An experiment using these panicle cDNA microarray slides is a case study presented involving hybridization of field-grown panicle samples from drought-tolerant and susceptible rice germplasm. The results indicated contrasting drought responses among both upland- and lowland-adapted cultivars and also between traditional and improved upland types (Kathiresan et al 2006).

Drought data integration

Quantitative trait locus (QTL) position data are also being integrated with gene data to assist researchers in managing and visualizing information from various experiments. Open-source software from the Generic Model Organism Database (GMOD, www.gmod.org) consortium, in particular the Genome Browser (GBrowse), is being used for this purpose. The Comparative Mapping Tool (CMAP) is also applied to the storage, comparison, and visualization of genetic and physical maps. These databases are currently hosted internally for IRRI only. An example of data integration is an anchoring of rice QTL maps to the rice physical map in order to find the candidate rice genes underlying a drought-tolerance QTL flanking RM212 to RM319 on chromosome 1 (Wang et al 2005). A set of candidate genes of known or inferred function were identified in this region using rice genome annotation. Published literature supports the candidacy of some of these genes in drought stress response.

Generation Challenge Program ortholog stress gene catalog

Comparative biology across multiple crop species is a key strategy for the identification of stress-responsive gene loci and their corresponding alleles of high agronomic value. Critical to the task of comparative biology is the elucidation of evolutionarily conserved gene orthology relationships across species and related paralogy relationships within a gene family. Such ortholog and paralog gene loci almost invariably share common gene functions; thus, important inferences of comparative biology may be possible once these relationships are clearly defined.

With the wealth of genome sequencing data, applications of bioinformatics regarding phylogenomic analysis have become possible. Phylogenomics is a methodology for assigning precise functional annotations to proteins encoded in a number of recently sequenced genomes using the evolutionary history of those proteins as captured by phylogenetic trees. This method combines phylogenetic tree construction, integration of experimental data, and differentiation of orthologs and paralogs (Sjölander 2004).

At present, several candidate genes for drought resistance have been identified and proven by transgenic methods (Parry et al 2005). Moreover, many drought stress–responsive genes were monitored by such a high-throughput method as microarray (Rabbani et al 2003, Seki et al 2002). However, the functional characterization of the genes responsible for drought and/or other abiotic stresses had been elucidated in a small range of organisms or only in model species such as rice and *Arabidopsis*. To transfer information among species, with a comparative genomics perspective, the phylogenomics method can be applied to help in the understanding of conserved gene functions and for application in breeding for stress resistance.

The Generation Challenge Program (GCP; www.generationcp.org) is funding a gene ortholog catalog project with the goal of developing a public comparative gene catalog, user interfaces, and data integration protocols meeting GCP research needs in comparative biology. The rationale is to identify stress-tolerant genes and their corresponding alleles of high agronomic value for application in breeding for stress tolerance. The task objectives are to (1) specify a suitable comparative gene catalog database using publicly available schema and content (ontology) standards, (2) curate, and add annotation to, a comprehensive set of plant stress gene information for the gene catalog, (3) develop Internet (Web services) cross-linkages to existing public plant database data sources, (4) develop suitable query and visualization interfaces for the resulting gene catalog, and (5) integrate with other GCP data sets, in particular, comparative gene expression data.

The project has established an online comparative stress gene catalog using the GMOD "Chado" database schema. The comparative gene catalog will be seamlessly connected to the tools developed for the GCP platform (http://pantheon.generationcp.org). The comparative gene catalog database was established and loaded with some initial data sets online at http://dayhoff.generationcp.org.

The data curation tasks behind the ortholog gene catalog are based on a phylogenomic analysis strategy as described by Sjölander (2004). The procedure consists of compiling genes, obtaining homologous sequences, and inferring the phylogenetic tree to discriminate between orthologous and paralogous genes.

So far, 500 genes collected from the literature and GCP partners have been compiled into the stress inventory tool. Based on this gene collection, 180 gene families have been annotated, validated, and loaded into the stress-gene catalog database.

The drought-responsive gene families currently available in the database include drought-responsive element binding factors (DREB) (Fig. 2); aquaporin; cell-wall and vacuolar invertases; sucrose synthase; and sucrose-phosphate synthase. These candidate genes have been well characterized by several experiments and their functions have been proven to be involved in pathways of drought-stress resistance.

Other bioinformatics data sets

The IRRI Plant Molecular Biology Laboratory undertakes proteomics experiments on germplasm subject to drought-resistance treatments. A pilot metabolomics experiment profiling the contents of pollen samples from two varieties (IR64 and Moroberekan) and two treatments (well-watered and drought-stressed) generated an interesting data set with some clues to the kinds of differential responses that account for pollen viability under stress.

Other pertinent tools and initiatives

Researchers wishing to apply genomics and bioinformatics to the problem of creating drought-resistant rice have various additional public resources to draw upon.

One of the key tasks of crop bioinformatics is the systematic documentation of germplasm and field data for germplasm. This is not a new challenge; therefore, not too surprisingly, suitable databases and software already exist to accommodate this need.

The International Crop Information System (ICIS; www.icis.cgiar.org; Portugal et al 2007, Fox and Skovmand 1994) is one example of such a computerized database system and suite of tools for general integrated management and use of genealogy, nomenclature, evaluation, and characterization data for a wide range of crops. One useful feature of ICIS is that it is a public open-source software collaboration involving several Consultative Group on International Agricultural Research (CGIAR) centers and non-CG partners in Australia (University of Queensland, University of Western Australia), Canada (Agriculture and Agri-Foods Canada), the Netherlands (a private seed company, Nunhems), and Singapore (Bayer hybrid rice seed division). Thus, the underlying computer code is free for the asking, hosted on a software project development site for agricultural projects called "CropForge" (http://cropforge.org), and it is being customized and enhanced by a wide public community of developers and end-users.

ICIS is successfully deployed for a number of nonrice crops: wheat, barley, maize, common beans, chickpea, cowpea, sugarcane, potato, sweet potato, and several vegetable crops. From the perspective of drought resistance for rice, the International Rice Information System (IRIS; www.iris.irri.org; McLaren et al 2005, Bruskiewich et al 2003) is of specific interest as the largest curated public rice installation of ICIS.

In addition to ICIS databases, many additional excellent public online plant and crop databases are available. A partial inventory of such resources is outlined in Table 1.

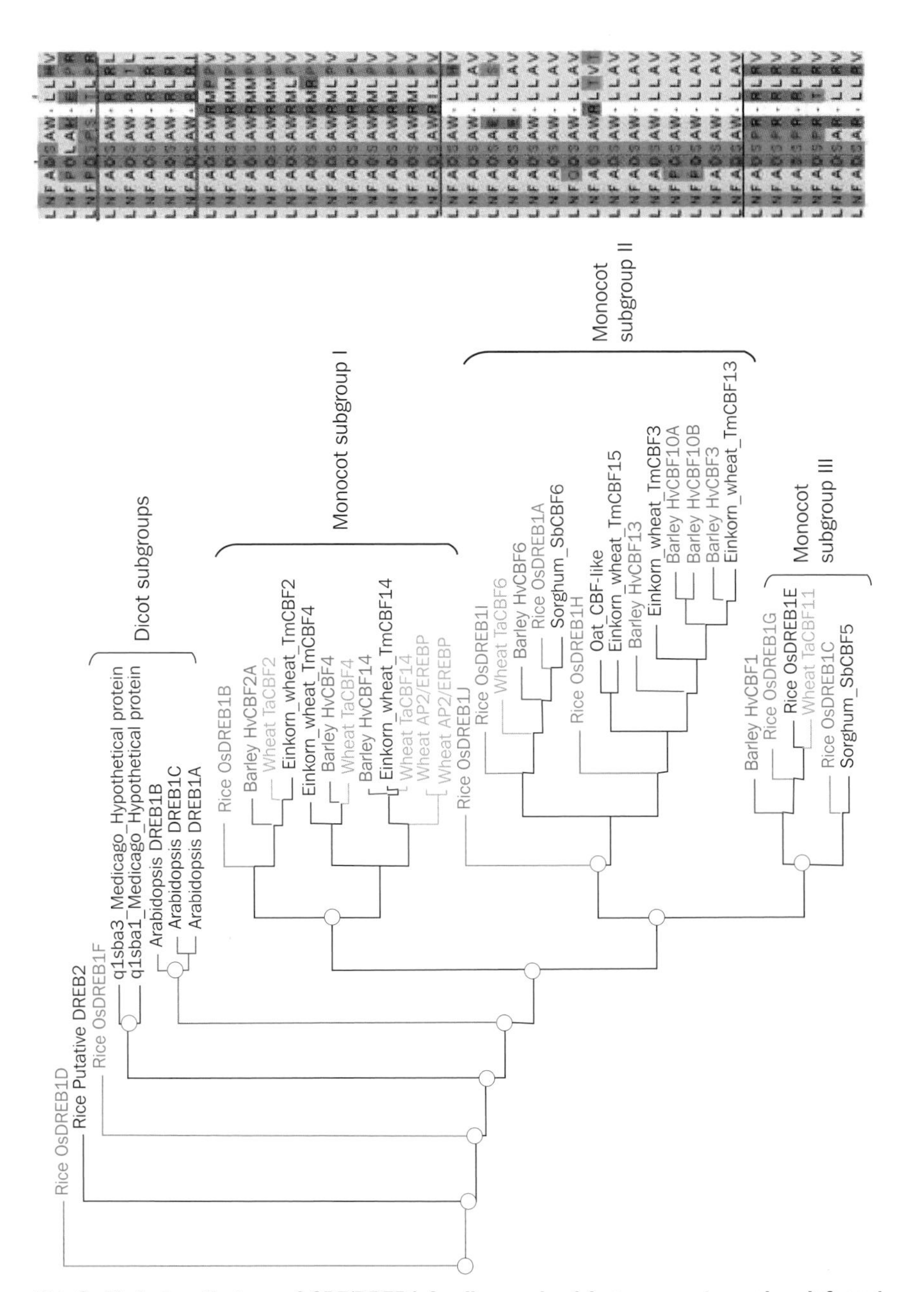

Fig. 2. Phylogenetic tree of CBF/DREB1 family, emphasizing monocot species, inferred by a phylogenomics analysis. CBF/DREB1 genes among monocots can be classified into three subgroups that were separated from the dicot subgroups and the three other rice putative DREBs, OsDREB1D, OsDREB1F, and putative OsDREB2. The multiple sequence alignment representing a portion of the conserved region in each subgroup is shown.

Table 1. Partial inventory of publicly accessible plant databases.

Database	Organism	URL
TAIR	*Arabidopsis thaliana*	www.arabidopsis.org
NASC	*Arabidopsis thaliana*	www.nasc.org
MATDB	*Arabidopsis thaliana*	www.mips.gsf.de
RAPDB	*Oryza sativa*	http://rapdb.lab.nig.ac.jp
Gramene	Comparative grasses	www.gramene.org
TIGR Plant Genomes	*Oryza sativa*	www.tigr.org/plantProjects.shtml
MaizeGDB	Maize	www.maizegdb.org
Barleybase	Barley (microarray data)	www.barleybase.org
SINGER	Plant genetic resources	www.singer.cgiar.org
GRIN	Plant genetic resources	www.grin.usda.gov

International Rice Functional Genomics Consortium

In the area of rice functional genomics, several resources are available. In January 2003, representatives of 18 institutions from 10 countries and two international agricultural research centers established an International Rice Functional Genomics Consortium (IRFGC; www.iris.irri.org/IRFGC). The proposed mandate of the IRFGC is to coordinate research in the postsequencing "functional genomics" era by exploring ways to consolidate international rice functional genomics resources and build common strategies.

The Consortium is striving to encourage sharing and consolidation of several useful resources for gene characterization of rice in the areas of genomics stocks, expression arrays, proteomics, and bioinformatics. For example, one specific bioinformatics resource under development is a "Rice MOBY Network," which will provide easier integrated access to globally distributed information of the Consortium.

Generation Challenge Program

A second pertinent international initiative is the Generation Challenge Program (GCP; www.generationcp.org), an international research consortium established in 2003, with a projected lifetime of 10 years, hosted by the CGIAR and involving global partners from international agricultural research centers, advanced research institutes, and national agricultural research and extension systems. The research themes of the GCP are directed to crop improvement through genomics and comparative biology across species, as well as molecular and phenotypic characterization of genetic resources to discover valuable alleles for crop improvement.

One major research and development subprogram of the GCP focuses on crop informatics. This bioinformatics subprogram is striving to develop global public standards for crop information management, including a comprehensive crop scientific domain model (see http://pantheon.generationcp.org; Bruskiewich et al 2006a, 2008) and a platform of tools for accessing and analyzing information available from the globally networked databases of GCP partners and other pertinent external data sources. All of the software technology of the GCP, such as ICIS, is open-source and will be available on CropForge. This platform and network is envisioned to specifically empower crop researchers to perform the kinds of integrated genomics and related data management and analyses discussed in this paper.

Of particular interest to drought research is a comparative gene ortholog catalog being developed to facilitate integration of knowledge about stress-responsive genes across crops (Wanchana et al 2008; see http://dayhoff.generationcp.org). An ICRISAT database of annotated orthologs from crop abiotic stress transcripts is also published (Balaji et al 2006; http://intranet.icrisat.org/gt1/tog/homepage.htm).

IRRI-CIMMYT Alliance in research informatics

Also pertinent to the task of engineering resistance into rice is an alliance between IRRI and CIMMYT in the area of research informatics. The resulting Crop Research Informatics Laboratory has as one of its goals the development of a comparative crop resource for rice, maize, and wheat. In collaboration with other international teams, this resource may be extended in the future to other cereal species such as barley, sorghum, finger millet, and others, which exhibit different degrees of drought resistance. We may be able to further learn from other plants what key strategies are required to achieve a drought-resistant rice. Even without identifying drought resistance in other species, comparative studies will help elucidate common pathways with greater clarity so that studies of stress processes will be more efficient.

Some brainstorming

What other (bio)informatics might prove useful for the task of developing drought resistance in rice? We will cast our net a bit wider here with several ideas (not presented in any particular order of precedence).

First, characterization of (all) drought-prone environments is an essential initial step in framing the problem of drought resistance; because not all environments are alike, recommendations for germplasm will not be monolithic (see http://drought.irri.org). Ideally, target environments should be geo-reference-indexed by toposequences, soil types and structure, cropping systems and technologies, climate profiles, and varieties being planted. Based on key criteria arising out of this classification, a kind of NARES-accessible decision-making system for drought could be developed to customize varieties and cropping systems to suit local conditions. An inventory of suitable germplasm and genotypes (genes) beneficial to particular environments or environmental characteristics could also be compiled. This characterization of environments is under way internationally with such efforts as the FAO "Climate Impact on Agriculture" initiative (www.fao.org/nr/climpag/index_en.asp) but these are not specific to rice and do not yet index production environments to more detailed information on germplasm or cropping strategies. Ideally, such information could be visualized within powerful graphical map tools such as Google maps (http://maps.google.com/).

A second idea is to leverage more nonrice genomics emphasizing host-pest interactions in, say, the roots. Nematode biology might be especially pertinent. Research could exploit available comparative nematode genomics project information from the sequenced genomes of *Caenorhabditis elegans* and other nematode species (Coghlan et al 2006). Information on several fungal genomes might also identify points for targeted research and intervention concerning pathogen complications relating to drought resistance.

Third, informatics tools now permit the dynamic compilation and maintenance of knowledge online. A kind of integrated encyclopedia for drought biology driven by "facilitated" community-wide expert curation could be compiled (Giles 2007, Salzberg 2007). This would serve as a longer-term community memory complementing scientific literature, for the current and next generation of researchers. Some of the encyclopedia's features could include

- Multilevel integration of information about phenotypes, genes/genotypes, germplasm, cropping systems, and associated drought environments.
- Online "expert editor" digests of drought literature for particular topics (e.g., gene loci, systems, etc.).
- Cross-linked drought data sets.
- A crop and simulation model repository.
- Thematic classification of information: for example, an integrated database of root biology, including developmental processes for root architecture, root-soil interactions, and root biotic stress (host-pest) interactions.

Fourth, the proposed use of infrared thermal (IRT) imaging for drought phenotyping suggests the application of high-throughput computer analysis of IRT images of fields and individual plants (Leinonen et al 2006).

Fifth, systems biology models and simulation could bear significant fruit in a complex trait such as drought resistance. The targets of study could include developmental (morphological) and hormonal signaling systems modeled in space and time. This kind of thing is starting to be tackled for specific systems such as roots (Sharp et al 2004). Most certainly, over time, more detailed modeling of gene product biological expression and activity based on molecular properties should be possible:

- **Gene expression:** Document the properties of transcription factor family members differing in DNA binding, protein-protein interactions, and the thermodynamics of complex formation.
- **Gene product activity:** Compile an inventory of control points for regulation, for each gene product, to highlight possible sites for the impact of genotype (sequence) variability.

Summary

Introducing drought resistance into rice is a significant challenge. Comparative genomics and bioinformatics can help identify molecular targets for modification or transfer into pertinent genetic backgrounds to achieve the biochemical and morphological changes required for the goal.

In addition, bioinformaticians could perhaps also help in a practical way by constructing consolidated online repositories of biological data and information about drought resistance to facilitate collaboration and sharing of knowledge. Thus, one final question is: What would a global rice drought resistance research community require from its informatics and communications team? Perhaps common community resources such as

- **Communication channels:** e-newsletter, list server, Web, videoconferences, etc.
- **Shared databases and networks:** Integrated online encyclopedia idea, with research digests and proceedings; cross-linked drought research databases managed by each member of the community.

References

Balaji J, Crouch JH, Petite PV, Hoisington DA. 2006. A database of annotated tentative orthologs from crop abiotic stress transcripts. Bioinformation 1:225-227.

Bedell JA, Budiman MA, Nunberg A, Citek RW, Robbins D, Jones J, Flick E, Rohlfing T, Fries J, Bradford K, McMenamy J, Smith M, Holeman H, Roe BA, Wiley G, Korf IF, Rabinowicz PD, Lakey N, McCombie WR, Jeddeloh JA, Martienssen RA. 2005. Sorghum genome sequencing by methylation filtration. PLoS Biol. 3(1):e13.

Bruskiewich R, Cosico A, Eusebio W, Portugal A, Ramos LR, Reyes T, Sallan MAB, Ulat VJM, Wang X, McNally KL, Sackville Hamilton R, McLaren CR. 2003. Linking genotype to phenotype: the International Rice Information System (IRIS). Bioinformatics 19(Suppl.1):i63-i65.

Bruskiewich R, Davenport G, Hazekamp T, Metz T, Ruiz M, Simon R, Takeya M, Lee J, Senger M, McLaren CG, van Hintum T. 2006a. The Generation Challenge Programme (GCP): standards for crop data. OMICS 10(2):215-219.

Bruskiewich R, Metz T, McLaren CG. 2006b. Bioinformatics and crop information systems in rice research. Int. Rice Res. Notes 31(1):5-12.

Bruskiewich R, Senger M, Davenport G, Ruiz M, Rouard M, et al. 2008. The Generation Challenge Programme platform: semantic standards and workbench for crop science. Int. J. Plant Genomics, vol. 2008, Article ID 369601, 6 pages. doi:10.1155/2008/369601.

Chan AP, Pertea G, Cheung F, Lee D, Zheng L, Whitelaw C, Pontaroli AC, SanMiguel P, Yuan Y, Bennetzen J, Barbazuk WB, Quackenbush J, Rabinowicz PD. 2006. The TIGR maize database. Nucl. Acids Res. 34(Database issue):D771-776.

Coghlan A, Stajich JE, Harris TW. 2006. Comparative genomics in *C. elegans, C. briggsae*, and other *Caenorhabditis* species. Methods Mol. Biol. 351:13-29.

Cordaux R, Udit S, Batzer MA, Feschotte C. 2006. Birth of a chimeric primate gene by capture of the transposase gene from a mobile element. Proc. Natl. Acad. Sci. USA 103(21):8101-8106.

Dias AP, Braun EL, McMullen MD, Grotewold E. 2003. Recently duplicated maize R2R3 *Myb* genes provide evidence for distinct mechanisms of evolutionary divergence after duplication. Plant Physiol. 131(2):610-620.

Elrouby N, Bureau TE. 2001. A novel hybrid open reading frame formed by multiple cellular gene transductions by a plant long terminal repeat retroelement. J. Biol. Chem. 276(45):41963-41968.

Elrouby N, Bureau TE. 2000. Molecular characterization of the *Abp1* 5.-flanking region in maize and the Teosintes. Plant Physiol. 124:369-377.

Feng Q, Zhang Y, Hao P, Wang S, Fu G, et al. 2002. Sequence and analysis of rice chromosome 4. Nature 420(6913):316-320.

Fox PN, Skovmand B. 1996. The International Crop Information System (ICIS): connects genebank to breeder to farmer's field. In: Cooper M, Hammer GL, editors. Plant adaptation and crop improvement. Wallingford (UK): CAB International. p 317-326.

Giles J. 2007. Key biology databases go wiki. Nature 445(7129):691.

Goff SA, Ricke D, Lan TH, Presting G, Wang R, et al. 2002. A draft sequence of the rice genome (*Oryza sativa* L. ssp. japonica). Science 296(5565):92-100.

IRGSP (International Rice Genome Sequencing Project). 2005. The map-based sequence of the rice genome. Nature 436:793-800.

Itoh T, Tanaka T, Barrero RA, Yamasaki C, Fujii Y, et al. 2007. Curated genome annotation of *Oryza sativa* ssp. japonica and comparative genome analysis with *Arabidopsis thaliana*. Genome Res. 17:175-183.

Jordan IK. 2006. Evolutionary tinkering with transposable elements. Proc. Natl. Acad. Sci. USA 103(21):7941-7942.

Jung KH, An G, Ronald PC. 2008. Towards a better bowl of rice: assigning function to tens of thousands of rice genes. Nat. Rev. Genet. 9(2):91-101.

Kathiresan A, Lafitte HR, Chen J, Mansueto L, Bruskiewich R, Bennett J. 2006. Gene expression microarrays and their application in drought stress research. Field Crops Res. 97:101-110.

Leinonen I, Grant OM, Tagliavia CPP, Chaves MM, Jones HG. 2006. Estimating stomatal conductance with thermal imagery. Plant Cell Environ. 29:1508-1518.

Liang C, Jaiswal P, Hebbard C, Avraham S, Buckler ES, Casstevens T, Hurwitz B, McCouch S, Ni J, Pujar A, Ravenscroft D, Ren L, Spooner W, Tecle I, Thomason J, Tung C-W, Wei X, Yap I, Youens-Clark K, Ware D, Stein L. 2008. Gramene: a growing plant comparative genomics resource. Nucl. Acids Res. 36:D947-953.

Mallory AC, Vaucheret H. 2004. MicroRNAs: something important between the genes. Curr. Opin. Plant Biol. 7:120-125

McLaren CG, Bruskiewich RM, Portugal AM, Cosico AB. 2005. The International Rice Information System: a platform for meta-analysis of rice crop data. Plant Physiol. 139(2):637-642.

McNally KL, Bruskiewich R, Mackill DC, Buell CR, Leach JE, Leung H. 2006. Sequencing multiple and diverse rice varieties: connecting whole-genome variation with phenotypes. Plant Physiol. 141:1-6.

Meinke DW, Cherry MJ, Dean C, Rounsley SD, Koornneef M. 1998. *Arabidopsis thaliana*: a model plant for genome analysis. Science 282:662-682.

Morgante M. 2006. Plant genome organisation and diversity: the year of the junk! Curr. Opin. Biotechnol. 17(2):168-173.

Ohyanagi H, Tanaka T, Sakai H, Shigemoto Y, Yamaguchi K, Habara T, Fujii Y, Antonio BA, Nagamura Y, Imanishi T, Ikeo K, Itoh T, Gojobori T, Sasaki T. 2006. The Rice Annotation Project Database (RAP-DB): hub for *Oryza sativa* ssp. japonica genome information. Nucl. Acids Res. 34(Database issue):D741-744.

Ouyang S, Zhu W, Hamilton J, Lin H, Campbell M, Childs K, Thibaud-Nissen F, Malek RL, Lee Y, Zheng L, Orvis J, Haas B, Wortman J, Buell CR. 2007. The TIGR Rice Genome Annotation Resource: improvements and new features. NAR 35 Database Issue:D846-851.

Parry MAJ, Flexas J, Medrano H. 2005. Prospects for crop production under drought: research priorities and future directions. Ann. Appl. Biol. 147:211-226.

Paterson AH, Freeling M, Sasaki T. 2005. Grains of knowledge: genomics of model cereals. Genome Res. 15(12):1643-1650.

Pennisi E. 2000. Plants join the genome sequencing bandwagon. Science 290:2054-2055.

Portugal A, Balachandra R, Metz T, Bruskiewich R, McLaren G. 2007. International Crop Information System for germplasm data management. In: Edwards D, editor. Plant bioinformatics: methods and protocols. Totowa, N.J. (USA): Humana Press. Methods Mol. Biol. 406:459-472.

Rabbani MA, Maruyama K, Abe H, Khan MA, Katsura K, Ito Y, Yoshiwara K, Seki M, Shinozaki K, Yamaguchi-Shinozaki K. 2003. Monitoring expression profiles of rice genes under cold, drought, and high-salinity stresses and abscisic acid application using cDNA microarray and RNA gel-blot analysis. Plant Physiol. 133:1755-1767.

Rabinowicz PD, Braun EL, Wolfe AD, Bowen B, Grotewold E. 1999. Maize R2R3 *Myb* genes: sequence analysis reveals amplification in the higher plants. Genetics 153(1):427-444.

Rice Annotation Project. 2008. The Rice Annotation Project Database (RAP-DB): 2008 update. Nucl. Acids Res. doi:10.1093/nar/gkm978.

Salzberg SL. 2007. Genome re-annotation: a wiki solution? Genome Biol. 8(1):102.

Sasaki T, Matsumoto T, Yamamoto K, Sakata K, Baba T, et al. 2002. The genome sequence and structure of rice chromosome 1. Nature 420(6913):312-316.

Seki M, Narusaka M, Ishida J, Nanjo T, Fujita M, Oono Y, Kamiya A, Nakajima M, Enju A, Sakurai T, Satou M, Akiyama K, Taji T, Yamaguchi-Shinozaki K, Carninci P, Kawai J, Hayashizaki Y, Shinosaki K. 2002. Monitoring the expression profiles of 7000 *Arabidopsis* genes under drought, cold and high-salinity stresses using a full-length cDNA microarray. Plant J. 31(3):279-292.

Sharp RE, Poroyko V, Hejlek LG, Spollen WG, Springer GK, Bohnert HJ, Nguyen HT. 2004. Root growth maintenance during water deficits: physiology to functional genomics. J. Exp. Bot. 55(407):2343-2351.

Sjölander K. 2004. Phylogenomic inference of protein molecular function: advances and challenges. Bioinformatics 20(2):170-179.

Wanchana S, Thongjuea S, Ulat VJ, Anacleto M, Mauleon R, Conte M, Rouard M, Ruiz M, Krishnamurthy N, Sjolander K, van Hintum T, Bruskiewich RM. 2008. The Generation Challenge Programme comparative plant stress-responsive gene catalogue. Nucl. Acids Res. 36:D943-946.

Wang X-S, Zhu J, Mansueto L, Bruskiewich R. 2005. Identification of candidate genes for drought stress tolerance in rice by the integration of a genetic (QTL) map with the rice genome physical map. J. Zhejiang Univ. Sci. B:382-388.

Wing RA, Ammiraju JS, Luo M, Kim H, Yu Y, Kudrna D, Goicoechea JL, Wang W, Nelson W, Rao K, Brar D, Mackill DJ, Han B, Soderlund C, Stein L, SanMiguel P, Jackson S. 2005. The *Oryza* map alignment project: the golden path to unlocking the genetic potential of wild rice species. Plant Mol. Biol. 59(1):53-62.
Wyrwicz LS, Grotthuss M, Pas J, Rychlewski, Leszek, Kikuchi S. 2004. How unique is the rice transcriptome? Science 303:168b.
Yu J, Hu S, Wang J, Wong GK, Li S, et al. 2002. A draft sequence of the rice genome (*Oryza sativa* L. ssp. indica). Science 296(5565):79-92.
Yu J, Wang J, Lin W, Li S, Li H, et al. 2005. The genomes of *Oryza sativa*: a history of duplications. PLoS Biol. 3(2):e38.
Yuan Q, Ouyang S, Liu J, Suh B, Cheung F, Sultana R, Lee D, Quackenbush J, Buell CR. 2003. The TIGR rice genome annotation resource: annotating the rice genome and creating resources for plant biologists. Nucl. Acids Res. 31(1):229-233.
Yue B, Xue W, Xiong L, Yu X, Luo L, Cui K, Jin D, Xing Y, Zhang Q. 2006. Genetic basis of drought resistance at reproductive stage in rice: separation of drought tolerance from drought avoidance. Genetics 172(2):1213-1228.

Notes

Authors' address: Crop Research Informatics Laboratory, International Rice Research Institute, DAPO Box 7777, Metro Manila, Philippines.
Acknowledgments: S. Wanchana and M. Mauleon are supported by postdoctoral fellowships from the Generation Challenge Program.

Phylogenomic analysis by Samart Wanchana was aided by analysis tools from K. Sjölander (University of California, Berkeley, USA; http://phylogenomics.berkeley.edu). Ideas were discussed relating to transposable element genome evolution and regulatory impact with T. Bureau (McGill University, Canada) and his student, D. Hoen.

Bioinformatics work relating to the annotation of the drought-stressed panicle library and QTL analysis was undertaken by various Rockefeller Foundation–funded scholars: Locedie Mansueto, Xusheng Wang, Jutharat Prayongsap, Ravindra Babu, Naveen Sharma, and Muthurajan Raveendran. IRRI national staff support for bioinformatics curation and software engineering was provided by Victor Jun Ulat, Ma. Teresa Ulat, Alexander Cosico, and Clarissa Pimentel.

Conclusions and recommendations

Drought-resistant rice for increased rainfed production and poverty alleviation: a concept note

R. Serraj and G. Atlin

Drought is the major constraint to rice production in rainfed areas across Asia and sub-Saharan Africa. Frequent droughts result in enormous economic losses and have long-term destabilizing socioeconomic effects on resource-poor farmers and communities. In the context of current and predicted water scarcity scenarios, irrigation is generally not a viable option to alleviate drought problems in the rainfed rice-growing systems. It is therefore critical that genetic management strategies of drought focus on maximum extraction of available soil moisture and its efficient use in crop establishment, growth, and maximum biomass and seed yield. Drought mitigation, through improved drought-resistant rice varieties and complementary management practices, represents an important exit pathway from poverty. Recent advances in drought physiology and genetics together with progress in cereal functional genomics have set the stage for an initiative focusing on the genetic enhancement of drought resistance in rice. Extensive genetic variation for drought resistance exists in the rice germplasm. However, the current challenge is to decipher the complexities of drought resistance in rice and exploit all available genetic resources to produce rice varieties combining drought adaptation with high yield potential, quality, and tolerance of biotic stresses. The aim will be to develop a pipeline for elite "prebred" varieties or hybrids in which drought-resistance genes can be effectively delivered to rice farmers. The Frontier Project on Drought-Resistant Rice will scale up gene detection and delivery for use in marker-aided breeding. The development of high-throughput, high-precision phenotyping systems will allow genes for component traits to be efficiently mapped, and their effects assessed on a range of drought-related traits, moving the most promising into widely grown rice mega-varieties.

The past success of the Green Revolution in the race between increasing world population and food production was principally achieved by increasing crop productivity in favorable zones. However, success has been limited in increasing rice productivity in rainfed systems, which are prone to frequent droughts and other abiotic stresses. Worldwide, drought affects approximately 23 million ha of rainfed rice. Drought is particularly frequent in bunded uplands and shallow rainfed lowland fields in many parts of South and Southeast Asia, sub-Saharan Africa, and Latin America. It also affects production on millions of hectares in water-short irrigated areas dependent on surface irrigation, where, in drought years, river flows and water in ponds and reservoirs may be insufficient to supply the crop (IRRI 2002).

The variation in rice production in areas dependent on rainfall and/or surface irrigation is closely related to total annual rainfall, but, even when the total is adequate, shortages at critical periods greatly reduce productivity, resulting in severe economic losses for some of the world's poorest communities. In the eastern Indian states of Jarkhand, Orissa, and Chhattisgarh alone, rice production losses in severe droughts (about one year in five) average about 40% of total production, with an estimated value of US$650 million (Pandey et al 2005). These losses affect the poorest farmers and their communities disproportionately. In drought years, food consumption decreases, indebtedness increases, assets are sold, children are withdrawn from school, and household members migrate. Droughts therefore have long-term destabilizing effects. Drought risk reduces productivity even in favorable years because farmers avoid investing in inputs when they fear crop loss. Risk-reducing technologies can therefore lead to increased investment and productivity in rainfed systems.

In the context of current and predicted water-scarcity scenarios, irrigation has only limited potential to alleviate drought problems in rainfed rice-growing systems (O'Toole 2004). It is therefore critical that both agronomic and genetic management strategies focus on the efficient use of available soil moisture for crop establishment, growth, and yield. Rice yields in drought-prone rainfed systems remain low at 1.0 to 2.5 t ha^{-1}, and tend to be unstable due to erratic and unpredictable rainfall. Drought mitigation through improved drought-resistant rice varieties and complementary practices, such as water conservation, represents an important exit pathway from poverty.

Most small farmers obtain rice seed from neighbors and relatives. Improved varieties spread rapidly through farmer-to-farmer exchange once farmers are convinced of their superiority. Varieties combining improved drought resistance with high yield under favorable conditions and quality characteristics preferred by farmers are therefore among the most promising and deliverable technologies for alleviating poverty in communities dependent on rainfed rice production.

Because of recent advances in our understanding of the physiology and genetics of drought resistance, and the integration of highly efficient breeding and genetic analysis techniques with the power of functional genomics, the time is now ripe for a major assault on the problem of improving rice drought resistance.

Rice toposequence, drought agroecology, and implications for breeding

Local watersheds in which rainfed rice is grown can often be characterized as a *toposequence*, or a series of terraced fields that drain into each other. Within distances of several hundred meters, the toposequence may include unbunded uplands, drought-prone upper fields that do not retain standing water, well-drained mid-toposequence fields, and poorly drained lower fields in which water accumulates to depths of 1 m or more during the rainy season. Water-related stresses are variable across years because of variability in the amount and distribution of rainfall, but occur with predictable frequency in a given field based on toposequence position and soil texture. They are most severe in unbunded uplands and shallow bunded fields at the top of the toposequence (Jongdee et al 2006). Asia has more than 20 million ha of such lands. Yield variability due to micro-geographic variation in field elevation and levelness can be great even within single fields, resulting in very large estimates of genotype by location by year interactions and residual error in the analysis of rainfed rice trials, thus complicating the selection of drought-resistant genotypes (Cooper et al 1999).

Rice yield is linearly related to the number of days in the growing season in which soil is saturated (Boling et al 2004, Haefele et al 2006). The ability to maintain biomass accumulation in relatively dry soils is therefore a key feature required in drought-resistant varieties. Intermittent soil drying substantially reduces biomass production, and therefore total yield potential. However, to a rice farmer, the word "drought" means not only physical water shortage that affects plant growth and development but also a lack of sufficient water to support land preparation, transplanting, fertilizer application, and weed control operations. All of these operations are dependent on the presence of a standing water layer in the paddy. If they are delayed or skipped, large yield losses often ensue, even though plants have not suffered physiological water stress. Losses from these management disruptions may be as great as those from direct drought damage.

Transplanting is the management step that is most vulnerable to water shortage. Farmers must often delay transplanting until sufficient water accumulates in fields to permit puddling (usually 400–500 mm of rainfall); this may result in transplanting seedlings that are 60 to 80 days old. Transplanting delays of this magnitude result in large yield losses because of reductions in both panicle number and weight. In experiments conducted at IRRI in 2005, transplanting 65-day-old seedlings as opposed to 22-day-old seedlings resulted in a yield reduction of more than 50%, averaged across 125 cultivars. Variability for the trait has not been systematically studied, but appears to be large, even in photoperiod-insensitive germplasm, and resistance to delayed transplanting has become an important target of the IRRI drought breeding program. The widespread eastern Indian weed management practice of *beushening* (also known as *beusani* or *biasi*, among other variants), which consists of uprooting the standing crop about 1 month after direct sowing by plowing, followed by re-rooting (Singh et al 1994), cannot be performed without standing water in the field and is also extremely sensitive to drought.

Understanding the complexity of rice drought in agroecological terms thus permits the identification of two distinct hydrological target environments for drought resistance breeding:

1. *Unbunded upland fields*

   Upland rice is established via direct seeding in unbunded, unpuddled fields that never accumulate standing water. The largest area of drought-prone upland rice is in the eastern Indian plateau region, where 3–5 million ha are grown, but large areas of upland rice production remain in West Africa,

Indonesia, and hilly Southeast Asia. In the unbunded upland environment, soil water content frequently falls to field capacity or below in the root zone, and the ability to root deeply, maintain biomass production in drying soil, and tolerate periods of severe stress throughout the growing season is critical. In direct-sown upland fields, a high degree of seedling vigor to compete with fast-growing upland weeds, as well as resistance to severe drought at any crop stage, is of paramount importance. Upland rice yields are closely related to total seasonal rainfall, and are very sensitive to short periods of stress during flowering.

2. *Bunded upper-toposequence fields*

   Bunded fields at the top of the lowland toposequence accumulate little or no standing water, but are usually either puddled and transplanted or established via beushening. In these fields, a compacted layer is often created by soil puddling that may reduce the ability of rice plants to root deeply. Because these fields can be highly productive in favorable years, combining high yield potential with drought resistance is a prerequisite for varietal adoption. Delayed transplanting, reduced biomass production due to soil drying, and sterility caused by stress around flowering are the major sources of yield loss in these fields. Drought risk could be reduced by developing cultivars that tolerate delayed transplanting, maintain vegetative growth when soil water content drops below field capacity, and tolerate severe stress during flowering and grain filling. Both transplanting and beushening risk could be avoided by developing cultivars adapted to direct seeding in dry soils, like rice or wheat. These *aerobic rice* cultivars are an important breeding target at IRRI. The bunded upper-field target environment occupies millions of hectares across South and Southeast Asia.

The two major target environments require distinctly different types of germplasm, and genetic solutions for one environment are not necessarily effective in the other. This was demonstrated in managed-stress trials at IRRI in the dry season of 2006, in which 64 lines were evaluated under both upland and lowland stress. The correlation between yields in the stress and nonstress environment was only 0.35, indicating that about 10% of the genetic variation for yield under stress was common to the two environments. However, gains made in one environment are likely to generate spillover effects and new options in the other. For example, high-yield, drought-resistant aerobic rice performs well when direct-sown in bunded upper fields, opening up new options for reducing transplanting risk.

The discussion above indicates that there are four critical entry points to the development of varieties with reduced risk and greater productivity in drought-prone fields:

- Improved maintenance of biomass production in drying soils (for both bunded and unbunded target environments).
- Resistance to severe stress during the critical reproductive stage (for both bunded and unbunded target environments).

- Adaptation to direct seeding in dry soil (for both bunded and unbunded target environments).
- Tolerance of delayed transplanting (for bunded target environments only).

Old challenges and recent breakthroughs

Progress in developing drought-resistant rice varieties has been slow, despite considerable past efforts. Drought resistance is a difficult trait to evaluate because of the unpredictable occurrence of natural droughts and lack of information on effective screening techniques. Consequently, few national programs have systematically incorporated drought resistance as a breeding objective. Biotechnological approaches have also had little impact on the development of improved cultivars, largely because the key mechanisms conferring improved drought resistance have not been clearly identified and few highly resistant donor varieties have been identified. Recently, however, research breakthroughs have revived interest in drought-resistance breeding and the use of new genomics tools to enhance crop water productivity. In rice, these breakthroughs include particularly

- Sequencing of the rice genome and the development of new genomics and postgenomics tools for detecting genetic polymorphism, gene discovery, and functional analysis of stress-related genes and mechanisms.
- Development of repeatable and predictive field screening methods suitable for use in breeding programs, and prospects for new high-throughput and precise phenotyping under drought (Campos et al 2004, Granier et al 2006, Serraj and Cairns 2006).
- Proof of concept that conventional breeding based on direct selection for yield under artificially imposed drought stress can result in actual gains in drought resistance (Venuprasad et al 2007, Kumar et al 2007, Lafitte et al 2006, Jongdee et al 2006).
- Identification of varieties and breeding lines with high levels of drought resistance for use as donors in breeding and gene discovery (Atlin 2005, Ouk et al, 2006, Toojinda et al 2005).
- Localization of QTLs with large effects on yield under drought stress that may be useful in marker-aided backcrossing (Bernier et al n.d., Venuprasad et al 2007, Yue et al 2005), and development of near-isogenic genetic stocks that differ in drought resistance (Li et al 2005, IRRI, unpublished data).
- Identification of genes and transcription factors from other plant species that may potentially improve drought resistance, if transferred to rice (see Yamaguchi-Shinozaki and Shinozaki K 2006 for a review).

These advances have set the stage for the development of varieties that can double[1] the yield of current widely grown rainfed rice varieties such as Swarna,

[1]The project aims at developing cultivars that can produce at least 2 t ha^{-1} under drought conditions in which the major rainfed lowland varieties Swarna, Sambha Mahsuri, and IR64 produce 1 t ha^{-1} or less.

IR36, Sambha Mahsuri, and Mahamaya under severe drought stress, but that have equivalent yield potential in high-rainfall years. Such varieties would have large effects on poverty by protecting food supply and income in drought years. Because of the reduction in risk of crop failure that would result from their adoption, they would also increase productivity and income in favorable years by encouraging farmers to increase investment in yield-enhancing inputs. In addition, any development of drought-adapted rice varieties for South Asia will also affect rice production in sub-Saharan Africa, where rice consumption is growing faster than anywhere else in the world, and where government policies are moving the continent as rapidly as possible away from imports and toward self-sufficiency. Rice cultivars combining improved drought resistance with responsiveness to favorable conditions are among the most promising and deliverable technologies for alleviating poverty. There is added value to drought resistance in rice: it will be a "carrier trait" for high-iron rice and golden rice. The importance of comprehensive linkage of farmer and nutritional traits is being increasingly recognized.

Critical research issues

Drought characterization and matching phenotype to the environment

Drought occurring at different points in crop development and in different soil types and management regimes affects crop growth differently; particular patterns of drought occurrence may require different sets of adaptive traits. For example, some drought-resistance traits and mechanisms affecting performance in upland systems may not be relevant under transplanted lowland systems. The timing, severity, and frequency of occurrence of drought stress must be modeled using available information on rainfall, soils, topography, and cropping patterns in the major rainfed rice-growing areas and used to identify combinations of traits and types of germplasm that fit the target environment.

The large genotype-by-environment (G × E) interactions observed in drought-prone environments have stymied progress in exploiting drought traits in the past because expression of genetic effects was unreliable. G × E for drought traits results to a large extent from the variation in the weather patterns among growing environments. The impact of any given drought trait is usually highly sensitive to the rate at which the drought develops, when in the season it develops, and its severity. A drought trait that might offer substantial benefit in one weather scenario might well result in a negative response in another (Sinclair and Muchow 2001). Simulation models provide a tool to combine mechanistic understanding of a drought trait with a range of weather scenarios. Given a historical record of weather for a location, the probability of a yield increase (and maybe decrease) resulting from the incorporation of any trait into the crop can be simulated. Combining the probabilities for yield change with understanding of growers' tolerance for risk will help breeders to assess the desirability of incorporating a particular drought trait in cultivars to be grown in a specific location.

High-throughput and precise phenotyping that predicts crop performance

Although the advances listed above have fulfilled some of the requirements for rapid progress toward the development and deployment of drought-resistant rice varieties, important bottlenecks remain. The rapid progress made in rice genomics must be matched with a better understanding of drought physiological mechanisms and their relationship to the performance of varieties in drought-affected farmers' fields. Identification of donors, mapping and use in marker-aided breeding (MAB) of QTLs and genes that affect performance under stress, and screening of improved varieties in national agricultural research and extension systems (NARES) breeding programs all depend on the development of repeatable, low-cost, high-throughput phenotyping procedures that reliably characterize genetic variation for drought resistance and its component traits. Therefore, a special effort is needed for the conceptualization, design, and management of phenotyping programs for drought resistance to maximize the chances of identifying donors, QTLs, and breeding lines that will be useful in the future improvement of resistance in the target environment.

Identifying donors of genes and QTLs for drought resistance

Extensive, unexploited variation for drought resistance exists within rice and its wild relatives. Because of the difficulty of screening for drought resistance in conventional breeding programs, MAB using QTLs with large effects on drought resistance is among the most promising routes to the development of cultivars with improved yield in drought-prone environments. Progress in breeding and identifying genes for drought resistance will result from the identification of existing traditional and improved varieties and wild species accessions that have unusually high levels of drought adaptation or traits that can contribute to it. Some potential donors have had their drought resistance confirmed by researchers at IRRI and other institutions, but many more are needed to serve as sources of useful alleles for breeding programs.

To date, several QTLs with large and consistent effects on grain yield under drought stress have been identified at IRRI. Many more such QTLs are likely to exist. They must be fine-mapped and their effects assessed in a range of backgrounds and environments for deployment in MAB.

Efficient integration of the various disciplines and approaches involved in drought research

It is now well accepted that the complexity of the drought syndrome can be tackled only with a holistic approach integrating plant breeding with physiological dissection of the resistance traits and molecular genetics tools together with agronomic practices that lead to better conservation and use of soil moisture and matching crop genotypes with the environment (Serraj et al 2003). Some important steps involved in this multidisciplinary approach are to

- Define the target drought-prone environment(s) and identify the predominant type(s) of drought stress and the rice varieties preferred by farmers.
- Define the phenological, morphological, and physiological traits that contribute substantially toward adaptation to drought stress(es) in the target environment(s).
- Use simulation modeling and systems analysis to evaluate crop response to the major drought patterns, and assess the value of candidate physiological traits in the target environment.
- Develop and refine appropriate screening methodologies for characterizing genetic stocks that could serve as donors, characterizing the effects of genes and QTLs, and selecting cultivars with resistance to specific stresses.
- Establish the relationship between putative drought-related traits and crop productivity under targeted drought stress.
- Harness functional genomics, transgenics, and reverse genetics tools to understand the genetic control of the relevant traits.
- Identify, map, and tag for marker-aided breeding QTLs within the rice germplasm that are critical for drought resistance, and incorporate them into locally adapted varieties.
- Evaluate the effect in rice of sequences known to affect drought resistance in other species by transforming important but susceptible varieties.
- Develop near-isogenic versions of important but drought-susceptible varieties containing genes affecting specific traits of interest to assess their value in improving drought adaptation of locally adapted varieties.
- Test the marker-assisted selection products under a well-managed screening facility and in farmer participatory multilocation trials.

IRRI's comparative advantage in drought research

Research commitment and capacity

IRRI has made a long-term and large-scale commitment to rice improvement for drought-prone areas. Seventeen senior IRRI staff members contribute directly to our drought research efforts. These efforts include varietal development, genomics and functional genomics, resource management, and research on social sciences and complementary practices. The IRRI drought research program takes full advantage of an excellent field screening environment, with a 5-month dry season in which water stress can be reliably imposed, yet the environmental conditions for rice production are otherwise nearly optimal. This has permitted the development of a large-scale field-based drought breeding and physiology research program; in the dry season of 2006, more than 7,000 genotypes were screened for yield in replicated field trials under managed-drought-stress conditions.

A record of success in drought breeding and gene detection

IRRI researchers have developed effective screening methods for identifying drought-resistant varieties. These methods are being applied in the IRRI program to develop drought-resistant cultivars, and are being extended to NARES. Low-cost, high-throughput QTL mapping approaches are also being used to detect and fine-map genes with large effects on drought resistance, with the result that several major QTLs have been recently identified in resistant by susceptible crosses. One major QTL located on chromosome 12 has been shown over three seasons to double yield (from 250 to 500 kg ha^{-1}) under severe upland stress in an already moderately resistant genetic background (Bernier et al n.d.). This QTL also affects yield to a slightly lesser degree under moderate stress in another cross. Several other QTLs affecting yield under severe upland drought stress have been identified in other crosses.

The world's largest collection of rice germplasm, breeding materials, and genetic stocks

The IRRI genebank, with its more than 108,712 accessions from 85 countries, is a prime potential source of useful genetic variation for drought resistance. IRRI also has a large collection of chemically induced mutants in the background of IR64, one of the world's most widely grown rice cultivars. Many of these mutants exhibit a differential response to drought stress and are being used to identify genes affecting resistance. IRRI has also developed a large set of near-isogenic stocks for genetic and functional genomics analysis. Identification, mapping, and deployment in MAB of genes conferring improved drought resistance depend on the development of stocks that are identical in most of their genes but that differ substantially in performance under drought stress. These near-isogenic stocks eliminate much of the background variability in gene expression patterns that exists between more distantly related materials, and allow the traits and genes associated with stress resistance to be clearly identified.

Additional comparative advantages of IRRI include

- Breeding networks that encompass the entire major rainfed rice-growing regions of Asia and are entering Africa (eastern and southern) aggressively.
- An unsurpassed track record of getting materials out to farmers in rice-growing countries.
- Increasing capacity in the technical and regulatory expertise needed to deliver transgenic products.

This effort will focus on the two key components of drought-resistant varieties:

1. The ability to maintain biomass production in dry soils through the preflowering period.
2. The ability to maintain pollination and seed set when stress occurs during the highly sensitive period around flowering.

The Frontier Project on Drought-Resistant Rice

Goal and objectives

The overall goal of the Frontier Project on Drought-Resistant Rice will be to develop new rice genotypes that can potentially double yield under drought stress in rainfed environments. The specific objectives are to

1. Explore rice genetic diversity and identify donors and parental lines for drought-resistance breeding.
2. Identify key traits and mechanisms affecting rice drought resistance, and develop relevant screening techniques.
3. Tag the alleles that increase rice productivity under drought stress for marker-aided breeding, validate them in farmers' fields, and use them widely in NARES breeding programs.
4. Deploy gene transformation systems for the identification and exploitation of drought-resistance genes in plant breeding.
5. Develop improved rice varieties and hybrids that double yield under drought under both lowland and upland conditions relative to current widely grown varieties, without constraining productivity in favorable years.

Project description

The Frontier Project on Drought-Resistant Rice aims to harness recent scientific breakthroughs in breeding and biotechnology and the wide genetic diversity in rice to develop and deliver improved drought-resistant rice varieties through the following approaches.

Gene discovery

This will be achieved through two complementary approaches: (1) mobilizing genes for resistance within the rice germplasm using high-throughput QTL analysis and allele mining leading to map-based cloning of rice alleles affecting growth and yield under drought stress, and (2) exploiting them in rice sequences demonstrated to enhance drought resistance in other species.

Mobilizing genes for drought resistance in rice germplasm. Rice germplasm has extensive unexploited variation for drought resistance. IRRI research has shown that variation for the trait can be strongly influenced by alleles of a relatively small set of QTLs with large effects. We have developed efficient "selective genotyping" strategies that allow many potential donors to be searched quickly and inexpensively for these valuable genes. This allows the hunt for useful alleles to be quickly narrowed down to small chromosome segments suitable for bioinformatics analysis. This approach, initiated at IRRI in 2003, has already led to the identification of three genes with large effects on yield under severe upland drought stress, one of which has been localized to a chromosome interval of less than 10 cM. The drought-resistance genes located by this approach are international public goods that can be made freely available to IRRI's clients and stakeholders. The Frontier Project will scale up the detection, tagging, and delivery of these genes for use in MAB.

Mobilizing genes for drought resistance from other species. Genes and their control sequences previously determined to improve drought resistance in other species are also of potential value in rice, and will be evaluated using high-throughput transformation and phenotyping systems. Sequences licensed from the commercial sector and already proven to enhance performance under drought stress in maize will be among the first to be evaluated using this approach. The Frontier Project will rapidly assess the effect of these sequences on a range of drought-related traits, and move the most promising into widely-grown, high-quality rice mega-varieties.

Bridge the "phenotype gap" through physiology and high-throughput phenotyping

To identify sources of drought resistance, it is necessary to develop screening methods that are simple, reproducible, and predictive of performance in the target environment. Therefore, managing drought-screening nurseries requires a careful analysis of likely sources of nongenetic variation among plots, replications, and repeated experiments, and establishing procedures for minimizing these factors. Several field and laboratory screening methods are being used and improved at IRRI for screening rice germplasm for drought resistance, including drip and line-source sprinkler irrigation systems, drained paddies, and the development of remote-sensing techniques for monitoring plant water status in the field. Research is needed to identify the physiological bases for differences in genotype performance in these screens, as well as their relationship to performance in farmers' fields under naturally occurring drought stress.

Development of productive, high-quality "platform" varieties

Farmers require much more than drought resistance from a variety. Only varieties that combine high yield potential in favorable seasons, high quality, and resistance to diseases and pests will be adopted by farmers. Highly stress-tolerant lines that are deficient in many other aspects will be rejected by farmers and are difficult for national programs to use as donors in breeding. IRRI will develop elite "prebred" varieties in which drought-resistance alleles identified through genetic analysis and functional genomics research can be effectively delivered to breeders and farmers. Hybrids have been shown to be more resistant to water and nutrient deficiency than pure lines; stress-resistant hybrids are therefore a promising "delivery platform" for drought-resistance genes.

Development of a system for delivering drought-resistant lines and hybrids to poor farmers in Asia and Africa

The products of the Frontier Project must be proven in the target environment and adopted by farmers in order to have impact. The project will link researchers in innovative partnerships with extension services and the commercial seed and input supply sectors to evaluate the drought-resistant varieties it produces under the real conditions faced by farmers, and deliver them to the poor. Breeding programs in drought-prone regions will be organized into networks of centers that collaborate in multienvironment and managed-stress testing to identify drought-resistant genotypes. Lines and hybrids proven to combine resistance and yield potential in on-station trials will be evaluated on-farm in participatory varietal selection programs managed by the partner centers, using the "mother-baby" trial model. Once varieties are released by national systems, they will be delivered via targeted dissemination to farmers who are facing water shortage, in collaboration with governmental and nongovernmental extension and community development partners, including private-sector seed companies.

Training of the next generation of rice scientists in Southeast Asia and sub-Saharan Africa

Capacity building of NARES has been one of the major reasons behind the historical successes achieved in crop improvement and agricultural research in Asia, especially in countries such as India and China. For drought research, a tremendous amount of work has been accomplished over the past decade in training rice scientists in biotechnology and its application in plant breeding, with support from donors such as the Rockefeller Foundation (RF). However, such efforts need to be sustained and extended to new areas in Asia and sub-Saharan Africa. The Frontier Project will closely collaborate with NARES in capacity building to increase the number of local rice scientists (breeders, agronomists, molecular biologists) with applied experience in drought breeding and foster the development of a community of practice in the use of upstream tools and best drought-resistant germplasm. This project will provide on-the-job training to new rice scientists joining NARES through regional, international, and South-South cooperation. Building on the past successes and achievements of the RF-funded Indian drought breeding network, NARES will be invited to contribute their breeding lines to the collaborative trial networks across South and Southeast Asia and sub-Saharan Africa. Annual project meetings will be organized with NARES in target countries to discuss and harmonize rice breeding approaches, review progress, and determine the technical backstopping needs of NARES breeding and graduate research projects.

Expected outputs

Phase I: 2007-09

- Rice germplasm (core collections, advanced breeding lines, hybrids, mutants, and wild relatives) collected, screened, and characterized for drought-resistance traits and yield performance under various water-deficit scenarios.
- Databases and detailed GIS maps of drought-prone rice production target areas in South Asia, sub-Saharan Africa, and Latin America.
- Control sequences and structural genes from rice and other plant species identified that, when incorporated into the rice genome, enhance drought resistance in specific drought-stress conditions and increase grain yield in the fields of drought-affected farmers.
- Methods developed for high-throughput and precise phenotyping of drought-resistance traits and field performance.
- QTLs that confer improved yield under drought stress fine-mapped and MAB systems developed for their deployment by NARES breeding programs.

- Gene transformation systems and "intermediary" gene products developed.
- Crop simulation modeling tools for knowledge integration and systems analysis of rice-adapted traits in target environments.

Phase II: 2010-12

- Physiological and structural traits involved in improved resistance to drought stress dissected and linked to candidate genes and gene networks.
- Localized, sequenced, and physiologically characterized alleles that result in improved grain yield under specific drought-stress conditions incorporated in widely grown rice varieties, and demonstrated to increase grain yield in the fields of drought-affected farmers.
- Capacity for high-throughput screening systems for drought resistance transferred to national breeding programs.
- NARES rice scientists trained in the various drought research disciplines, technology transferred, and networks strengthened.
- MAB programs implemented with NARES and products tested with farmers for multilocation participatory varietal selection (PVS).
- Gene transformation products with putative drought-resistance genes tested in controlled water-deficit conditions.
- Large-scale multilocation trials of breeding products for PVS in farmers' fields.

Phase III: 2013-15

- MAS and transformation products with putative drought-resistance traits tested in drought-prone field conditions.
- Widely adapted, pest-resistant pure lines and hybrids with high yield potential and high grain quality with alleles for drought resistance incorporated, for direct release, and seed increased for distribution.
- Dissemination of improved varieties on at least 1 million ha in collaboration with government and nongovernment extension organizations and private-sector seed companies.

Partnerships and linkages: consortium building

The implementation of the Frontier Project will require close and efficient integration among the various disciplines and approaches involved in drought research, including

- Germplasm use: identifying donors of genes for drought resistance among traditional and improved varieties, wild species, and mutants.
- Breeding: yield- and trait-based selection for drought resistance, detection, mapping, and mobilization of alleles for drought resistance by MAS.
- Understanding physiology and development of high-throughput precision phenotyping.

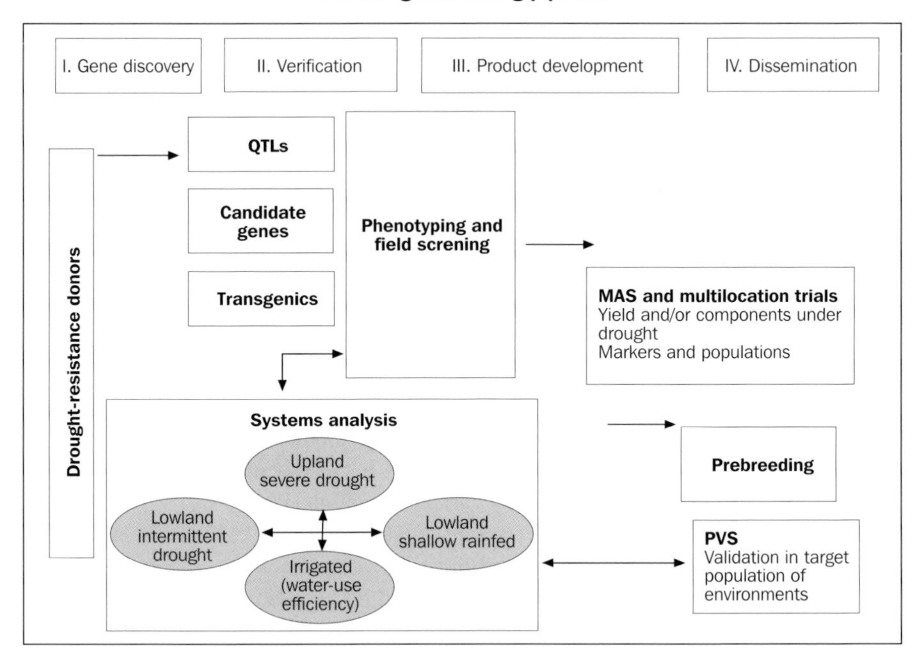

- Transgenic technology and functional genomics.
- Comparative genomics and bioinformatics.
- Systems analysis, modeling, and GIS.
- Socioeconomics, dissemination strategy and delivery mechanisms, and networks, including private-sector seed companies for hybrid dissemination.

In addition to IRRI's own resources and capacities in these various scientific fields, the Frontier Project will build a scientific consortium with experts and advanced research institutes with expertise in drought and rice research. Research and extension institutions from the NARES in Asia and Africa will also play a central role in carrying out the field research and disseminating its outputs to farmers and end-users. Collaboration with private-sector companies will mainly aim at the exploitation in rice of genetic sequences and products with demonstrated drought-resistance characteristics. The regional research networks (Consortium for Unfavorable Rice Environments, drought breeding networks, etc.) will be directly and actively involved in the planning and implementation of all aspects of the project.

References

Atlin GN. 2005. Detecting alleles conferring improved reproductive-stage drought tolerance in rainfed rice. Rockefeller Report. International Rice Research Institute, Manila, Philippines.

Bernier J, Kumar A, Venuprasad R, Spaner D, Atlin G. n.d. A large-effect QTL for grain yield under reproductive-stage drought stress in upland rice. Crop Sci. (Accepted.)

Boling A, Tuong TP, Jatmiko SY, Burac M. 2004. Yield constraints of rainfed lowland rice in Central Java, Indonesia. Field Crops Res. 90:351-360.

Campos H, Cooper A, Habben JE, Edmeades GO, Schussler JR. 2004. Improving drought tolerance in maize: a view from industry. Field Crops Res. 90:19-34.

Cooper M, Rajatasereekul S, Immark S, Fukai S, Basnayake J. 1999. Rainfed lowland rice breeding strategies for Northeast Thailand. I. Genotypic variation and genotype by environment interactions for grain yield. Field Crops Res. 64:131-151.

Granier C, Aguirrezabal L, Chenu K, Cookson SJ, Dauzat M, Hamard P, Thioux, J, Rolland G, Bouchier-Combaud S, Lebaudy A, Muller B, Simonneau T, Tardieu F. 2006. PHENOPSIS, an automated platform for reproducible phenotyping of plant responses to soil water deficit in *Arabidopsis thaliana* permitted the identification of an accession with low sensitivity to soil water deficit. New Phytol. 169:623-635.

Haefele SM, Konboon Y, Patil S, Mishra VN, Mazid MA, Tuong TP. 2006. Water by nutrient interactions in rainfed lowland rice: mechanisms and implications for improved nutrient management. Presented at the CURE Resource Management Workshop in Dhaka, Bangladesh, March 2006.

IRRI (International Rice Research Institute). 2002. Annual report. Manila (Philippines): International Rice Research Institute.

Jongdee B, Pantuwan G, Fukai S, Fischer K. 2006. Improving drought tolerance in rainfed lowland rice: an example from Thailand. Agric. Water Manage. 80:225-240.

Kumar R, Venuprasad R, Atlin GN. 2007. Genetic analysis of rainfed lowland rice drought tolerance under naturally-occurring stress in eastern India: heritability and QTL effects. Field Crops Res. 103:42-52.

Lafitte HR, Li ZK, Vijayakumar CHM, Gao YM, Shi Y, Xu JL, Fu BY, Ali AJ, Domingo J, Maghirang R, Torres R, Mackill D. 2006. Improvement of rice drought tolerance through backcross breeding: evaluation of donors and selection in drought nurseries. Field Crops Res. 97:77-86.

Li ZK, Fu BY, Gao YM, Xu JL, Ali J, Lafitte HR, Jiang YZ, Domingo-Rey J, Vijayakumar CHM, Dwivedi D, Maghirang R, Zheng TQ, Zhu LH. 2005. Genome-wide introgression lines and a forward genetics strategy for genetic and molecular dissection of complex phenotypes in rice (*Oryza sativa* L.). Plant Mol. Biol. 59:33-52.

O'Toole JC. 2004. Rice and water: the final frontier. First International Conference on Rice for the Future, 31 August-2 Sept. 2004, Bangkok, Thailand.

Ouk M, Basnayake J, Tsubo M, Fukai S, Fischer KS, Cooper M, Nesbitt H. 2006. Use of drought response index for identification of drought tolerant genotypes in rainfed lowland rice. Field Crops Res. 99:48-58.

Pandey S, Bhandari H, Sharan R, Naik D, Taunk SK, Sastri ADRAS. 2005. Economic costs of drought and rainfed rice farmers' coping mechanisms in eastern India. Final project report. International Rice Research Institute, Manila, Philippines.

Serraj R, Bidinger FR, Chauhan YS, Seetharama N, Nigam SN, Saxena NP. 2003. Management of drought in ICRISAT's cereal and legume mandate crops. In: Kijne JW, Barker R, Molden D, editors.Water productivity in agriculture: limits and opportunities for improvement. Wallingford (UK): CABI Publishing.
Serraj R, Cairns, JE. 2006. Diagnosing drought. Rice Today 5(3):32-33.
Sinclair TR, Muchow RC. 2001. System analysis of plant traits to increase grain yield on limited water supplies. Agron. J. 93:263-270.
Singh RK, Singh VP, Singh CV. 1994. Agronomic assessment of beushening in rainfed lowland rice cultivation in Bihar, India. Agric. Ecosyst. Environ. 51:271-280.
Toojinda T, Tragoonrung, S, Vanavichit A, Siangliw JL, Pa-In N, Jantaboon J, Siangliw M, Fukai S. 2005. Molecular breeding for rainfed lowland rice in the Mekong region. Plant Prod. Sci. 8:300-333.
Venuprasad R, Lafitte HR, Atlin G. 2007. Response to direct selection for grain yield under drought stress in rice. Crop Sci. 47:285-293.
Yamaguchi-Shinozaki K, Shinozaki K. 2006. Transcriptional regulatory networks in cellular responses and tolerance to dehydration and cold stresses. Annu. Rev. Plant Biol. 57:781-803.
Yue B, Xiong L, Xue W, Xing Y, Luo, L, Xu C. 2005. Genetic analysis for drought resistance of rice at reproductive stage in field with different types of soil. Theor. Appl. Genet. 111:1127-1136.

Notes

Authors' addresses: R. Serraj, International Rice Research Institute, Los Baños, Philippines; G. Atlin, International Maize and Wheat Improvement Center, Mexico.